WONDER-2015 - 4th International Workshop on Nuclear Data Evaluation for Reactor Applications

EPJ Web of Conferences Volume 111 (2016)

Aix-en-Provence, France
5 - 8 October 2015

Editors:

Olivier Serot

ISBN: 978-1-5108-2205-4

Printed from e-media with permission by:

Curran Associates, Inc.
57 Morehouse Lane
Red Hook, NY 12571

Some format issues inherent in the e-media version may also appear in this print version.

Printed by Curran Associates, Inc. (2016)

For additional information, please contact EDP Sciences – Web of Conferences at the address below.

EDP Sciences – Web of Conferences
17, Avenue du Hoggar
Parc d'Activité de Courtabœuf
BP 112
F-91944 Les Ulis Cedex A
France

Phone: +33 (0) 1 69 18 75 75
Fax: +33 (0) 1 69 28 84 91

contact-edps@webofconferences.org

Additional copies of this publication are available from:

Curran Associates, Inc.
57 Morehouse Lane
Red Hook, NY 12571 USA
Phone: 845-758-0400
Fax: 845-758-2633
Email: curran@proceedings.com
Web: www.proceedings.com

TABLE OF CONTENTS

EPJ Web of Conferences **111**, 00001 (2016)
DOI: 10.1051/epjconf/201611100001

Preface

WONDER-2015 is the fourth International Workshop on Nuclear Data Evaluation for Reactor Applications (the first three workshops were organised respectively in 2006, 2009 and 2012). It was held on October 5th to 8th, 2015 at the Hotel Aquabella in Aix-en-Provence (France).

About 70 experts in the field of nuclear data were participating at this workshop. The main objective was to review the current modelling and evaluation methods of nuclear data for reactor applications (GEN. II, III, IV) and to discuss possible areas of improvements. Presentations were organised in sessions on Nuclear Data needs for reactor applications, Microscopic nuclear data measurements, Evaluation of nuclear data, Thermal scattering laws, Processing and benchmarking, Integral nuclear data measurements, Decay data and Fission Yields, Uncertainties and covariance matrices, Fission Fragments Obervables.
Several posters covering these items were also presented.

We thank the French Alternative Energies and Atomic Energy Commission (CEA), the Nuclear Energy Agency (NEA) and La Communauté du Pays d'Aix (CPA) for their active participation in the organisation of this workshop. Lastly, we would like to warmly thank Mrs Alexandra Herrenschmidt who assumed with enthusiasm and dynamism the secretariat of the workshop.

Taking into account the undeniable success of WONDER-2015, the 5th edition could be born in 2018!

January 2016,
The Organizing Committee

edp sciences

EPJ Web of Conferences

Statement of Peer review

In submitting conference proceedings to EPJ Web of Conferences, the editors of the proceedings certify to the Publisher that they adhere to the **Policy on Publishing Integrity** of the journal in order to safeguard good scientific practice in publishing.

1. All articles have been subjected to peer review administered by the proceedings editors.
2. Reviews have been conducted by expert referees, who have been requested to provide unbiased and constructive comments aimed, whenever possible, at improving the work.
3. Proceedings editors have taken all reasonable steps to ensure the quality of the materials they publish and their decision to accept or reject a paper for publication has been based only on the merits of the work and the relevance to the journal.

Title, date and place of the conference

WONDER 2015, October 5th to 8th 2015
Aix-en-Pce

Proceedings editor(s):

SEROT OLIVIER . CEA CADARACHE

Date and editor's signature

8/02/2016

EPJ Web of Conferences **111**, 01001 (2016)
DOI: 10.1051/epjconf/201611101001

The nuclear data, A key component for reactor studies, Overview of AREVA NP needs and applications

Simon Ravaux[1], Pierre-Marie Demy[1], and Clément Rechatin[2]

[1]*AREVA NP, PEPDV-F, Tour AREVA – 1 Place Jean Millier, 92084 Paris La Défense Cedex, France*
[2]*AREVA NP, FE/FL, 10 rue Juliette Recamier, 69456 Lyon Cedex, France*

Abstract. The quality of the nuclear data is essential for AREVA NP. Indeed, many AREVA NP activities such as reactor design, safety studies or reactor instrumentation use them as input data. So, the nuclear data can be considered as a key element for AREVA NP. AREVA NP's contribution in the improvement of the nuclear data consists in a joint effort with the CEA. It means a financing and a sharing of information which can give an orientation to the future research axis. The aim of this article is to present the industrial point of view from AREVA NP on the research on nuclear data. Several examples of collaborations with the CEA which have resulted in an improvement of the nuclear data are presented.

1 Short overview of AREVA NP France neutronic activities

1.1 Core analysis studies

AREVA NP performs many core analysis studies. These studies mainly concern core designs, refuelling studies and safety demonstration. They consist in modelling as accurately as possible in a reasonable period the core of a nuclear reactor in steady-state or in transient conditions.

Two core calculation chains are currently available inside AREVA NP France: SCIENCE [1] and ARCADIA [2]. These chains couple many physics fields such as thermal-hydraulic, mechanic and neutronic.

SCIENCE-V2 is the chain which is currently in production inside AREVA NP France. The neutronic part is based on the JEF-2.2 library [3], the spectral code APOLLO2.5 [4] and the core code SMART [1]. This chain has been validated for many applications and is currently used in France for most of the core studies. ARCADIA is a "new" chain which starts its industrial phase. The neutronic part is based on the JEFF-3.1.1 library [5], the spectral code APOLLO2.8 [4] and the core code ARTEMIS [2]. ARCADIA has already been licensed in the U.S [2]. It is currently used in AREVA NP France for internal studies in the perspective of its licensing in France.

1.2 Other activities

In addition to the core analysis studies, AREVA NP is also involved in other neutronic activities such as radiation protection studies, fuel assembly conception, residual power calculations… For these activities, many other codes are available inside AREVA NP such as the Monte Carlo code MCNP5[6] and the evolution codes SCALE/ORIGEN-S[7] and ALEPH[8]…

Article available at http://www.epj-conferences.org or http://dx.doi.org/10.1051/epjconf/201611101001

MCNP5 is considered as a reference code for the criticity and for the neutron propagation. It is used for radiation protection studies, and also to qualify the deterministic codes (including the core calculation chains) through benchmarks and to evaluate uncertainties.

The evolution codes are also essential for AREVA NP. They are used for many applications such as: inventory of used fuel, corium criticity, estimation of delayed neutron or gamma sources, residual power...

2 Research on nuclear data within AREVA NP France

All the AREVA NP activities mentioned in the previous chapter make use of nuclear data libraries such as punctual or multigroup cross sections, fission yields, decay data. AREVA NP is involved in the improvement and consequently the research on the nuclear data. This involvement is mainly performed in the frame of the collaboration with the CEA through the so-call "Institut tri-partite" or "I3P".

2.1 AREVA NP actions inside the I3P

Inside the I3P, the "direct" research on the nuclear data is mainly performed by the CEA which get a funding from CEA, EDF and AREVA NP. AREVA NP is responsible of the physical validation of its own calculation codes which make use of the nuclear data libraries provided by the CEA. This validation is based on a lot of experimental and power reactor measurements: reactivity, fission rate distribution, depleted pin inventory, boron let-down curve, temperature coefficient, flux map, rod worth… AREVA NP then provides to the CEA a feedback on the basis of the EDF and the worldwide nuclear fleet which can give an orientation to the future research axis on the nuclear data.

2.2 CEA actions inside the I3P

The CEA is an important actor inside the European project JEFF. It proposes many new nuclear data evaluations which are then integrated in the new releases of the JEFF library. The new libraries are then transmitted to AREVA NP in binary file readable by AREVA NP codes and more precisely by ARCADIA.

The CEA also designs critical mock-up in experimental reactors. It performs measurements and provides the results to AREVA NP and EDF. These experiments can then be added to the physical validation base of AREVA NP codes.

The last but not least action from the CEA is an expert advice. It can give an efficient support to the industrial needs on the nuclear data.

3 Example of joint AREVA/EDF/CEA activities

3.1 Implementation of the ^{238}U resonant up-scattering

In the frame of the development of the ARCADIA calculation chain, AREVA NP has noticed some calculation-measurement discrepancies on power reactors. The Reactivity difference (difference on the bore concentration in the moderator) calculated by ARCADIA between the hot zero power (HZP) and the hot full power (HFP) core states, which is also called the HZP/HFP swing, was too low in comparison with the measurements.

In agreement with the CEA, it has been decided to implement the resonant up-scattering model inside the spectral code APOLLO2. The CEA has processed a new multigroup library for APOLLO2 and transmitted it to AREVA NP.

Thanks to these new developments, the discrepancies on the swing have been significantly decreased and the predictability of ARCADIA has been increased.

3.2 Feedback on the JEFF-3.1 library

The CEA delivered to AREVA NP the new application library based on JEFF-3.1 in 2006. AREVA NP quickly observed that, with this new library, the fuel cycle length was slightly under-estimated. This discrepancy was signalled to the CEA.

This problem has been presented by the CEA to the JEFF group. Seven new fission product evaluations were proposed and inserted by the CEA in a new version of the library called JEFF-3.1.1[5]. A new application library based on JEFF-3.1.1 was delivered to AREVA NP and allowed to consequently decrease the under-estimation of the cycle length.

4 Conclusion

The quality of the neutronic calculations is linked to the quality of the nuclear data libraries. Thus AREVA NP, as a major actor of the nuclear industry, is involved in the research on the nuclear data through the collaboration with the CEA inside the I3P. This collaboration is valuable for both organisms.

References

1. P. Girieux, proc. of ENC'94 (1994)
2. ARCADIA topical report, http://pbadupws.nrc.gov/docs/ML1015/ML101550189.pdf
3. JEF report 17, https://www.oecd-nea.org/dbdata/nds_jefreports/jefreport-17/Table_of_content.pdf
4. S. Loubiere, R. Sanchez, M. Coste, A. Hebert, Z. Stankovski, C. Van Der Gucht, and I. Zmijarevic, Proc. Int. Conf. Mathematics and Computation, Reactor Physics and Environmental Analysis in Nuclear Applications (1999)
5. JEF report 22, https://www.oecd-nea.org/dbdata/nds_jefreports/jefreport-22/nea6807-jeff22.pdf
6. X-5 Monte Carlo Team, LA-UR-03-1987 (2003)
7. S. M. Bowman and I. C. Gauld, ORNL/TM-2010/43 (April 2010)
8. W Haeck, B. Verboomen, SCK CEN-BLG-1003 Rev. 0 (2006)

EPJ Web of Conferences **111**, 01002 (2016)
DOI: 10.1051/epjconf/201611101002

Review of nuclear data improvement needs for nuclear radiation measurement techniques used at the CEA experimental reactor facilities

Christophe Destouches[1,a]

[1]*CEA, DEN, DER, Service of Experimental Physic - CEA Cadarache, F-13108 Saint-Paul-Lez-Durance, France*

Abstract. The constant improvement of the neutron and gamma calculation codes used in experimental nuclear reactors goes hand in hand with that of the associated nuclear data libraries. The validation of these calculation schemes always requires the confrontation with integral experiments performed in experimental reactors to be completed. Nuclear data of interest, straight as cross sections, or elaborated ones such as reactivity, are always derived from a reaction rate measurement which is the only measurable parameter in a nuclear sensor. So, in order to derive physical parameters from the electric signal of the sensor, one needs specific nuclear data libraries. This paper presents successively the main features of the measurement techniques used in the CEA experimental reactor facilities for the on-line and off-line neutron/gamma flux characterizations: reactor dosimetry, neutron flux measurements with miniature fission chambers and Self Power Neutron Detector (SPND) and gamma flux measurements with chamber ionization and TLD. For each technique, the nuclear data necessary for their interpretation will be presented, the main identified needs for improvement identified and an analysis of their impact on the quality of the measurement. Finally, a synthesis of the study will be done.

1 Introduction

The constant improvement of the neutron and gamma calculation codes used in the experimental nuclear reactors goes hand in hand with that of the associated nuclear data libraries such as JEFF-3.1.1 [1], ENDF/B-VII.1 [2], JENDL4.0 [3]...The validation of these calculation schemes always requires the confrontation with integral experiments performed in experimental reactors to be completed. Nuclear data of interest, straight as cross sections, or elaborated ones such as reactivity, are always derived from a reaction rate measurement which is the only measurable parameter in a nuclear sensor. So, in order to derive physical parameters from the rough electric signal of the sensor, one needs specific nuclear data libraries. These nuclear data libraries are either directly extracted from the general libraries when the data exist, or re-evaluated in dedicated libraries such as IRDFF. However, the reactions put in games being generally of not much interest for neutron calculation codes, these data are often old and associated to high level of uncertainty when existing. As efforts needed to proceed to new evaluation or new measurements are rather consequents, a motivation for each ones

[a] Corresponding author: christophe.destouches@cea.fr

should be established. This paper is thus going to present successively the main features of the measurement techniques used in the CEA experimental reactor facilities for the on-line and off-line neutron/gamma flux characterizations: reactor dosimetry, neutron flux measurements with miniature fission chambers and Self Power Neutron Detector (SPND) and gamma flux measurements with ionization chamber and TLD. For each technique, the nuclear data necessary for their interpretation is presented, the main needs for improvement identified as well as an analysis of their impact on the quality of the measurement. Finally, a synthesis of the study will be done.

2 Nuclear data used in the main measurement techniques

2.1 Activation measurement techniques

2.1.1 Neutron measurements

In the reactor dosimetry process applied at the CEA Cadarache [4, 5], reaction rates and neutron fluences are derived from activity measurements of irradiated dosimeters using the inversion of the Bateman equation. These dosimeters (figure 1) are selected according to their response functions on the studied neutron spectrum (neutron cross section) and to their characteristics with respect to activity measurement process (nature and energy of the emitted particles, decay constant). Typical precision estimates for these fluence monitors are estimated to range from 3 to 10%.

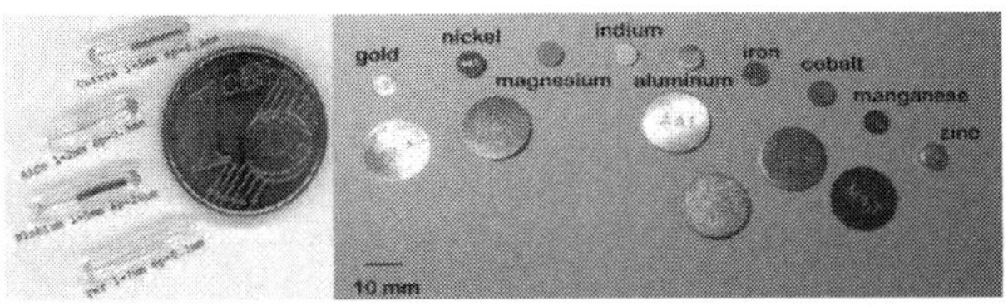

Figure 1. Typical set of dosimeters

Decay data used to derive the dosimeter's activity from the rough experimental counting spectra are extracted from specific libraries, for instance, Nucleides [6] or NUDAT [7]. For deriving neutron fluences from these activity measurements activation cross sections are taken from international dosimetry activation files, IRDFF [8], themselves extractions from the main international nuclear libraries ENDF-B/VII , JEFF3.1.1, JENDL4.0, or EAF2010 [9].

These nuclear data are not only used for activity measurements and reaction rate derivation (mean values) but also to assess output uncertainties (covariance). In addition the latter could also be used in an unfolding spectrum analysis combing several dosimeters results, each one covering different ranges of the studied neutron spectrum.

Thus, nuclear data accuracies impact both measurement and interpretation processes. Although the reactor dosimetry process is used as a reference for absolute calibration of other methods and nuclear sensors, dosimetry nuclear reactions are not of much interest for neutron calculation codes; they could present poor or discrepant evaluations in general purpose data library due mainly to a lack of experimental evaluations [4]. We shall note here the constant effort realized by the nuclear data service NDS of the IAEA to propose and maintain compilations as up to date as possible of these data with covariance matrix evaluations when possible, for example, the IRDFF validation project CRP [8]. Focusing on dosimetry measurements performed at the CEA Cadarache, particular nuclear data improvement needs could be identified. The $^{117}Sn(n,n')^{117}Sn^{m}$ reaction is an interesting estimator for upper part of the epithermal and for the fast neutron flux evaluations in short duration (minutes up to few days) and low to intermediate level power irradiations because of its unique characteristics combination: $E_{Threshold} = 0.314$ MeV, $T_{1/2} = 14$ d,
$E_{g1} = 158$ keV $I_{\gamma 1} = 86\%$.

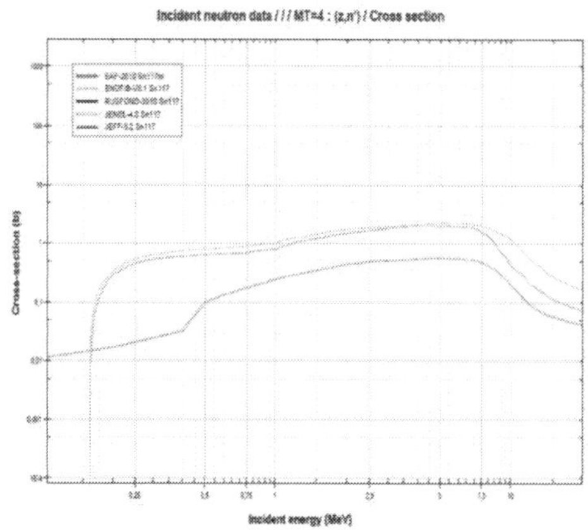

Figure 2. Comparison of the $^{117}Sn(n,n')^{117}Sn^{m}$ cross section evaluations.

Enriched Tin dosimeters (93% at. ^{117}Sn) have been tested in different spectra from Mockup Reactor (EOLE) to MTR facilities (OSIRIS). The associated measurement activity method has been upgraded at MADERE facility (CEA/Cadarache) up to reach 5% uncertainty (1σ). However, figure 2 extracted from JANIS 4.0 [10], shows that nuclear data still need upgrades to allow this reaction to be used in absolute way: discrepancies between the different library evaluations and lack of uncertainties.

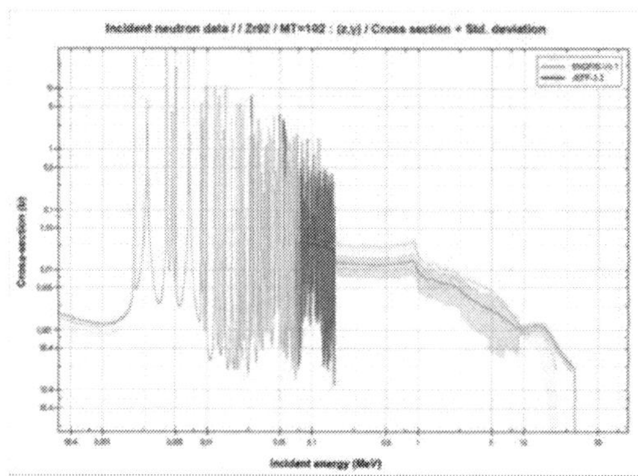

Figure 3. Comparison of the $^{92}Zr(n,\gamma)^{93}Zr$ cross section.

Continuing the investigation for reactions able to give information between 10keV and 1MeV, the $^{92}Zr(n,\gamma)$ reaction has been identified as very promising [11]. ^{92}Zr enriched Zirconium dosimeters have been irradiated under BN cover at OSIRIS giving promising results. Here again nuclear data need to be upgraded due to large discrepancies between evaluations (figure 3).

In addition, the status of the following reactions needs to be revised:

- $^{93m}Nb(n,\gamma)^{94}Nb$ cross section is unknown leading to possible errors in ^{93m}Nb activity analysis in the MTR high level thermal flux conditions. The $^{93}Nb(n,n')^{93m}Nb$ reaction is one of the most important estimator for fast neutron fluence in MTR and Reactor Vessel Survey.
- $^{103}Rh(n,n')^{103}Rh^{m}$ reaction, used in many benchmarks used for ZPR neutron calculation scheme validation, presents C/M discrepancies over 10%. Mass attenuation coefficient μ/ρ, I_x $^{103}Rh^{m}$ and cross section should to be reevaluated.
- Calculation/Measurement discrepancies (around 10%) are still observed for $^{55}Mn(n,\gamma)^{56}Mn$ reaction [12].
- $^{237}Np(n,f)$ cross section used for interpretation of the Vessel Surveillance capsules shows high level of uncertainties and evaluation discrepancies in the 1keV – 100keV range.

More generally speaking, experimental benchmarks for integral cross sections should be revisited in order to take into account the effect of the decay data evolution since the measurements were made.

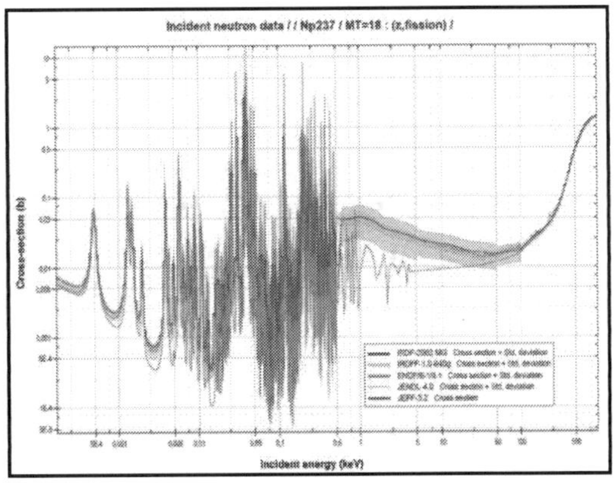

Figure 4. Comparison of the ^{237}Np(n,f) cross section.

2.1.2 Gamma measurements

Two types of dosimeters are commonly used for the photon heating measurements in the CEA mockup reactors where gamma flux level: TLD (CaF2:Mn) and OSLD (Al2O3:C). The dosimeters are made of ionic crystal containing point defects generated by the addition of impurities (Mn or C) which creates localized energy states in the band gap. If an external radiation deposits energy with the slowing down of a charged particle in the crystal lattice, an electron can go up from the valence band to the conduction band. This electron which moves in the conduction band tends to go down to the valence band and can be trapped (energy level in the band gap - Figure 4 a). The energy stored in the traps is proportional to the deposited dose in the crystal. For measurement, dosimeters are thermally (TLD) or optically (OSLD) stimulated. Trapped electrons break free from their traps and move freely in the conduction band until their recombination with a hole. The latter, is associated with a photon emission of a specific energy if the center of the hole is a luminescent center (case of a radiative recombination). The measured light intensity is detected by a photomultiplier tube in the reader and is directly proportional to the energy deposited by electrons/positrons in the crystal. Measurement Method is described in detail in [13, 14].

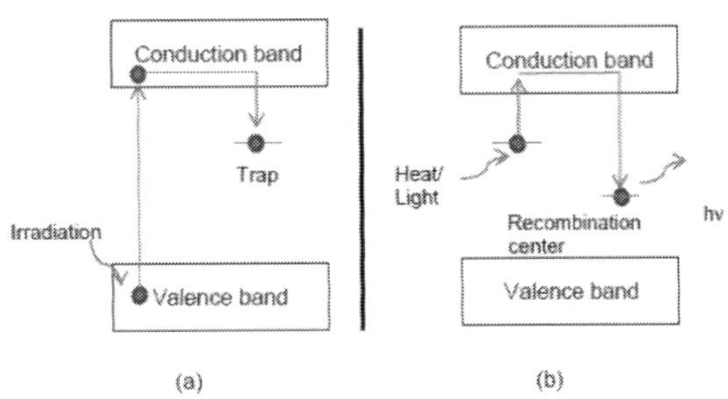

Figure 5. TLD and OSL schematic process during the irradiation (a) and the measurement step (b) after irradiation.

The measurement analysis is based on Monte Carlo TRIPOLI4© [15] calculations modelling the core exact three-dimensional geometry. The JEFF nuclear data library is used for the calculation of the neutron transport and the photon emission. The photon transport is made on the basis of the EPDL97 photo-atomic library [16]. The prompt and delayed doses deposited in dosimeters are estimated separately. The transport of 4 (neutron, photon, electron and positron) or 3 particles (photon, electron and positron) is simulated in the calculations depending whether the prompt or delayed dose is calculated. The TRIPOLI4© calculations enable to model the electromagnetic cascade shower with both electrons and positrons using EEDL (Evaluated Electron Data Library) [17] and the atomic relaxation with the EADL (Evaluated Atomic Data Library) library [17].

The evaluation of the contributions of the gamma radiation to the total and local peak power is an important stake for nuclear core physic of NPPs but also for designing irradiation devices in MTRs. Since the end of the 90s, CEA has upgraded this method for in EOLE and MINERVE facilities [18] for experimental validation of neutron/gamma calculation schemes used for reactor core modelling.

Latest experimental CEA results show a delayed dose of about 25% of the total photon energy deposition in the dosimeters. A systematic underestimation of the calculated global photon energy deposition is still remaining but has been reduced drastically from about -20% ± 13 % to -8 % ± 4.5 % (1σ) [18]. This underestimation could be partially explained by an underestimation in the JEFF3 prompt photon emission ($^{235}U(n,f)$) and for ^{27}Al neutron interactions in the sensor surrounding [18].

However, concerning measurement process itself, low level of covariance data information and lack of recent reevaluated data in EEDL, EPDL, EADL and EXDL limits uncertainty derivation. However, these files have been recently updated by D.E Cullen (August 2015) and renamed EPICS2014 [19]. Improvements brought by this version will be evaluated on existing measurements data.

2.2 Online neutron measurement techniques

2.2.1 Self Power Neutron Detector (SPND)

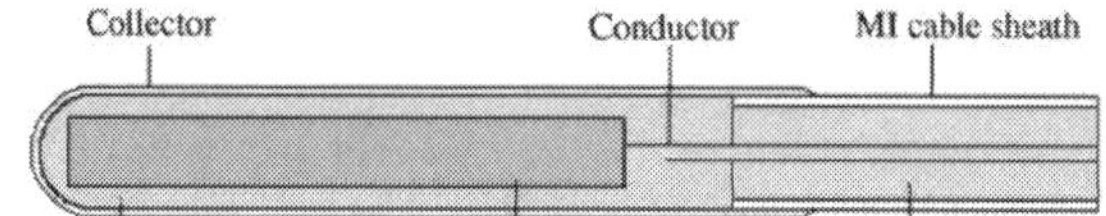

Figure 6. Schematic diagram of a SPND.

Since the 60s SPNDs are widely used in MTRs and NPPS to monitor thermal neutron flux. The SPND (Figure 6) consists of three main parts: the emitter (V, Rh, Co, Pt), the insulator (Al_2O_3 or MgO) and the collector (Inconel). Electrons are emitted when exposed to radiation then they penetrate the thin insulation around the emitter and reach the outer sheath. Some electrons are also emitted from the insulator and sheath. The net flow of electrons from the emitter gives rise to a DC signal between the emitter and sheath, which is proportional to the incident neutron flux. Rh and V SPNDs work on the basis of (n, β) reactions and are used for flux mapping, while Co and Pt SPNDs work on the basis of a (n, γ-e) prompt reactions and are used for reactor control and safety. Specific SPND with Bismuth emitter (SPGD) have also been developed to be exclusively sensible to gamma rays measurement via (γ, e) interactions. Due to different constant time in concerned reaction, SPND total time response can be generalized as follow:

$$R_{total} = R_{n(prompt)} + R_{n(delayed)} + R_{\gamma(prompt)} + R_{\gamma(rdelayed)}$$

Those reaction contributions are evaluated through the MATISSe code [20] which is a multistep Monte Carlo calculations using the JEFF3.1 neutron data library, the ENDF/B-VII.1 gamma data library and libraries associated to charged particles (EEDL,EADL). Knowledge of the shapes of the neutron and gamma spectra (figure 7) is needed for this SPND modelling. However, to give a precise absolute value of the measured thermal neutron (<5%, k=1), SPND have to be calibrated in situ with dosimetry. As the Nuclear data libraries associated to gamma and charged particles are the same as those used for TLDs, analysis performed in the previous paragraph is valid, in particular concerning covariance matrices.

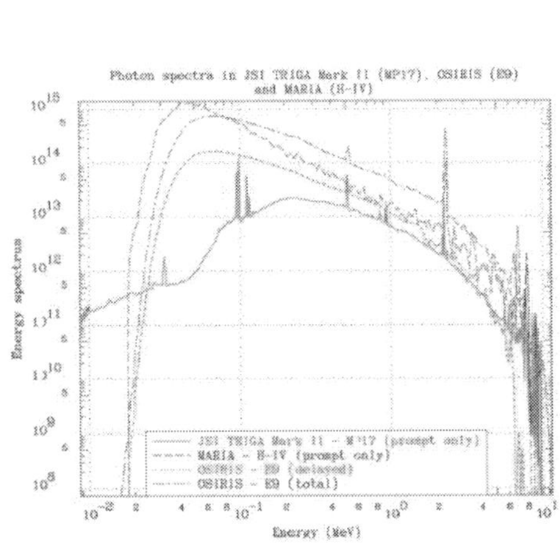

Figure 7. Typical MTR gamma spectra [20].

2.2.2 Fission chambers

Fission chambers are used for real time and online flux monitoring in experimental devices in MTRs and for spectral indexes measurements in mock-up reactor facilities. The neutron energy range of interest drives the choice of the fissile deposit (^{232}Th, ^{233}U, ^{235}U, ^{238}U, ^{238}Pu, ^{239}Pu, ^{240}Pu, ^{242}PU, ^{237}Np, ^{241}Am, ^{243}Am). One notes, that the CEA Cadarache LDCI is one of the last supplier for subminiature fission chambers with these deposits.

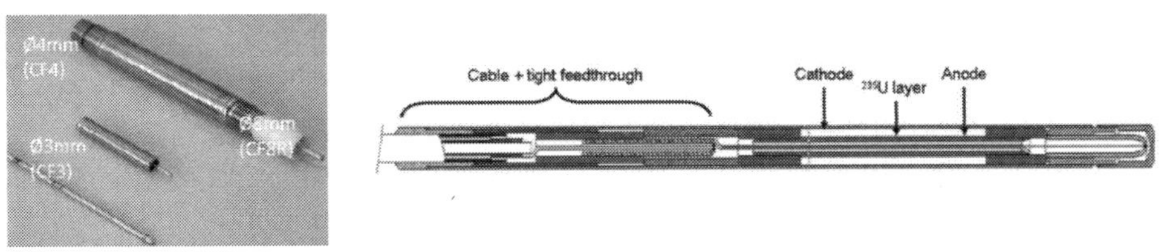

Figure 8. Fission chamber manufactured at CEA-Cadarache.

A typical fission chamber is composed of two coaxial cylindrical electrodes, one of which is covered with fissile material (Figure 8). The inter-electrode gap is filled with gas, often argon. After a neutron-induced fission in the deposit, one of the fission products is ejected into the gas, creating a large number of charge pairs. These charges are collected by a polarization voltage applied between the electrodes, leading to a current pulse. A system dedicated to fast neutron measurement, FNDS [21], has been developed at the CEA in collaboration with SCK.CEN. This device is at the moment the only system available for online fast neutron flux measurement in MTRs with an acceptable uncertainty (>10%). FNDS uses a ^{242}Pu deposit fission chamber for fast neutron flux measurement because ^{242}Pu has been shown to be the best choice for measuring the fast component in the case of a high flux with a significant thermal component [21]. Unfortunately, fission and capture cross sections of ^{242}Pu and moreover ^{243}Pu are badly known [22]. A new evaluation of these data is strongly welcomed, especially for the future use in the JHR facility.

2.2.3 Miniature Ionization Chamber

The miniature ionisation is a fission chamber without fissile coating. Recently this device has shown that the current level is sufficient to be measured and that the model used to analyse the signal has a good accordance with independent nuclear heating measurements. In particular, MCNP modelling has shown a good selectivity to the photon flux in MTR conditions with a calculated neutron contribution to the signal of the ionization chamber of less than 3% [FOURMENTEL].These innovative experimental results highlight the interest to measure the photon flux to improve the assessment of the nuclear heating in MTRs. But, nuclear data associated to gamma and charged particles need to be re-evaluated as in the case of TLD and SPND at least in terms of covariance matrices.

3 Conclusions

This paper has focused on the nuclear data aspects of the main features of the measurement techniques used in the CEA experimental reactor facilities for the on-line and off-line neutron/gamma flux characterizations. For each technique, the nuclear data necessary for their interpretation have been presented, the main needs for improvement listed and an analysis of their impact on the quality of the measurement. Beside demands for specific isotopes and reactions identified for neutron and gamma measurements, a generic one exists for having a coherent set of data with improved uncertainties. One should note here the constant effort made by the IAEA – Nuclear Data Service and by the NEA. In addition, nuclear data for charged particles improvements become necessary to allow instrumentation modelling to give results applicable for instrumentation calibration. Indeed, measurement processes developed for nuclear measurements take into account the lack of knowledge of some nuclear data into the calibration process such as for example dosimetry. Progress in nuclear data dedicated to instrumentation is propagated through more accurate measurements, leading to a better validation of the calculation schemes used for physic phenomena modelling. Finally, having a better experimentation validation, these codes will increase safety and operational margins for industrial using this instrumentation.

References

1. The JEFF-3.1.1 Nuclear Data Library; NEA Data Bank (May 2009) - JEFF Report 22
2. M.B. Chadwick et al., Nucl. data Sheets 107, 2931(2006).
3. K. Shibata et al., Journal of the Korean Physical Society, Vol. 59, No. 2, August 2011
4. C. Destouches et al., Int. Conf. on Nuclear Data for Science and Technology (2007)
5. C. Destouches et al., Physor 2006, Vancouver (2006)
6. M.-M. Be et al., "Nucleide. Table of Radionuclides" www.laraweb.free.fr
7. NuDat 2.0: Nuclear Structure and Decay Data http://www.nndc.bnl.gov/nudat2
8. AIEA-Coordinated Research Project (CRP - F41031)
9. J.-Ch. Sublet et al., EASY Documentation Series CCFE-R (10) 05
10. JANIS4.0 web nuclear data library available at http://www.oecd-nea.org/janis/
11. V. Sergeyeva, 4th Int. conf. ANIMMA 2015, 20-24 April 2015, Lisbon, Portugal
12. N. Thiollay et al., 15th Int. Symp. on Reactor Dosimetry Aix en Provence France May 2014
13. Standard practice for application of CaF2(Mn) thermos-luminescence dosimeters in mixed neutron-photon environments, ASTM International, E 2450-06
14. H. Amharrak et al., IEEE transaction on nuclear science, 59, 4, 1360-1368 (2012)
15. J.-C. Trama, ISFNT-11, Barcelona, Spain, September16-20, 2013
16. D.E. Cullen, J.H. Hubbell, L. Kissel, Lawrence Livermore National Laboratory, UCRL-LR-50400, Vol. 6, (1997)
17. S. T. Perkins and D. E. Cullen, UCRL-ID-117796 (1994)
18. C. Vaglio-Gaudard, Nuclear Science, IEEE Transac Volume:61 Issue1 Feb 2014
19. D. E. Cullen, IAEA-NDS-218, rev.1 – 2015
20. L. Barbot et al., 4th Int. conf. ANIMMA 2015, 20-24 April 2015, Lisbon, Portugal
21. B. Geslot et al., Rev. Sci. Instrum. March 2011 vol 82(3)
22. O. Cabellos et al., Nucl. Instr. and Meth. in Physics Research A 618 248–259 (2010)

EPJ Web of Conferences **111**, 02001 (2016)
DOI: 10.1051/epjconf/201611102001

Recent Developments in Nuclear Data Measurement capabilities at the Gaerttner LINAC Center at RPI

Y. Danon[1,a], A. Daskalakis[1], B. McDermott[1], N. Thompson[1], A. Youmans[1], R. Block[2], D. Barry[2], B. Epping[2], G. Leinweber[2], M. Rapp[2], and T. Donovan[2]

[1]*Gaerttner LINAC Center Rensselaer Polytechnic Institute, Troy, NY, 12180, USA*
[2]*Bechtel Marine Propulsion Corporation, Knolls Atomic Power Laboratory, P.O. Box 1072, Schenectady, NY 12301, USA*

Abstract. The Gaerttner LINAC Center at RPI uses a 60 MeV electron linear accelerator to produce short pulses of neutrons with duration of 5-5000 ns. The main research thrust at the Center is nuclear data for nuclear reactors and criticality safety applications. The Center includes several setups for time-of-flight measurements including neutron transmission, capture and scattering detectors, and a lead slowing-down spectrometer. Experiments were designed to produce neutron interaction cross sections that cover the energy range of 0.01 eV to 20 MeV. Recently added experiments include: setups for keV and fast neutron transmission, a C_6D_6 detector array for keV neutron capture measurements, and a fast neutron scattering system. Results discussed here include fast neutron scattering and angular distributions for ^{nat}Fe, iron capture measurements for incident neutrons from 1 keV to 2 MeV, fast neutron transmission through W and H_2O samples, and keV transmission through Mo isotopes.

1 Introduction

The Gaerttner LINAC Center at RPI utilizes a 60 MeV electron Linear Accelerator (LINAC) to produce short pulses of neutrons used for time-of-flight measurements of nuclear data [1]. Recent developments include measurement setups for keV and fast neutron transmission, fast neutron scattering, and keV neutron capture. Each setup is a combination of a neutron production target with a detector array and the LINAC operated at appropriate conditions. These new capabilities provide new accurate experimental data that can help improve current evaluations.

2 Experimental Setups

Table 1 gives a list of the recently added experimental capabilities discussed here. Each setup uses its own data acquisition system, and some experiments on different flight paths can be performed simultaneously and thus better utilize the LINAC operation time.

[a]e-mail: danony@rpi.edu

Table 1. Summary of recently added capabilities.

Setup	Target	Detector	Flight path
keV Transmission	Pacman [2]	Li-Glass Array [3]	100 m
Fast Transmission	Bare bounce [2]	EJ-301 array [4]	250 m
Fast Scattering	Bare Bounce [2]	EJ-301 array [4]	30 m
keV Capture	Pacman [2]	C_6D_6 array [5]	45 m

3 Results

Examples of recent experimental results utilizing the configurations shown in Table 1 are reviewed next:

3.1 Fast neutron transmission

The setup of this detector was described in detail in reference [6] where the detector was located at the 100m flight station. For the measurements described here the detector was located at a 250m flight station utilizing two detector modules. For this type of experiment, the LINAC typically operates with a pulse width of 6-8ns and repetition rate of 400 pulses/s. Recent experiments include transmission of ^{nat}Pb, ^{nat}W and H_2O. The measured transmission of ^{nat}W is shown in Figure 1 left, and provides a good example of the capabilities of the system to help resolve differences between evaluations and provide new experimental data where none existed before. Above 6 MeV the experimental data agree well with the experimental data of reference [7] and with the ENDF/B-VII.1 [8] evaluation. Below 6 MeV there is a discrepancy between the evaluations and also with the RPI experimental data.

One of the materials evaluated in the CIELO [9] project is ^{16}O. Previously, Cerjacks measured the total cross section of oxygen in two experiments; the first was published in 1968 (CI68) [10]. The second paper 1980 (CI80) [11] mentions that the experiment was normalized to the 1968 experiments at 3.5 MeV. However, the EXFOR data do not seem to include this normalization. The CI80 experiment is the highest energy resolution experiment for ^{16}O, and thus it is important to clarify its normalization. A high accuracy transmission experiment was designed at RPI to resolved this normalization issue using water samples and using the ENDF/B-VII.1 evaluation to correct for the H_2 in the sample. The experiment was designed such that at the minima of the ^{16}O cross section near 2.35 MeV, the cross section of H_2 is measured. This serves as a verification of the normalization of the current transmission experiment. Figure 1 right shows the results of this measurement. It was found that for the energy range of 3.2 MeV $< E <$ 6 MeV the ratio of the CI80 data to the RPI experiment was 0.968 ± 0.002 where the ratio of the CI68 data to the RPI experiment was 1.009 ± 0.002. At the ^{16}O minima near 2.35 MeV the ratio of the H_2O to H_2 transmission was 0.996 ± 0.003. The energy resolution of the new RPI measurement is lower than the CI80 experiment but higher than the CI68 experiment. The finding for the CI80 experiment supports the normalization used in reference [12] (reduction of 3.5%) to agree with the Johnson experiment [13].

3.2 Mid energy (keV) transmission

The Mid-energy transmission detector setup was described in detail elsewhere [14]. This detector uses an array of four Li-glass detector modules located at 100m flight path. It was designed to cover the energy range from a couple of keV to 0.5 MeV, which includes resolved and unresolved regions. The results of measurements of Mo isotopes can be found in references [14] and [15]. An example of the

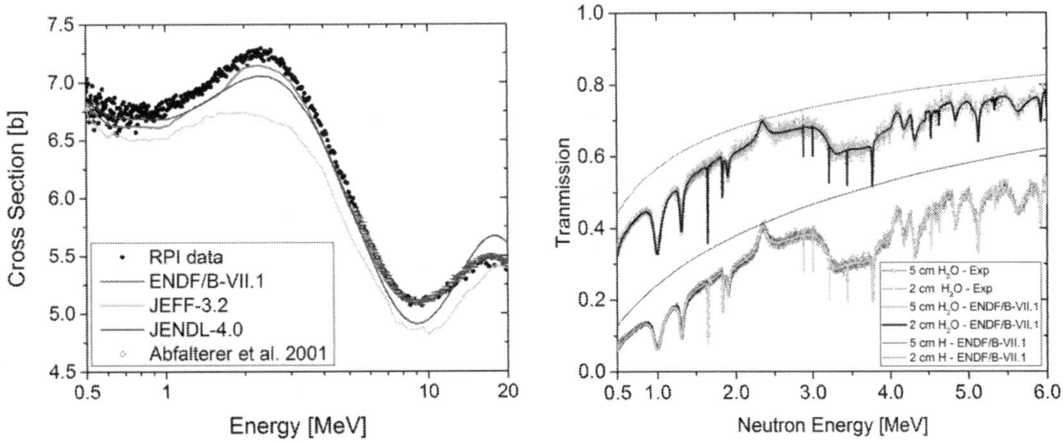

Figure 1. Left-Comparison of the total cross section of ^{nat}W with evaluations and with the data from reference [7]. Right- transmission through H_2O samples and the ENDF/B-VII.1 evaluation (constructed for H_2O). The measured transmission near 2.35 MeV matches the evaluated transmission of H_2 which is used as validation of the normalization.

transmission of Mo isotopes is shown in Figure 2 with the ENDF/B-VII.0 evaluation. The new data show resonance structure that is not in the current evaluation and can be used to revise the evaluation. The data were used for unresolved resonance analysis in reference [15].

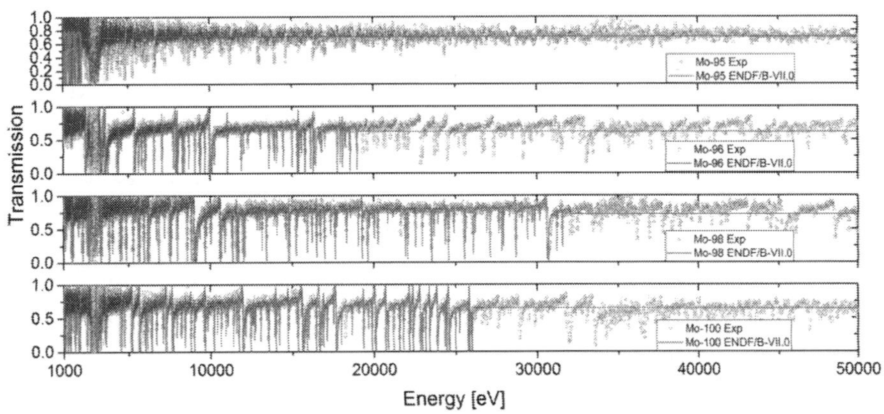

Figure 2. Measured transmission of $^{95,96,98,100}Mo$ and the ENDF/B-VII.0 evaluation.

3.3 Mid energy (keV) capture

An array of four C_6D_6 detectors was constructed at a dedicated 45m flight station. This setup was designed to extend the upper energy of capture measurements from ≈2 keV to ≈2 MeV. The C_6D_6 detector concept is well established, see for example [16], and was chosen because it provides low sensitivity to scattered neutrons. However, in order to reduce the sensitivity to variations in the capture gamma cascade spectra and to account for the detection efficiency, a weighting function is needed. To provide all the data necessary for applying a weighting function and for flexibility of the data analysis, all 4 detectors were digitized with 10 bit pulse height resolution and 0.8 ns timing. To process the data a new multithreaded data acquisition system and data processing codes were written. The first measurements with this system were for ^{nat}Fe and Ta samples. A measurement of ^{56}Fe was also recently completed. Fe has a capture cascade that is dominated by high energy gamma rays ($\approx$7 MeV) which allows discrimination of the gammas from first inelastic state. Previous experimental data are available up to 200 keV, and one set by Spencer (measured at ORELA) that was not published is available up to 600 keV. Above this energy only a few experimental points exist. An experiment with ^{nat}Fe was performed, and a dedicated weighing function was used for the analysis. The neutron flux shape was measured using a B_4C sample. The experimental results, evaluations, and other experimental data are shown in Figure 3. For comparison with other experiments both the RPI experimental capture yield and MCNP simulations were converted to cross sections using the thin sample approximation. Above 847 keV inelastic gammas were measured but can be discriminated as illustrated in Figure 3. Multiple scattering was estimated using MCNP. Above 850 keV the new data is in good agreement with the evaluations and other available experiments. Above the first inelastic state of ^{54}Fe at about 1.4 keV the new data is substantially lower than the current evaluations. A capture measurement of ^{56}Fe was performed (analysis in progress) which might help resolve this issue.

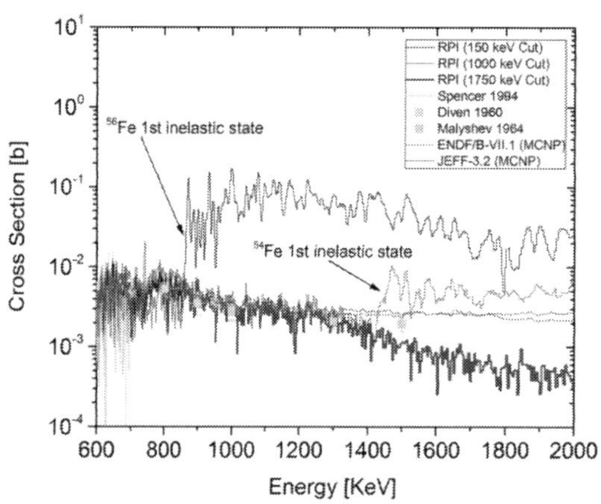

Figure 3. Neutron capture on 20 mm thick ^{nat}Fe sample. Above 600 keV the plots also illustrate the increase in count rate due to inelastic gammas that was eliminated by increasing the pulse integral discriminator settings.

3.4 Fast neutron scattering

The neutron scattering detector array was previously described in detail [4]. Recently the data analysis codes and methodology were improved [17] . This includes better pulse shape classification for gamma rejection and implementation of a gamma rejection technique for high gamma rates. These developments enable neutron scattering measurements in the presence of high gamma background such as for ^{238}U and ^{235}U where fission is occurring and producing multiple gammas. Although this system is referred to as a neutron scattering system, it measures all the neutrons scattered or emitted from the sample. The experimental data were then compared to MCNP simulations that included detailed experimental parameters such as geometry, incident neutron flux, and detector efficiencies. This method provides a benchmark to compare how different nuclear data evaluations perform. For the case of Fe a method was developed to separate elastic from inelastic scattering. This is possible because of the relatively high energy levels for inelastic scattering which results in a large difference between elastically and inelastically scattered neutron. The scattered neutron energy can thus be inferred from the detected pulse shape. As an example the results for a scattering angle of 153 deg are shown in Figure 4. A comparison of the experiment and evaluations in Figure 4 indicates that for some time-of-flight regions the JEFF-3.2 [18] and JENDL-4.4 [19] evaluations are closer to the experiment compared to ENDF/B-VII.1. The ratio of inelastic to elastic scattering is in good agreement with the evaluations. For some of the points ENDF/B-VII.1 seems slightly higher than the other evaluations and the experimental data.

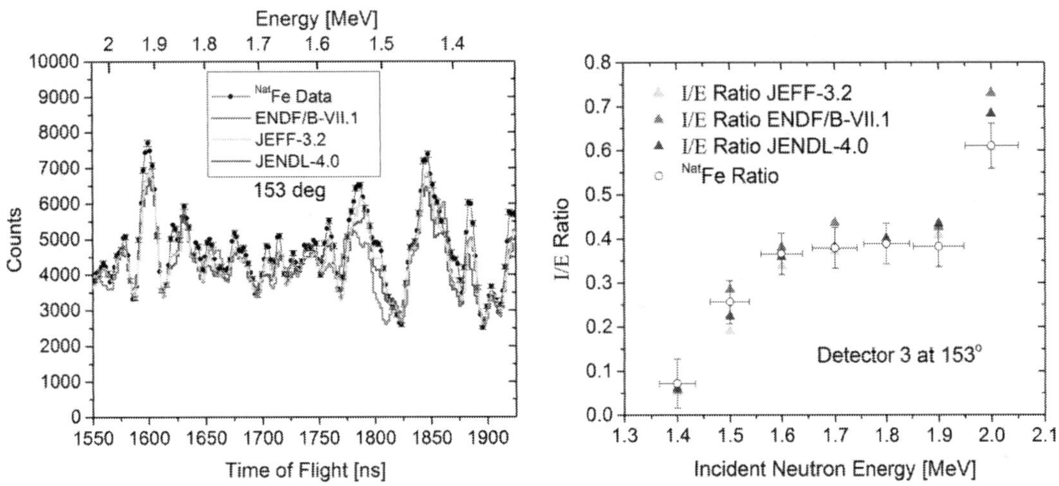

Figure 4. Neutron scattering from ^{nat}Fe sample. The left plot shows the data in TOF (energy scale is on the top x-axis) compared with MCNP simulations using different evaluations. The right plot shows the ratio of inelastic to elastic scattering.

4 Summary and Conclusions

The Gaerttner LINAC Center at RPI added new capabilities, and some of them are presented here. Fast neutron transmission of W shows large deviations between the evaluations and the RPI experiment near 2.5 MeV. Fast neutron transmission through H_2O provides new high resolution data for ^{16}O below 3.2 MeV where the CI80 starts. The measurement also established that the CI80 [11] is about 3.2% lower than the CI68 [10]. The current experiment uses the unique minima in the ^{16}O cross section as an anchor for the normalization of the data. A mid energy Li-Glass detector array enabled new transmission measurements on several Mo isotopes. The data showed resolved and unresolved resonance structure that can help improve the evaluations. New capture measurements of Fe in the keV and MeV incident neutron energy regions were enabled by a new C_6D_6 detector array.

References

[1] Y. Danon, R. Block, R. Bahran, M. Rapp, F. Saglime, C. Romano, J. Thompson, D. Barry, N. Drindak, J. Hoole et al., Proceedings of The Eighth International Topical Meeting on Nuclear Applications and Utilization of Accelerators(AccApp 07) (2007)

[2] M. Overberg, B. Moretti, R. Slovacek, R. Block, Nuclear Instruments and Methods in Physics Research Section A: Accelerators, Spectrometers, Detectors and Associated Equipment **438**, 253 (1999)

[3] R. Bahran, Y. Danon, M. Rapp, D. Williams, D. Barry, G. Leinweber, R. Block, J. Hoole, Proceedings of the Tenth International Topical Meeting on Nuclear Applications of Accelerators (AccApp 2011) (2011)

[4] F. Saglime, Y. Danon, R. Block, M. Rapp, R. Bahran, G. Leinweber, D. Barry, N. Drindak, Nuclear Instruments and Methods in Physics Research Section A: Accelerators, Spectrometers, Detectors and Associated Equipment **620**, 401 (2010)

[5] Y. Danon, L. Liu, E. Blain, A. Daskalakis, B. McDermott, K. Ramic, C. Wendorff, D. Barry, R. Block, B. Epping et al., Transactions of the American Nuclear Society, Vol. 109, p. 897-900, Washington, D.C (2013)

[6] M.J. Rapp, Y. Danon, F.J. Saglime, R.M. Bahran, D.G. Williams, G. Leinweber, D.P. Barry, R.C. Block, Nuclear Science and Engineering **172**, 268 (2012)

[7] W.P. Abfalterer, F.B. Bateman, F.S. Dietrich, R.W. Finlay, R.C. Haight, G.L. Morgan, Phys. Rev. C **63**, 044608 (2001)

[8] M. Chadwick, M. Herman, P. Obložinský, M. Dunn, Y. Danon, A. Kahler, D. Smith, B. Pritychenko, G. Arbanas, R. Arcilla et al., Nuclear Data Sheets **112**, 2887 (2011), special Issue on ENDF/B-VII.1 Library

[9] M. Chadwick, E. Dupont, E. Bauge, A. Blokhin, O. Bouland, D. Brown, R. Capote, A. Carlson, Y. Danon, C.D.S. Jean et al., Nuclear Data Sheets **118**, 1 (2014)

[10] S. Cierjacks, P. Forti, D. Kopsch, L. Kropp, J. Nebe, H. Unseld, *High Resolution Total Neutron Cross Sections Between 0.5 TO 30 MeV.* (1968)

[11] S. Cierjacks, F. Hinterberger, G. Schmalz, D. Erbe, P. Rossen, B. Leugers, Nuclear Instruments and Methods **169**, 185 (1980)

[12] R. Sayer, L. Leal, N. Larson, R. Spencer, R. Wright, Journal of Nuclear Science and Technology **39**, 88 (2002)

[13] C.H. Johnson, J.L. Fowler, L.A. Galloway, N.W. Hill, ORNL-4937 (1973)

[14] R. Bahran, PhD thesis, Rensselaer Polytechnic Institute (2013)

[15] R. Bahran, D. Barry, R. Block, G. Leinweber, M. Rapp, A. Daskalakis, E. Blain, D. Williams, B. McDermott, L. Leal et al., Phys. Rev. C **92**, 024601 (2015)

[16] A. Borella, G. Aerts, F. Gunsing, M. Moxon, P. Schillebeeckx, R. Wynants, Nuclear Instruments and Methods in Physics Research Section A: Accelerators, Spectrometers, Detectors and Associated Equipment **577**, 626 (2007)

[17] A. Daskalakis, R. Bahran, E. Blain, B. McDermott, S. Piela, Y. Danon, D. Barry, G. Leinweber, R. Block, M. Rapp et al., Annals of Nuclear Energy **73**, 455 (2014)

[18] JEFF-3.2 NEA Data Bank (2012)

[19] K. Shibata, O. Iwamoto, T. Nakagawa, N. Iwamoto, A. Ichihara, S. Kunieda, S. Chiba, K. Furutaka, N. Otuka, T. Ohasawa et al., Journal of Nuclear Science and Technology **48**, 1 (2011)

EPJ Web of Conferences **111**, 02002 (2016)
DOI: 10.1051/epjconf/201611102002

Towards the high-accuracy determination of the ^{238}U fission cross section at the threshold region at CERN – n_TOF

M.Diakaki[1,a], L.Audouin[2], E.Berthoumieux[1], M.Calviani[3], N.Colonna[4], E.Dupont[1], I.Duran[5], F.Gunsing[1,3], E.Leal-Cidoncha[5], C.Le Naour[2], L.S.Leong[2,6], M.Mastromarco[4], C.Paradela[5,7], D.Tarrio[5,8], L.Tassan-Got[2], G.Aerts[1], S.Altstadt[9], H.Alvarez[5], F.Alvarez-Velarde[10], S.Andriamonje[1], J.Andrzejewski[11], G.Badurek[12], M.Barbagallo[4], P.Baumann[13], V.Becares[10], F.Becvar[14], F.Belloni[7], B.Berthier[2], J.Billowes[15], V.Boccone[3], D.Bosnar[16], M.Brugger[3], F.Calvino[17], D.Cano-Ott[10], R.Capote[18], C.Carrapiço[19], P.Cennini[3], F.Cerutti[3], E.Chiaveri[3], M.Chin[3], G.Cortes[20], M.A.Cortes-Giraldo[21], L.Cosentino[22], A.Couture[23], J.Cox[23], S.David[2], I.Dillmann[24], C.Domingo-Pardo[25], R.Dressler[26], W.Dridi[1], C.Eleftheriadis[27], M.Embid-Segura[10], L.Ferrant[2], A.Ferrari[3], P.Finocchiaro[22], K.Fraval[1], K.Fujii[28], W.Furman[29], S.Ganesan[30], A.R.Garcia[10], G.Giubrone[25], M.B.Gomez-Hornillos[20], I.F.Goncalves[19], E.Gonzalez-Romero[10], A.Goverdovski[31], F.Gramegna[32], E.Griesmayer[12], C.Guerrero[3], P.Gurusamy[30], R.Haight[33], M.Heil[24], S.Heinitz[26], M.Igashira[34], S.Isaev[2], D.G.Jenkins[35], E.Jericha[12], Y.Kadi[3], F.Kaeppeler[24], D.Karadimos[36], D.Karamanis[36], M.Kerveno[13], V.Ketlerov[31], N.Kivel[26], M.Kokkoris[37], V.Konovalov[31], M.Krticka[14], J.Kroll[14], C.Lampoudis[27], C.Langer[9], C.Lederer[9], H.Leeb[12], S.Lo Meo[38], R.Losito[3], M.Lozano[21], A.Manousos[27], J.Marganiec[11], T.Martinez[10], S.Marrone[4], C.Massimi[39], P.Mastinu[32], E.Mendoza[10], A.Mengoni[38], P.M.Milazzo[28], F.Mingrone[39], M.Mirea[40], W.Mondelaers[7], C.Moreau[28], M.Mosconi[24], A.Musumarra[41], S.O'Brien[23], J.Pancin[1], N.Patronis[36], A.Pavlik[42], P.Pavlopoulos[3], J.Perkowski[11], L.Perrot[1], M.T.Pigni[12], R.Plag[24], A.Plompen[7], L.Plukis[1], A.Poch[17], C.Pretel[17], J.Praena[21], J.Quesada[21], T.Rauscher[43,44], R.Reifarth[9], A.Riego[20], F.Roman[3], G.Rudolf[13], C.Rubbia[3], P.Rullhusen[7], J.Salgado[19], C.Santos[19], L.Sarchiapone[3], R.Sarmento[19], A.Saxena[30], P.Schillebeeckx[7], S.Schmidt[9], D.Schumann[26], C.Stephan[2], G.Tagliente[4], J.L.Tain[25], L.Tavora[19], R.Terlizzi[4], A.Tsinganis[3], S.Valenta[14], G.Vannini[39], V.Variale[4], P.Vaz[19], A.Ventura[39], R.Versaci[3], M.J.Vermeulen[35], D.Villamarin[10], M.C.Vincente[10], V.Vlachoudis[3], R.Vlastou[37], F.Voss[24], A.Wallner[45], S.Walter[24], T.Ware[15], M.Weigand[9], C.Weiß[3], M.Wiesher[23], K.Wisshak[24], T.Wright[15], P.Zugec[16]

[1]CEA, Saclay, Irfu/SPhN, Gif-sur-Yvette, France ; [2]CNRS/IN2P3 – IPN, Orsay, France ; [3]CERN, Geneva, Switzerland ; [4]INFN, Bari, Italy ; [5]Universidade de Santiago de Compostela, Spain ; [6]JAEA, Japan ; [7]EC JRC, IRMM, Geel, Belgium ; [8]Department of Physics and Astronomy, Uppsala University, Sweden ; [9]Johann-Wolfgang-Goethe Universität, Frankfurt, Germany ; [10]CIEMAT, Madrid, Spain ; [11]University of Lodz, Poland ; [12]Atominstitut der Österreichischen Universitäten, Technische Universität Wien, Austria ; [13]CNRS/IN2P3 – IPHC, Strasbourg, France ; [14]Charles University, Prague, Czech Republic ; [15]University of Manchester, UK ; [16]Department of Physics, Faculty of Science, University of Zagreb, Croatia ; [17]Universidad Politécnica de Madrid, Spain ; [18]IAEA Nuclear Data Section, Vienna, Austria ; [19]Instituto Superior Técnico/CTN, Universidade de Lisboa, Portugal ; [20]Universitat Politécnica de Catalunya, Barcelona, Spain ; [21]Universidad de Sevilla, Spain ; [22]INFN - Laboratori Nazionali del Sud, Catania, Italy ; [23]University of Notre Dame, Indiana, USA ; [24]KIT, Karlsruhe, Germany ; [25]Instituto de Física Corpuscular, CSIC-Universidad de Valencia, Spain ; [26]PSI, Villigen, Switzerland ; [27]Aristotle University of Thessaloniki, Greece ; [28]INFN, Trieste, Italy ; [29]JINR, Dubna, Russia ; [30]BARC, Mumbai, India ; [31]IPPE, Obninsk, Russia ; [32]INFN - Laboratori Nazionali di Legnaro, Italy ; [33]LANL, New Mexico, USA ; [34]Tokyo Institute of Technology, Japan ; [35]University of York, United Kingdom ; [36]University of Ioannina, Greece ; [37]National Technical University of Athens, Greece ; [38]ENEA, Bologna, Italy ; [39]Dipartimento di Fisica, Università di Bologna, and Sezione INFN di Bologna, Italy ; [40]Horia Hulubei National Institute of Physics and Nuclear Engineering, Bucharest - Magurele, Romania ; [41]Dipartimento di Fisica e Astronomia, Università di Catania and INFN-Laboratori Nazionali del Sud, Catania, Italy ; [42]University of Vienna, Faculty of Physics, Austria ; [43]Centre for Astrophysics Research, School of Physics, Astronomy, Mathematics, University of Hertfordshire, UK ; [44]Department of Physics, University of Basel, Switzerland ; [45]Research School of Physics and Engineering, Australian National University, Australia

[a] Corresponding author: maria.diakaki@cea.fr

Abstract. The ^{238}U fission cross section is an international standard beyond 2 MeV where the fission plateau starts. However, due to its importance in fission reactors, this cross-section should be very accurately known also in the threshold region below 2 MeV. The ^{238}U fission cross section has been measured relative to the ^{235}U fission cross section at CERN – n_TOF with different detection systems. These datasets have been collected and suitably combined to increase the counting statistics in the threshold region from about 300 keV up to 3 MeV. The results are compared with other experimental data, evaluated libraries, and the IAEA standards.

1 Introduction

The fission cross sections of ^{235}U and ^{238}U are of fundamental importance in the field of nuclear technology, as well as for other fields of basic and applied nuclear physics. In particular, fission cross section in the hundreds-of-keV region are of paramount importance for the development of innovative fast reactors. The ^{235}U(n,f) cross section is an international standard at 0.0253 eV and above 0.15 MeV, and ^{238}U(n,f) is a standard beyond 2 MeV [1]. While the ^{235}U(n,f) standard is commonly used for neutron flux measurements from thermal to high energy, the ^{238}U(n,f) threshold cross section can be more conveniently used in the presence of a low energy neutron background.

To address the need of new and accurate data for future improvements of these standards a series of measurements of the ^{238}U(n,f)/^{235}U(n,f) cross section ratio were performed at the CERN n_TOF facility up to 1 GeV. Some of these measurements are described in detail in Ref. [2]. In the present contribution, the results of one more dataset are included, and the possibility to obtain high accuracy ^{238}U(n,f) cross section results at the threshold (0.3 MeV - 3 MeV) is examined.

2 Experimental setups

The measurements were performed at the CERN n_TOF facility [3–7], which is based on the spallation of 20 GeV/c protons on a lead target. Data were collected in different campaigns from the experimental area 1 (EAR1) approximately 190 m downstream of the neutron source. Two detection setups were used: a fast fission ionization chamber (FIC), in which a single fission fragment (FF) is detected, and an array of parallel plate avalanche counters (PPAC), in which the two FFs are detected in coincidence.

2.1 The Parallel Plate Avalanche Counters

The PPAC are gaseous detectors with very thin windows, small gaps between electrodes and low gas pressure [8-11]. These features make the detector practically insensitive to the prompt γ-flash. The actinide samples were deposited on very thin backings and positioned between two PPACs for measuring FFs in coincidence. The main advantage of the coincidence technique is the very high efficiency for rejecting the α-particle background from the sample, as well as for discriminating fission against competing reactions. Another advantage of PPACs is that they can also measure the angular distribution of FFs. The main drawback of the system is the limited angular acceptance and the angular dependence of the efficiency.

In the first experimental campaign the PPAC detectors were mounted perpendicular to the neutron beam direction, hereafter referred to as "PPAC perpendicular". In this configuration the setup is affected by a loss of efficiency for FFs emitted at angles larger than ~60°. To overcome this problem, a new geometrical configuration was adopted in a second campaign, with the detectors and the samples tilted by 45°. Two measurements were performed with the tilted setup, hereafter referred to as "PPAC tilted 1" and "PPAC tilted 2". Details on the analysis of the data in each configuration can be found in [2,8-14]. The PPAC samples were prepared at IPN (Orsay, France), and most of the targets used were characterised by means of alpha spectroscopy and Rutherford Backscattering Spectrometry (RBS).

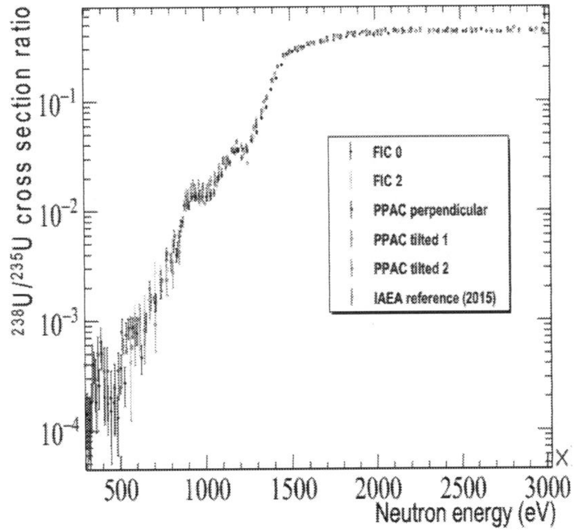

Figure 1. The $^{238}U/^{235}U$ fission cross-section ratio measured at n_TOF with the different detection systems, in the neutron energy range 0.3-3MeV. The statistical uncertainties are plotted. The ratio calculated based on the IAEA references (2015) is shown for comparison.

2.2 The Fast Ionization Chamber

The fast ionization chamber [15-16] is made of a stack of ionization cells consisting of two external electrodes and a central electrode plated on both sides with the fissile deposit. The FIC0 and FIC1 versions of the detector were specifically built as "sealed sources" for measurements of highly radioactive samples. The FIC2 was used as a neutron flux monitor, thus it was much lighter and directly coupled to the vacuum tube. The samples were prepared using the painting technique at the IPPE (Obninsk, Russia) and the JINR (Dubna, Russia), and most of the targets used were characterised by means of alpha spectroscopy and the RBS technique.

For the extraction of the $^{238}U(n,f)/^{235}U(n,f)$ cross section ratio, one ^{238}U and two ^{235}U samples were used from the FIC0 measurements, two ^{235}U and four ^{238}U samples from FIC1, and one ^{238}U and two ^{235}U samples from FIC2. Different analysis procedures and corrections were applied for the data from the different detectors as described in [2,15-18]. The data collected with FIC2 were normalised to the results obtained with the FIC1 chamber in the 1–10 MeV energy region, due to the large uncertainty of the ^{238}U sample mass.

3 Results for the $^{238}U(n,f)/^{235}U(n,f)$ cross section ratio measurement

The systematic uncertainties from the different datasets are summarised in Table 1 (assuming a uniform distribution). Fig. 1 shows the results of the different measurements at n_TOF of the $^{238}U/^{235}U$ fission cross-section ratio for neutron energies in the range 0.3–3 MeV. A good agreement among the various datasets was observed, within their uncertainties, especially at the first chance fission plateau. Very few differences bigger than 1σ are noticed at the threshold at ~1.2 MeV where the systematic uncertainty due to the FF emission anisotropy becomes larger.

Each of the five datasets collected at n_TOF represents a new result by itself, and should be considered independently from each other, for re-evaluating the fission cross section ratio. Nevertheless, in order to compare the n_TOF data with previous measurements and current evaluations, the n_TOF $^{238}U/^{235}U$ fission cross section ratio has been calculated as the weighted average value of all the datasets, taking into account the uncorrelated statistical uncertainties. Firstly, all the datasets were normalised to the ratio calculated with the IAEA standards (2006) in the energy range 2.2-2.6 MeV, and the normalisation factor did not exceed 3%. An energy shift was applied to all the datasets in order for them to match in the energy region of 1.34-1.48 MeV where the slope of the ^{238}U(n,f) cross section is largest. This energy-matching shift was less than 0.7%. A binning of 50 bins/decade was chosen for E>800 keV and 25 bins/decade for E<800keV, and a linear interpolation was applied between consecutive points in order to match this binning.

Table 1. Systematic uncertainties (in %) on the data collected in the measurements of the ^{238}U /^{235}U fission cross-section ratio. Since the same samples were used in the "perpendicular" and "tilted 1" PPAC measurements, the corresponding uncertainties on the sample mass are fully correlated. For the last dataset, normalized to ENDF/B-VII.1 (see [2]), the uncertainty in the mass is replaced by the one in the evaluated cross sections.

Setup	Samples	Efficiency	Dead-time
FIC 0	2	1-2	<0.5
FIC 1, 2	1.5	1	<3
PPAC perpendicular	1.1	3	< 1
PPAC tilted 1	1.1	2	< 1
PPAC tilted 2	(~1)	1	< 1

The extracted n_TOF ^{238}U/^{235}U fission cross section ratio is shown in Fig. 2, along with experimental data from the EXFOR database [19] and major evaluations, as well as the associated residuals divided by 1σ (statistical uncertainty). The systematic uncertainty of the weighted average ratio, taking into account the normalisation to the IAEA standards, can be calculated from the energy-dependent uncertainties given in Table 1 and is roughly estimated to be less than 1.5% (1σ).

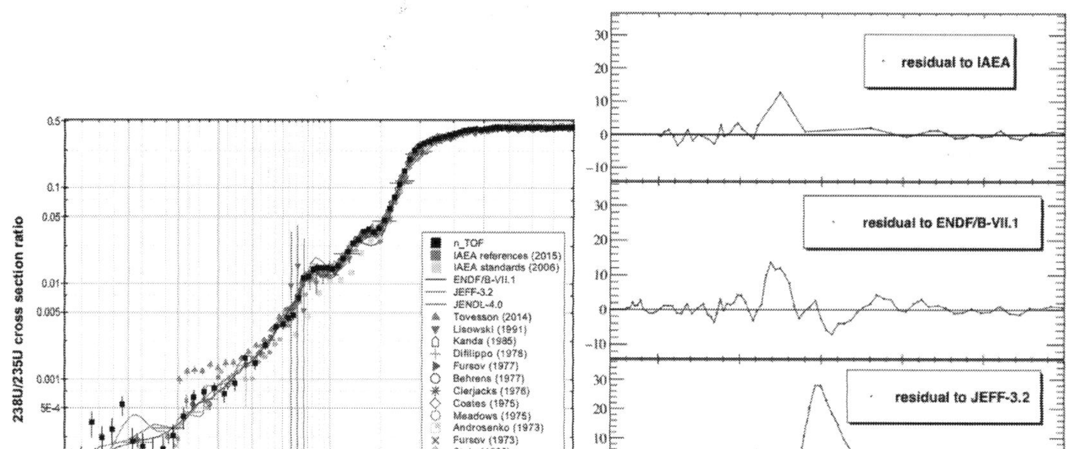

Figure 2. a) The ^{238}U/^{235}U(n,f) cross section ratio obtained from the weighted average of the n_TOF datasets compared with the IAEA standard (2006) and reference (2015) and major evaluations, as well as data available in EXFOR. b) The residual of the n_TOF dataset with selected references and evaluations, divided by 1σ (statistical) of the n_TOF dataset.

It can be noted that at the threshold region the evaluations present differences >5σ from the n_TOF data, and that below ~750keV the latest data of Tovesson (2014) are systematically above the n_TOF data.

4 Determination of the $^{238}U(n,f)$ cross section

Based on the merged n_TOF dataset and the IAEA (2006) standard $^{235}U(n,f)$ cross section, the $^{238}U(n,f)$ cross section was extracted at the threshold region. It is shown in Fig.3, along with selected previous data available in the EXFOR database.

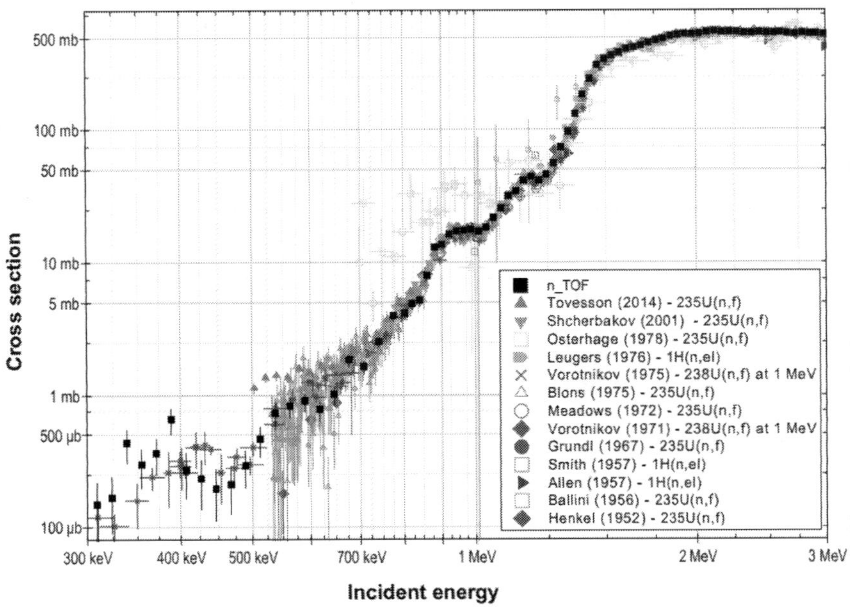

Figure 3. The $^{238}U(n,f)$ reaction cross section in the energy range 0.3-3MeV measured with reference to the $^{235}U(n,f)$ cross section at n_TOF (black squares), along with experimental data available in the EXFOR database. Next to the label the reference reaction used is noted.

The n_TOF data extracted with $^{235}U(n,f)$ as a reference agree within 1σ with most other datasets extracted with the same reference. as well as with the data from Allen (1957) using $^{1}H(n,el)$ as reference. Only a few other datasets exist at the threshold region with reference to the $^{1}H(n,el)$, (such as Smith (1957) and Leugers (1976)) but for E<1.5 MeV they present differences bigger than 1σ with respect to all available data using the $^{235}U(n,f)$ as reference.

Despite the statistical uncertainties, interesting shoulders can be noticed in the threshold region at ~0.4, ~0.6, ~0.8, ~1, and ~1.2MeV, most of them not properly described in the latest evaluations and worth being studied in the context of the fission barrier parameters related to different fission modes, resonant sub-threshold structure, etc. [20] as well as to the competition with the inelastic neutron channel.

5 Outlook

The $^{238}U/^{235}U$ fission cross section ratio has been measured at n_TOF with two different detection systems, one of which was used in two different geometrical configurations. The results of the measurements are consistent within their combined uncertainties. The datasets have been combined in order to obtain a unique dataset and the $^{238}U(n,f)$ cross section was derived. The extracted ratio could be used to improve the accuracy of current libraries, and in particular of the IAEA standard and of reference datasets used in a variety of applications. New measurements of the $^{238}U/^{235}U$ fission cross section ratio are foreseen using the new vertical neutron beam line of the n_TOF facility [21] with much higher statistics in the region of the threshold, especially for neutron energies below 0.5 MeV.

References

1. A. Carlson et al., Nuclear Data Sheets **110**, 3215 (2009).
2. C. Paradela et al., Phys. Rev. C **91**, 024602 (2015).
3. http://www.cern.ch/ntof.
4. U. Abbondanno et al., CERN/INTC-O-011 INTC-2002-037, CERN-SL-2002-053 ECT (2003).
5. C. Guerrero et al., Eur. Phys. J. A **49**, 27 (2013).
6. M. Barbagallo et al., Eur. Phys. J. A **49**, 156 (2013).
7. N. Colonna et al., Energy Environ. Sci. **3**, 1910 (2010).
8. C. Paradela et al., Phys. Rev. C **82**, 034601 (2010).
9. D. Tarrio et al., Nucl. Instr. Meth. A **743**, 79 (2014).
10. C. Paradela, I. Duran, D. Tarrio, L. Audouin, L. Tassan-Got, and C. Stephan, *Proceedings of the 2nd International Conference on Advancements in Nuclear Instrumentation Measurement Methods and their Applications (ANIMMA)*, June, 2011, Ghent, Belgium, pp. 6–9 (IEEE, 2011).
11. D. Tarrio et al., Nucl. Data Sheets **119**, 35 (2014).
12. D. Tarrio et al., Phys. Rev. C **83**, 044620 (2011).
13. D. Tarrio, PhD thesis, Universidade de Santiago de Compostela, Spain (2012).
14. L.S. Leong, PhD thesis, University of Paris-Sud, France (2013).
15. M. Calviani et al., Nucl. Instr. Meth. A **594**, 220 (2008).
16. M. Calviani et al., Phys. Rev. C **80**, 044604 (2009).
17. D. Karadimos et al., Nucl. Instr. Meth. B **268**, 2556 (2010).
18. D. Karadimos et al., Phys. Rev. C **89**, 044606 (2014).
19. N. Otuka et al, Nuclear Data Sheets **120**, 272 (2014).
20. E. Birgesson et al., Nucl. Phys. A **817**, 1 (2009).
21. C. Weiss et al., Nucl. Instr. Meth. A **799**, 90 (2015).

EPJ Web of Conferences **111**, 02003 (2016)
DOI: 10.1051/epjconf/201611102003

High accuracy ^{235}U(n,f) data in the resonance energy region

C. Paradela[1,2,a], I. Duran[2], L.Tassan-Got[3], L.Audouin[3], B. Berthier[3], S. Isaev[3], C. Le Naour[3], C. Stephan[3], D. Tarrío[2,17], U. Abbondanno[4], G. Aerts[5], H. Álvarez-Pol[2], F. Álvarez-Velarde[6], S. Andriamonje[5], J. Andrzejewski[7], G. Badurek[8], P. Baumann[9], F. Becvar[10], E. Berthoumieux[5], F. Calviño[11], M. Calviani[12], D. Cano-Ott[6], R. Capote[13], C. Carrapiço[14], P. Cennini[12], V. Chepel[15], E. Chiaveri[12], N. Colonna[16], G. Cortes[11], A. Couture[17], J. Cox[17], M. Dahlfors[12], S. David[3], I. Dillmann[18], C. Domingo-Pardo[19], W. Dridi[5], C. Eleftheriadis[20], M. Embid-Segura[10], L. Ferrant[3], A. Ferrari[12], R. Ferreira-Marques[15], K. Fujii[21], W. Furman[22], I.F. Gonçalves[14], E. Gonzalez-Romero[6], A. Goverdovski[23], F. Gramegna[24], C. Guerrero[12.13], F.Gunsing[5], R.Haight[25], M. Heil[18], M. Igashira[26], E. Jericha[8], Y. Kadi[12], F. Kaeppeler[18], D. Karadimos[27], M. Kerveno[9], V. Ketlerov[23], P.Koehler[25], V.Konovalov[22], M. Krticka[10], C.Lampoudis[20], C. Lederer[8], H. Leeb[8], A. Lindote[15], S. Lukic[9], J. Marganiec[7], T. Martinez[6], S. Marrone[16], C. Massimi[28], P. Mastinu[24], A. Mengoni[12,2], P.M. Milazzo[21], C. Moreau[21], M. Mosconi[18], S, J. Pancin[5], A. Pavlik[42], P. Pavlopoulos[28], L. Perrot[5], R. Plag[18], A. Plompen[1], A. Plukis[5], A. Poch[11], C. Pretel[11], J. Praena[24], J. Quesada[13], T. Rauscher[30], R. Reifarth[25], C. Rubbia[12], G. Rudolf[9], P. Rullhusen[1], J. Salgado[14], C. Santos[14], L. Sarchiapone[12], I. Savvidis[20], G. Tagliente[4], J.L. Tain[19], L. Tavora[14], R. Terlizzi[4], P. Vaz[14], A. Ventura[29], D. Villamarin[6], M.C.Vincente[10], V. Vlachoudis[12], R. Vlastou[27], F. Voss[18], S. Walter[18], C. Weiss[12], M. Wiesher[17], and K. Wisshak[18] (n_TOF Collaboration)

[1]*EC-JRC-IRMM, Retieseweg 111, 2440 Geel, Belgium*
[2]*Universidad de Santiago de Compostela, Santiago de Compostela, Spain*
[3]*CNRS/IN2P3 – IPN, Orsay, France*
[4]*INFN, Bari, Italy*
[5]*CEA, Saclay, Irfu/SPhN, Gif-sur-Yvette, France*
[6]*CIEMAT, Madrid, Spain*
[7]*University of Lodz,Lodz, Poland*
[8]*Atominstitut der Österreichischen Universitäten, Technische Universität Wien, Austria*
[9]*CNRS/IN2P3 – IPHC, Strasbourg, France*
[10]*Charles University, Prague, Czech Republic*
[11]*Universitat Politécnica de Catalunya, Barcelona, Spain*
[12]*CERN, Geneva, Switzerland*
[13]*Universidad de Sevilla, Sevilla, Spain*
[14]*Instituto Superior Técnico/CTN, Universidade de Lisboa, Portugal*
[15]*LIP-Coimbra,Coimbra, Portugal*
[16]*Instituto Nazionale di Fisica Nucleare, Bari, Italy*
[17]*Department of Physics and Astronomy, Uppsala University, Sweden*
[18]*Karlsruhe Institute of Technology, Institut für Kernphysik, Campus North, Karlsruhe, Germany*
[19]*Instituto de Física Corpuscular, CSIC-Universidad de Valencia, Spain*
[20]*Aristotle University of Thessaloniki, Greece*
[21]*Instituto Nazionale di Fisica Nucleare, Trieste, Italy*
[22]*Joint Institut for Nuclear Research, Frank Laboratory of Neutron Physics, Dubna, Russia*
[23]*IPPE, Obninsk, Russia*
[24]*Instituto Nazionale di Fisica Nucleare, Laboratori Nazionali di Legnaro, Italy*
[25]*Los Alamos National Laboratory, New Mexico, USA*
[26]*Tokyo Institute of Technology, Tokyo, Japan*
[27]*National Technical University of Athens (NTUA), Greece*
[28]*Dipartimento di Fisica, Università di Bologna, Italy*
[29]*ENEA, Bologna, Italy*
[30]*Department of Physics and Astronomy, University of Basel, Switzerland*

[a] Corresponding author: carlos.PARADELA-DOBARRO@ec.europa.eu

Abstract. The ^{235}U neutron-induced cross section is widely used as reference cross section for measuring other fission cross sections, but in the resonance region it is not considered as an IAEA standard because of the scarce experimental data covering the full region. In this work, we deal with a new analysis of the experimental data obtained with a detection setup based on parallel plate ionization chambers (PPACs) at the CERN n_TOF facility in the range from 1 eV to 10 keV. The relative cross section has been normalised to the IAEA value in the region between 7.8 and 11 eV, which is claimed as well-known. Comparison with the ENDF/B-VII evaluation and the IAEA reference file from 100 eV to 10 keV are provided.

1 Introduction

The ^{235}U neutron-induced fission cross-section is of greatest importance for reactor physics, and has been extensively studied since the earliest nuclear research. Hence, it has become one of the main references for neutron-induced measurements, being used in a large number of applications involving neutron dosimetry or monitoring. Nevertheless, it is only considered as a standard at the thermal point and in the fast energy range (150 keV-30 MeV), leaving certain room to improve its knowledge in the resonance region.

On the other hand, different fission cross sections measured at the CERN - n_TOF facility have been released in the last years, using as reference the ^{235}U cross section [1,2], or even providing the ratio to it [3]. Moreover, it has been employed to obtain the n_TOF neutron flux at high energies [4]. However, the ^{235}U(n,f) cross section data from n_TOF has been released only for a limited range of the resonance region, that are presently in EXFOR [5]. In this work, we report on an improved release of the fission cross section in the range between 1 eV and 10 keV, where the n_TOF flux shape is smooth and very well known. These data are compared to the ENDF/B-VII evaluated dataset and to the integral values given by the IAEA reference file.

2 Experimental setup

We will focus on the data obtained with the detection setup based on Parallel Plate Avalanche Counters (PPAC) used during the n_TOF experiment Phase-I, when the detection setup consisted of ten detectors and nine targets interleaved, placed perpendicularly to the neutron beam [1]. The data provided in this work are the sum of the three different measurements taken during the 2003 fission campaign. Between the different measurements, the chamber was removed from the Experimental Area to replace part of the targets. However, the first and the second target in the beam direction, ^{238}U and ^{235}U, respectively, were not changed.

The targets were made very thin and the detector layers were minimised, being operated at a very low gas pressure. Furthermore, special attention was devoted to minimise too the amount of material in the experimental area, using thin Kapton windows to separate the chamber gas from the evacuated beam pipe. Altogether, the amount of in-beam scattered neutrons was reduced and, moreover, the upstream position of the ^{235}U target in the experimental setup minimises any possible attenuation from the in-beam materials.

The low background achieved at the CERN - n_TOF facility is one of the main characteristics of these experiments. Contrary to other time-of-flight facilities such as WNR in LANL or GELINA at JRC-IRMM, n_TOF has a very long flight-path with the experimental area placed at around 185 m from the spallation target, which reduces the time-independent neutron background. In addition, the duty factor is very low, in the order of seconds, which eliminates completely the background due to in-beam neutrons from the previous pulses (overlap neutrons) and reduces significantly the contribution from fissions events induced by thermal neutrons backscattered in the surrounding walls.

The use of the same detector configuration and very similar experimental conditions resulted in three data sets which agree within their statistical uncertainties.

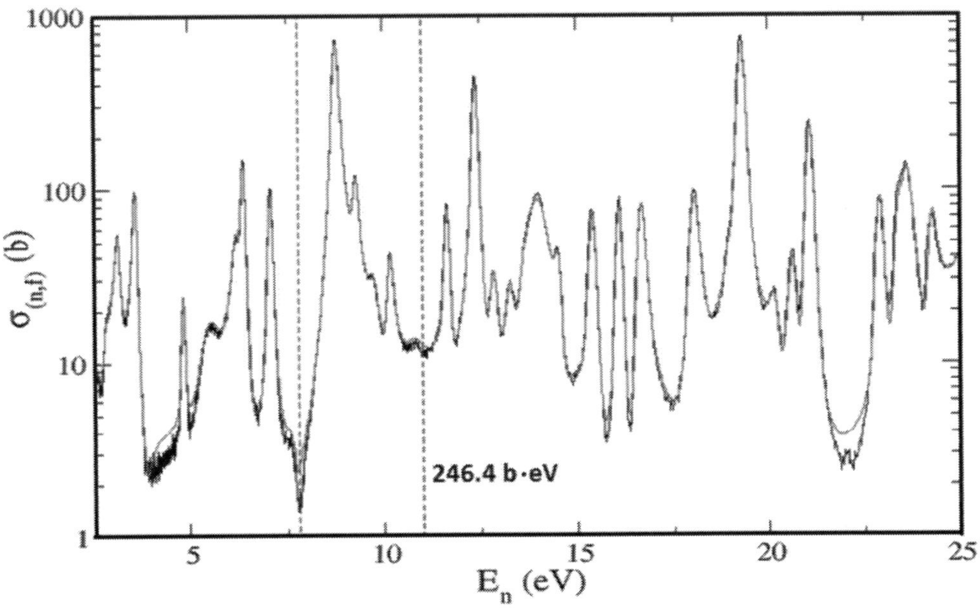

Figure 1. $^{235}U(n,f)$ cross section in the energy range from 2.5 to 25 eV. The n_TOF PPACs result (2000 bins per decade) is compared with the ENDF/B-VII evaluation. Vertical lines indicate the IAEA reference interval.

3 Data analysis and results

The analysis of the $^{235}U(n,f)$ cross section in the low energy region is based on the procedure described in Ref. [1], where the cross sections of ^{234}U and ^{237}Np were obtained using ^{235}U as reference isotope. For this analysis we have included all results taken during the 2003 fission campaign, by directly adding the fission yields obtained in the three data taking periods. Once the final yield is produced, the fission cross section shape is determined by dividing this yield by the official neutron flux provided by measurements with different neutron monitors based on the reference cross sections of $^{6}Li(n,t)$ and $^{10}B(n,\alpha)$ during the capture campaign in Phase-I. A comparison between measurements taken with a ^{10}B-based chamber using both the fission and capture collimators shows that the shape of the neutron flux in the region of interest is conserved below 1% through the collimator change.

In order to provide an accurate normalisation for the obtained fission cross section shape, we have used the well-known 7.8 to 11.0 eV fission resonance integral. With this method, it is neither necessary to know the mass of the ^{235}U and ^{10}B deposits nor the efficiency of the detection setup, only assuming that it is constant in the considered energy range. A normalisation value for this integral of 246.4 ±1.2 b · eV has been taken from the IAEA standards library [6]. In Figures 1 and 2, the n_TOF $^{235}U(n,f)$ cross section is compared to the current ENDF/B-VII evaluation [7] for two different neutron energy regions. It is worth to mention that the ENDF integral value for the 7.8-11 eV interval (241.6 b · eV) differs by 2 % from the IAEA reference value used in this work for the cross section normalisation.

An excellent reproduction of the resonance structure is achieved thanks to our good energy resolution, while the significant statistics allow us to show the different behaviour in the dips. The excellent background conditions led to reduced counts in the dips between resonances (see in Fig. 1). On the other hand, it is worth to note a problem with the energy calibration of the ENDF data above 100 eV. The complete n_TOF dataset, including uncertainties, will be available in EXFOR.

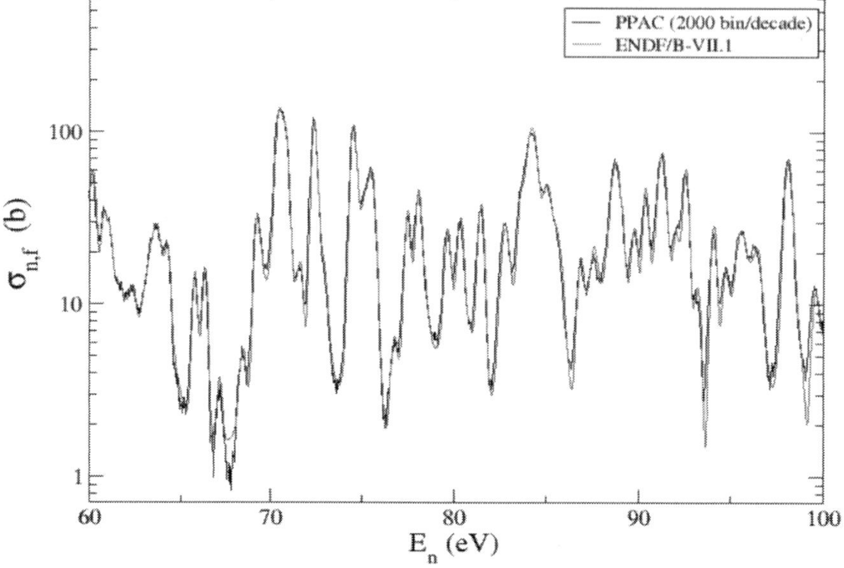

Figure 2. $^{235}U(n,f)$ cross section in the energy range from 60 to 100 eV.

4 Averaged integral values

In Table 1 the n_TOF values are compared with those provided in the IAEA reference file, as well as with the ones taken from the ENDF/B-VII data base. All these values are averaged cross-sections (in barn) centred in the same energy intervals as given by IAEA in Ref. [6]. Those from n_TOF were normalised as previously mentioned but those from ENDF maintain its original normalisation.

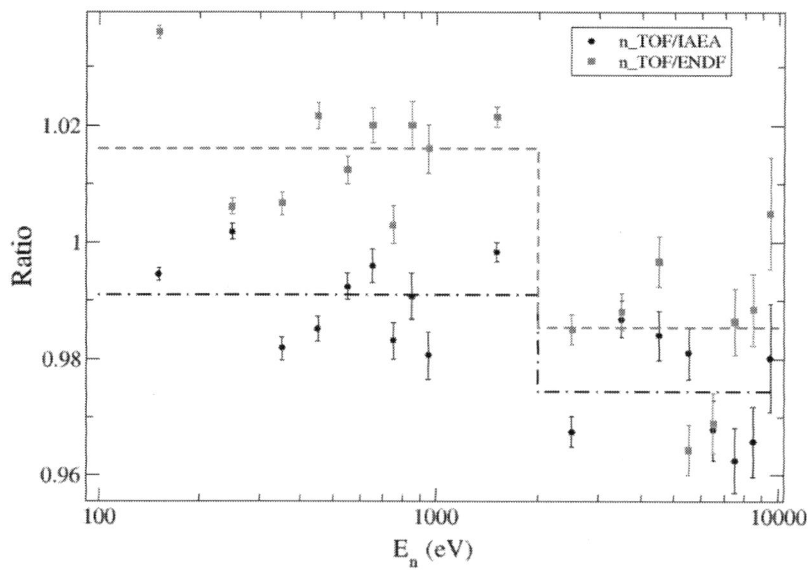

Figure 3. Ratios of the n_TOF data to IAEA reference values (black dots) and to ENDF/B-VII (red squares).

In order to clearly show the existing differences between the IAEA recommended values and those obtained from the ENDF evaluation, we have calculated their ratio with respect to the present results.

These ratios are shown in Fig. 3, centred in the intervals as in Table 1, with error bars corresponding to their statistical uncertainties. The dashed lines represent the average values below and above 2 keV. The abrupt drop in the ratio with the ENDF/B-VII evaluation denotes the influence of different experimental sources above and below 2 keV, corresponding to the transition at around 2.25 keV from the resolved to the unresolved resonance regions in the ENDF evaluation.

On the other hand, an offset appears in Table 1 as a systematic deviation of our dataset from the IAEA one. To some extent this offset could be due to the low background in the n_TOF experiment. Fig. 4 shows the new ratios after subtracting 0.09 b from every value in the IAEA file (the error bars also include the IAEA reported uncertainties). The ratio distribution along the whole energy range is now extremely flat, its mean value is 0.998, and the spread is compatible with the low statistical uncertainties of the n_TOF data, endorsing so the goodness of both n_TOF and IAEA datasets.

Table 1. Averaged integral cross-sections (in barn) compared to the IAEA reference file [6] and to ENDF/B-VII. Uncertainties included between brackets refer to the last significant digits.

E_n [eV]	n_TOF [b]	IAEA [b]	ENDF/B-VII [b]
100-200	21.054(22)	21.17(11)	20.321
200-300	20.726(28)	20.69(11)	20.601
300-400	12.891(26)	13.135(72)	12.806
400-500	13.575(30)	13.781(76)	13.288
500-600	15.055(35)	15.174(86)	14.870
600-700	11.463(33)	11.513(65)	11.238
700-800	10.912(35)	11.101(64)	10.880
800-900	8.137(32)	8.213(48)	7.977
900-1000	7.356(30)	7.502(44)	7.240
1000-2000	7.291(12)	7.303(40)	7.138
2000-3000	5.211(14)	5.386(33)	5.290
3000-4000	4.721(15)	4.784(30)	4.778
4000-5000	4.193(16)	4.261(25)	4.207
5000-6000	3.766(17)	3.838(24)	3.905
6000-7000	3.185(17)	3.291(21)	3.287
7000-8000	3.115(18)	3.236(19)	3.158
8000-9000	2.906(18)	3.009(18)	2.940
9000-10000	3.058(29)	3.120(19)	3.043

5 Conclusions

A new analysis of the ^{235}U(n,f) data taken with PPAC detectors at the CERN-n_TOF facility are presented in the energy range 100 eV to 10 keV. The data show extremely low background and very high energy resolution and have been normalised to the IAEA recommendation for the integral value in the range 7.8 to 11.0 eV. Its comparison with the last IAEA reference files and with the present version of the ENDF evaluation leads to the following conclusions:

a) There is very good agreement with the shape of the ENDF cross-section in the RRR, while showing a lower background.
b) The ENDF integral values, apart from a 2% difference in the normalisation value at 7.8-11.0 eV, show a sharp drop at the transition from the resolved to the unresolved resonance energy regions.

c) There is very good agreement with the IAEA integral-data set, provided that an offset of 0.09 barn is applied in the whole energy range.

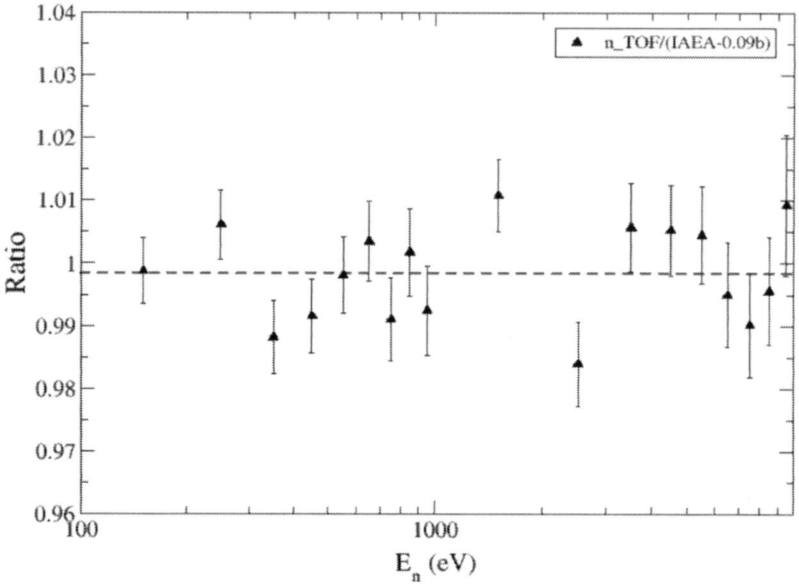

Figure 4. New ratio of the n_TOF data to the IAEA reference values, after applying a 0.09 b offset. The mean value is shown by the blue line.

References

1. C. Paradela *et al.*, Phys. Rev. C **82**, 034601 (2010)
2. D. Tarrío *et al*, Phys Rev. C. **83**, 044620 (2011)
3. C. Paradela *et al.*, Phys. Rev. C **91**, 024602 (2015)
4. M. Barbagallo *et al.*, Eur. Phys. J. A (2013) **49:** 12 - 24
5. C. Guerrero *et al.*, Eur. Phys. J. A. (2013) **49:** 27- 42
6. L. Audouin *et al.*, Proc. of Conf. on Nucl. Data for Sci. and Tech., Nice 2007, Vol. 1, p. 421.
7. https://www-nds.iaea.org/standards/
8. M.B. Chadwick *et al.*, Nucl. Data Sheets **112**, 2887 (2011)

EPJ Web of Conferences **111**, 02004 (2016)
DOI: 10.1051/epjconf/201611102004

Isomeric ratio measurements for the radiative neutron capture ^{176}Lu(n,γ) at DANCE

D. Denis-Petit[1,a], O. Roig[1], V. Méot[1], M. Jandel[2], D.J. Vieira[2], E.M. Bond[2], T.A. Bredeweg[2], A.J. Couture[2], R.C Haight[2], A.L. Keksis[2], R.S. Rundberg[2], and J.L. Ullmann[2]

[1]*CEA DAM DIF, F-91297 Arpajon, FRANCE*
[2]*Los Alamos National Laboratory, Los Alamos, New Mexico 87545, USA*

Abstract. The isomeric ratio for the neutron capture reaction ^{176}Lu(n,γ) on the $J^\pi = 5/2^-$, 761.7 keV, $T_{1/2}$=32.8 ns level of ^{177m}Lu, has been determined in the neutron energy range 8.5 eV-100 keV for the first time using the DANCE array at the Los Alamos National Laboratory.

1 Introduction

Neutron capture cross sections are of high interest in nuclear astrophysics to investigate the s-process in which the synthesis of heavy elements is dominated by neutron induced reactions. Some isotopes known as s-only nuclei are only produced by the s-process [1] and then give an access to specific characteristics of this nucleosynthesis process. Many investigations were performed on Lutetium isotopes [2–5] due to their great importance in the s-process. The s-only ^{176}Lu exhibits a thermally enhanced beta decay rate [6], making it a sensitive branch point, for estimating neutron density and temperature at the nucleosynthesis site [7]. Macklin [8, 9], Beer [10, 11], Bokhovko [12] and Wisshak [5] have performed various measurements on lutetium isotopes in the neutron energy region of astrophysical interest between 3 keV and 200 keV. Partial cross sections feeding the ground states or isomers are nevertheless particularly crucial in certain cases of the s-process nucleosynthesis [2, 7, 13], or of some astrophysical environments such as neutron stars or supernovae where reactions on the isomeric states can occur [14, 15]. Few measurements of these cross section branchings or of isomeric ratios, defined by the ratio of the isomeric/total cross sections, were performed by combining TOF (Time Of Flight) technic and a 4π array. Three methods were relevant up to now: activation method when lifetime are sufficiently long [16, 17], calorimetry method when excitation energy of isomeric states are sufficiently high compared to the detector energy resolution [18, 19] and combining time and calorimetry for short lifetimes and high excitation energy of isomeric states [20, 21].

In this paper we describe a new method using the high efficiency and granularity of the Detector for Advanced Neutron Capture Experiments (DANCE) array (Los Alamos National Laboratory) working with a digital data acquisition system. We have measured the isomeric ratio for a short lifetime isomer of ^{177}Lu formed by radiative capture reactions on ^{176}Lu.

[a]e-mail: david.denis-petit@cea.fr

2 Experimental setup

2.1 The DANCE array

The experiment was conducted at the Los Alamos Neutron Science Center (LANSCE) of the Los Alamos National Laboratory. The neutron beam is produced by spallation reactions of the 800-MeV pulsed proton beam impinging on a moderated tungsten target [22]. The ^{176}Lu target is placed on the center of the DANCE detector which is located on the 20.25 m long flight path FP14 of the LANSCE [23]. This 4π array is composed of 160 barium fluoride (BaF_2) detectors and uses a digitizer acquisition system to record waveforms coming from theses ones. The DANCE array provides for each γ-cascade following the neutron capture, the γ-sum energy, the multiplicity of crystal hit and the γ-rays energies from each crystal. To take into account the Compton effect inside the ball array, a cluster of hits is defined as where at least one neighboring crystal is hit. Information is provided as cluster multiplicty and cluster γ-ray energies determined for each γ-cascade. A cluster time is also defined as a mean of all the crystal times which compose the cluster. In our case, the continuous mode of the DANCE acquisition system is used for neutron energy ranging from 8.5 eV up to 100 keV. More details about the DANCE detector and data analysis can be found in [24].

2.2 The targets

The ^{176}Lu target is unique as the isotopic enrichment reaches 99.95 % for a mass of 0.5 mg/cm^2 and a deposit diameter of 7 mm on a 1 μm aluminized Mylar foil. Enrichment was achieved using a mass spectrometer, SIDONIE available at the CSNSM laboratory (Orsay, France).

To evaluate the background due to neutron scattering on target backing, we have used a blank target which is composed of the same backing material than the Lu target.

3 Isomeric ratio measurement

The isomeric cross-section ratio $R_{iso}(E_{iso})$ of the isomer located at an exitation energy of E_{iso} is written as:

$$R_{iso}(E_{iso}) = \frac{N_{iso}(E_{iso})\,\epsilon_{casc}}{N_{casc}\,\epsilon_{iso}(E_{iso})} \quad (1)$$

where $N_{iso}(E_{iso})$ is the number of detected isomers, N_{casc} is the number of detected γ-cascades, ϵ_{casc} is the cascade detection efficiency and $\epsilon_{iso}(E_{iso})$ is the isomer detection efficiency. The different terms of the equation (1) are described below.

3.1 Selection of the isomers

Our method to determine isomeric cross section ratios consists in a careful study of γ-cascades using the time information of each γ-ray of the cascade. In a large coincidence window of 250 ns, all the γ rays following the neutron capture are recorded. All the γ-rays of this window located in the first 20 ns define the prompt γ-cascade. The others γ-rays are assigned as delayed γ-rays. The time of the prompt cascade is defined as a mean of the cluster times associated to the γ-rays which compose this cascade. The selection of events with particular number of delayed γ-rays highlights particular isomer decays. This selection is illustrated in figure 1.

In this figure, the following γ-ray spectra for the Lu target are presented: the spectrum with no time selection (energies of the prompt and delayed γ-rays), the spectrum with only one γ-ray for the Lu and blank target and the spectrum with two delayed γ-rays.

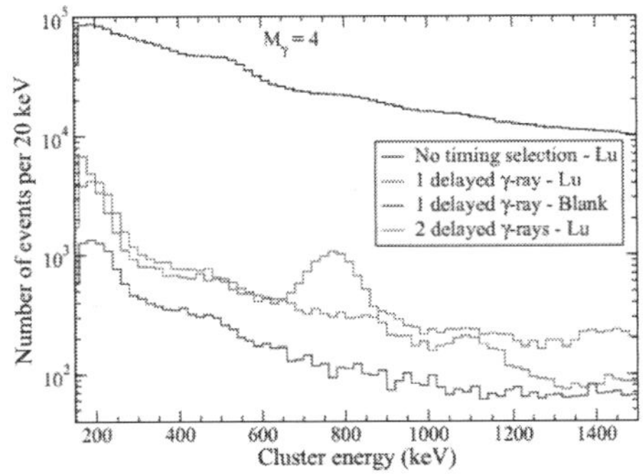

Figure 1: γ-ray spectra obtained with the DANCE array after a $M_\gamma = 4$ γ-cascades including all the prompt and delayed γ-rays (no timing selection), only one delayed γ-ray for the Lu and the blank targets and two delayed γ-rays for the Lu target.

The spectrum which represents the events with only one delayed γ-ray (three prompt γ-rays for $M_\gamma = 4$) exhibits two peaks located at 762 keV and at 1.1 MeV. In this paper, we will focus on the first one. This peak is not present in the case of the blank target which indicates that it is associated to the ^{177}Lu decay. This peak is related to the de-excitation of the $T_{1/2} = 32.8$ ns isomer to the ground state. It proceeds mainly by the emission of only one γ-ray at 761.7 keV (γ intensity of $I_\gamma(761.7) = 70.5 \pm 2.8$ %). That why It disappears when two delayed γ-rays are selected.

On the spectrum with one delayed γ-ray, energy gates, which take into account the BaF_2 energy resolution, are defined around the full energy peak at 761.7 keV to obtain the time decay of the isomers. In figure 2, the spectra which represent the difference between the time of the prompt γ-cascade and the delayed γ-ray are plotted for $M_\gamma = 4$ and in the cases of the Lu and blank targets.

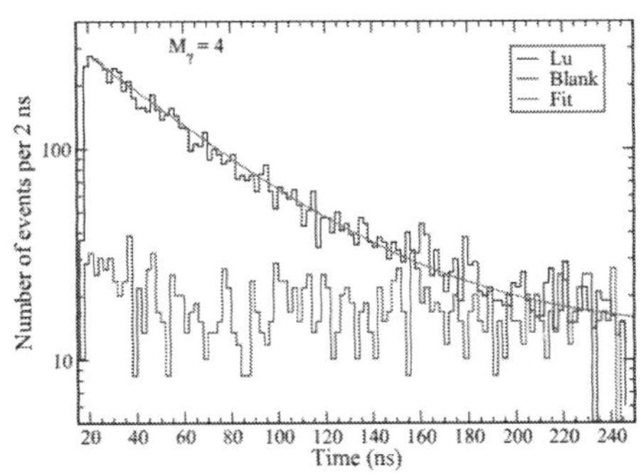

Figure 2: Time spectra for $M_\gamma = 4$ and for the $T_{1/2} = 32.8$ ns isomer obtained by applying energy gates on energy spectra with one delayed γ-ray for the Lu and blank targets. The exponential fits used to obtain the detected isomer number is also plotted.

The isomer exponential decay is clearly visible for the Lu target whereas the time spectrum is flat for the blank target. An exponential fit, $f(t) = A_0 \exp\left(-\frac{\ln 2}{T} t\right) + B$ is performed to extract the number of formed isomers. The fit parameters are reported in table 1 for multiplicities ranging from 3 to 6.

Table 1: Results of the fits performed for the $T_{1/2}$ = 32.8 ns isomer and for different cluster multiplicities.

M_γ	A_0 (counts/ns)	T (ns)	B (counts/ns)
3	173.9 (77)	39.0 (22)	66.7 (24)
4	199.3 (67)	33.8 (11)	13.5 (11)
5	106.8 (45)	34.6 (14)	2.7 (7)
6	38.8 (33)	28.4 (19)	1.1 (3)

The half-life obtained by the fits are compatible at 2σ with the previous measured ones: $T_{1/2}$ = 32.8 ± 2.4 ns. The number of detected isomer $N_{iso}(E_{iso})$ is deduced from the fits by:

$$N_{iso}(E_{iso}) = \sum_{M_\gamma} \frac{A_0(E_{iso}, M_\gamma)T(E_{iso}, M_\gamma)}{\ln 2\, I_{1\gamma}(E_{iso})} \tag{2}$$

where E_{iso} is the energy of the considered isomer and $I_{1\gamma}(E_{iso})$ is the probability that this isomer is detected by only on delayed γ-ray by taking into account the DANCE energy threshold. In the case of the 761.7 keV isomer, we have: $I_{1\gamma}(761.7) = I_\gamma(761.7) = 70.5 \pm 2.8$ %.

3.2 Number of detected cascades

The number of detected cascades N_{casc} is summed for cluster multiplicities ranging from 3 to 6. It is determined, for each multiplicity, by integrating the total energy spectrum between 3.5 and 7.5 MeV. Background obtained with the blank target is subtracted before the integration.

3.3 Detection efficiencies

The cascade detection efficiency ϵ_{casc} is calculated with GEANT4 simulations of the DANCE detector [25] thanks to a γ-cascade obtained with the Monte-Carlo code EVITA, which is based on Hauser-Feshbach formalism [26]. This code is able to reproduce a γ-cascade event using all available information for each nucleus involved in a capture reaction. Ingredients used by EVITA are based on the TALYS code [27] and optimized on the specified available information on the desired isotope and reaction. ϵ_{casc} is defined as the ratio of the number of detected cascade for $M_\gamma \in [\![3, 6]\!]$, which is calculated by integrating the total energy spectrum in the energy range [3.5 – 7.5] MeV, and the total number of initial cascades obtained with EVITA. In our case, we obtain ϵ_{casc} = 60.2 %.

The isomer detection efficiency $\epsilon_{iso}(E_{iso})$ is also calculated with GEANT4 simulations using EVITA. It is defined as the ratio of the full energy peak integral of the γ-rays used to evaluate $N_{iso}(E_{iso})$ summed over the multiplicities indicated in the table 1 and the total number of the same γ-rays in the initial cascades calculated with EVITA. $\epsilon_{iso}(E_{iso})$ takes into account the geometric efficiency, the photopeak efficiency and the multiplicity cut. We have: $\epsilon_{iso}(761.7)$ = 32.9 %.

4 Results and discussions

Experimentally, we obtain: $R_{iso}(761.7) = 11.3 \pm 0.8$ %. This experimental value can be compared with a TALYS calculation. In this calculation, we have used the Gilbert and Cameron model for the level density, the Enhanced Generalised LOrentzian (EGLO) model for the E1 strength function and the dispersive and global spherical optical potential developed by Morillon and Romain [28]. The calculated isomeric ratios are: $R_{iso}(761.7)$ = 4.1 %. Large discrepancies are observable between experimental and evaluated isomeric ratios. They may indicate that the γ-ray cascades which lead to the isomers are not properly modelled in our Hauser-Feshbach calculations. Further investigations must be done on level scheme and γ-strength function models.

5 Conclusion

In conclusion, we have obtained for the first time an isomeric ratio of the radiative capture reaction ^{176}Lu(n,γ) for an isomer of the ^{177}Lu compound nucleus, at 761.7 keV ($J^{\pi} = 5/2^{-}$, $T_{1/2}$ = 32.8 ns), integrated over all neutron energies. This isomeric ratio was extracted from TOF experiment at DANCE. Comparisons with Hauser-Feshbach prediction have shown large discrepancies.

Acknowledgments

The authors thank B. Morillon, P. Romain for their precious help on TALYS and EVITA codes. This work has benefited from the use of the Lujan Center at the Los Alamos Neutron Science Center, funded by the DOE Office of Basic Energy Sciences and Los Alamos National Laboratory funded by the Department of Energy under contract W-7405-ENG-36.

References

[1] E. Anders, N. Grevesse, Geochim. Cosmochim. Acta **53**, 197 (1989)
[2] H. Beer, F. Käppeler, K. Wisshak, R. Ward, Astrophys. J. Suppl. Series **46**, 295 (1981)
[3] M.A. Garner, D.G. Gardner, and R.W. Hoff, J. Phys. G: Nucl. Phys. **14**, S315 (1988)
[4] J.J. Carroll, J.A. Anderson, J.W. Glesener, C.D. Eberhard, and C.B. Collins, Astrophys. J. **344**, 454 (1989)
[5] K. Wisshak, F. Voss, F. Käppeler, and L. Kazakov, Phys. Rev. C **73**, 015807 (2006)
[6] K. Takahashi, K. Yokoi, At. Data Nucl. Data Tables **36**, 375 (1987)
[7] N. Klay, F. Käppeler, H. Beer, and G. Shatz, Phys. Rev. C **44**, 2839 (1991)
[8] R. Macklin, J. Gibbons, Phys. Rev. **159**, 1007 (1967)
[9] R.L. Macklin, D.M. Drake, and J.J. Malanify, Los Alamos Scientific Lab. Reports 7479, Los Alamos National Laboratory (1978)
[10] H. Beer, F. Käppeler, Phys. Rev. C **21**, 534 (1980)
[11] H. Beer, G. Walter, R.L. Macklin, and P.J. Patchett, Phys. Rev. C **30**, 464 (1984)
[12] M.V. Bokhovko, V.N. Kononov, N.S. Rabotnov, A.A. Voevodskiy, G.N. Manturov, and V.M. Timokhov, Obninsk Reports FEI 2169, Fiziko Energetscheskij Institut (1991)
[13] D.N.S. J. Audouze, W. A. Fowler, Nature Physical Science **238**, 8 (1972)
[14] F. Käppeler, A. Mengoni and R. Gallino, Nucl. Phys. A **718**, 173c (2003)
[15] T. Kawano, P. Talou, J.E. Lynn, M.B. Chadwick, and D.G. Madland , Phys. Rev. C **80**, 024611 (2009)
[16] C. Heiser und K.F. Alexander, Nucl. Phys. **70**, 415 (1965)
[17] X. Ledoux, J. Sigaud, T. Granier, J-P. Lochard, Y. Patin, P. Pras, C. Varignon, J-M. Laborie, Y. Boulin, and F. Gunsing, Eur. Phys. J. A **27**, 59 (2006)
[18] K. Wisshak, F. Voss, F. Käppeler, and L. Kazakov, Phys. Rev. C **61**, 065801 (2000)

[19] K. Wisshak, F. Voss, F. Käppeler, and L. Kazakov and F. Bečvàr and M. Krtička and R. Gallino and M. Pignatari, Phys. Rev. C **73**, 045807 (2006)

[20] M. Jandel, T.A. Bredeweg, E.M. Bond, M. Chadwick, A.J. Couture, J.M. O'Donnell, M.M. Fowler, R.C. Haight *et al.*, Los Alamos Scientific Lab. Reports LA-UR-13-20355, Los Alamos National Laboratory (2013)

[21] M. Jandel, T.A. Bredeweg, E.M. Bond, M. Chadwick, A.J. Couture, J.M. O'Donnell, M.M. Fowler, R.C. Haight *et al.*, *Neutron-induced fission measurements at the DANCE and LSDS facilities at LANL*, in *Proceedings of the Fifth International Conference on Fission and Properties of Neutron-Rich Nuclei* (World Scientific, 2013), pp. 416–423

[22] M. Mocko, G. Muhrer, Nuclear Instruments and Methods in Physics Research Section A: Accelerators, Spectrometers, Detectors and Associated Equipment **704**, 27 (2013)

[23] M. Heil, R. Reifarth, M.M. Fowler, R.C. Haight, F. Käppeler, R.S. Rundberg, E.H. Seabury, J.L. Ullmann, J.B. Wilhelmy, K. Wisshak, Nucl. Instr. Meth. A **459**, 229 (2001)

[24] J. M. Wouters, A.A. Vicente, T.A. Bredeweg, E. Esch, R.C. Haight, R. Hatarik, J.M. O'Donnell, R. Reifarth, R.S. Rundberg, J.M. Schwantes *et al.*, IEEE Transactions on nuclear science **53**, 880 (2006)

[25] M. Jandel, T.A. Bredeweg, A. Couture, M.M. Fowler, E.M. Bond, M.B. Chadwick, R.R.C. Clement, E.-I. Esch, J.M. O'Donnell, R. Reifarth *et al.*, Nucl. Instr. and Meth. B **261**, 1117 (2007)

[26] B. Morillon, P. Romain (2008), private communication

[27] A.J. Koning, S. Hilaire and M.C. Duijvestijn, *TALYS-1.0*, in *Int. Conf. on Nuclear Data for Science and Technolgy*, edited by O. Bersillon, F. Gunsing, E. Bauge, R. Jacqmin and S. Leray (EDP Sciences, 2008), 2, pp. 211–214

[28] B. Morillon, P. Romain, Phys. Rev. C **70**, 014601 (2004)

EPJ Web of Conferences **111**, 02005 (2016)
DOI: 10.1051/epjconf/201611102005

New measurement of the ^{242}Pu(n,γ) cross section at n_TOF

J. Lerendegui-Marco[1,a], C. Guerrero[1], M. A. Cortés-Giraldo[1], J. M. Quesada[1], E. Mendoza[2], D. Cano-Ott[2], K. Eberhardt[3], A. Junghans[4], and the n_TOF Collaboration[5]

[1] *Dpto. de Física Atómica, Molecular y Nuclear, Universidad de Sevilla, 41012 Sevilla, Spain*
[2] *Centro de Investigaciones Energéticas, Medioambientales y Tecnológicas (CIEMAT), 28040 Madrid, Spain*
[3] *Johannes Gutenberg-Universität Mainz, 55128 Mainz, Germany*
[4] *Helmholtz-Zentrum Dresden-Rossendorf, D-01314 Dresden, Germany*
[5] *European Center for Nuclear Research (CERN),CH-1211 Geneva, Switzerland*

Abstract. The use of MOX fuel (mixed-oxide fuel made of UO_2 and PuO_2) in nuclear reactors allows substituting a large fraction of the enriched Uranium by Plutonium reprocessed from spent fuel. With the use of such new fuel composition rich in Pu, a better knowledge of the capture and fission cross sections of the Pu isotopes becomes very important. In particular, a new series of cross section evaluations have been recently carried out jointly by the European (JEFF) and United States (ENDF) nuclear data agencies. For the case of ^{242}Pu, the two only neutron capture time-of-flight measurements available, from 1973 and 1976, are not consistent with each other, which calls for a new time-of flight capture cross section measurement. In order to contribute to a new evaluation, we have perfomed a neutron capture cross section measurement at the n_TOF-EAR1 facility at CERN using four C_6D_6 detectors, using a high purity target of 95 mg. The preliminary results assessing the quality and limitations (background, statistics and γ-flash effects) of this new experimental data are presented and discussed, taking into account that the aimed accuracy of the measurement ranges between 7% and 12% depending on the neutron energy region.

1 Introduction and motivation

The measurement of accurate capture and fission cross sections is essential for the design and operation of current and innovative nuclear systems aimed at the reduction of the nuclear waste [1]. The spent fuel from current nuclear reactors contains a significant fraction of plutonium, which can be separated from the fuel matrix. This plutonium contains 66% of the fissile ^{239}Pu and ^{241}Pu, that can be combined with depleted uranium (^{238}U) to make what is known as mixed oxide (*MOX*) fuel. In this way the Pu from spent fuel and the depleted uranium, otherwise considered as waste, are used in a new reactor cycle, contributing in this way to the long-term sustainability of nuclear energy. Currently, the use of MOX fuel in thermal power reactors has been established in several countries. However, a much more efficient use of plutonium will ultimately be made in fast reactors, where multiple recycling is possible and has been demonstrated. The reader is referred to Ref.[2] for more details on the status and progress in MOX fuel technologies.

[a] e-mail: jlerendegui@us.es

Table 1. Current and required accuracy for the capture cross section of ^{242}Pu for different nuclear reactor concepts, including as well the neutron energy range of interest. See Refs. [7, 8]

System	Neutron energy range	Present accuracy (%)	Required accuracy (%)
SFR	2-500 keV	35	8
EFR	2-67 keV	35	25
GFR	2-183 keV	35	7-8
LFR	9-183 keV	35	11-12
ADMAB (ADS)	9-25 keV	35	10
PHENIX	0.5-2 keV	14	7
NEA/HPRL	0.5-2 keV	14	8

The extensive use of MOX fuels in fast reactors calls for a revision of the neutron cross sections that play a role in the neutronics of such reactors and are not known with enough accuracy yet. For the particular case of ^{242}Pu, the first attempts to measure its neutron capture cross section were made in 1973 and 1976, when Portmans et al. [3] and Hockenbury et al. [4] used the time-of-flight technique for measurements at low (below 1.3 keV) and high (6-87 keV) neutron energies, respectively. A few years later, Wisshak and Kaeppeler [5, 6] measured the capture cross section in the 10-90 and 50-250 keV energy intervals. The comparison of the different results indicates an uncertainty of about 35% in the capture cross section in the keV region. In this context, the Nuclear Energy Agency recommends in its "High Priority Request List" [7] and its report WPEC-26 [8] that the capture cross section of ^{242}Pu should be measured with an accuracy of at least 7-12% in the neutron energy range between 500 eV and 500 keV (see Table 1). Furthermore, interpretations with JEFF-3.1 of the PROFIL and PROFIL-2 experiments carried out in the fast reactor PHENIX have shown an overestimation of about 14% of the ^{242}Pu capture cross section [9, 10]. In addition to the direct measurement in the fast energy region (E_n>2 keV), accurate average radiation width and strength function are required to solve some ambiguous results obtained between optical model calculations and the statistical analysis of the s-wave resonance parameters [11]. This calls as well for an accurate measurement of the resonance region (1 to 1000 eV) with enough resolution and statistics to determine accurately the corresponding average resonance parameters.

For all of the above, a new measurement of the ^{242}Pu cross section at the n_TOF facility was proposed to the ISOLDE and Neutron Time-of-Flight Committee (*INTC*) [12] and was approved in September 2013. The material production and sample preparation took place during the first semester of 2015 in collaboration with the JGU Mainz and the HZ Dresden-Rossendorf research centers. Finally, the experiment was succesfuly performed in summer 2015. In the following section we briefly describe the n_TOF facility and the experimental setup. Section 3 provides some details on the measuring technique and the preliminary data and some preliminary results are presented in Section 4.

2 Experimental setup

The measurement was performed at the n_TOF-EAR1 facility at CERN [13]. The pulsed neutron beam at n_TOF is generated through spallation of 20 GeV/c protons from the CERN Proton Synchroton (CPS) impinging on a thick lead target. Each proton bunch contains, in average, $7 \cdot 10^{12}$ protons with a time distribution of σ=7 ns and an average repetition rate of 0.17 Hz. The spallation neutrons, with energies in the MeV-GeV range, are partially moderated in the water cooling and moderation layers that surround the lead target and travel towards the experimental areas along two beam lines: EAR1 at 185 m (horizontal) and EAR2 at 19 m (vertical). A complete description and experimental

Figure 1. Experimental setup for the detection of γ-rays from capture on ^{242}Pu: (Left) 4 C_6D_6 detectors looking at the ^{242}Pu sample assembly (center), composed of by several thin targets (right).

characterization of EAR1 can be found in Ref. [13], while the description of the EAR2 beam line, still in its commissioning phase, is given in Ref. [14]. The measurement presented in this work was performed in EAR1, featuring a better time resolution and consequently a better capability to resolve resonances up to higher neutron energies than in EAR2.

At n_TOF, the radiative capture measurements can be performed using two different detection systems available: the Total Absorption Calorimeter TAC [15] (a 4π BaF_2 array) and the Total Energy detectors (deuterized benzene (C_6D_6) scintillators [16]). The latter have been chosen for this measurement mainly because they suffer significantly less from the so-called γ-flash, thus allowing us to measure up to the required neutron energy (see Table 1). In addition, this detection setup features a much lower neutron sensitivity than the TAC. This advantage is counterbalanced by a more complicated analysis technique, discussed in Section 3.

Neutrons coming from the spallation target, travel 185 m in vacuum along the beam line until they reach the ^{242}Pu sample, with a diameter larger than the beam, inducing radiative capture reactions. The γ-ray cascades are detected with 4 *BICRON* C_6D_6 scintillators placed at $\sim$10 cm of the sample (Figure 1). The gain of the C_6D_6 detectors was adjusted to record signals up to 14 MeV in the flash-ADC cards. The detectors were calibrated in energy using five γ-ray energies from ^{137}Cs, ^{88}Y, ^{241}Am-^{9}Be and ^{244}Cm-^{13}C and the energy calibration was validated with the comparison between experiment and Geant4 simulations of the detector response to this sources. Upstream from the sample, the neutron beam is monitored with the SiMon monitor which consists of four silicon detectors looking at a thin lithium foil for detecting ^{6}Li(n,α) reactions [17]. The signals registered in each detector are recorded by our digital data acquisition system. The new 12-bit cards with a maximum sampling rate of 900MSamples/s enhance the amplitude (deposited energy) resolution, timing and signal reconstruction capabilities.

The sample preparation was carried out within the CHANDA project [18] by the University of Mainz and the HZDR research center. A total of 95 mg of 99% pure ^{242}Pu, a series of eight thin ^{242}Pu samples 45mm in diameter were prepared by electrodeposition of ^{242}Pu on a thin (10 μm) Al backing covered with 50 nm Ti coating, reaching a maximum density of 0.8 mg/cm^2 of ^{242}Pu per sample. The thickness of the backings was minimized to a level where the associated background becomes negligible. The right panel of Figure 1 shows schematic view of the structure of each thin target and the final assembly as a stack of targets.

3 Total Energy Detection Technique

Neutron capture measurements with C_6D_6 detectors are analyzed following the *Total Energy Detection Technique* [19, 20], based in two principles. First,the efficiency of the detectors is low enough so that just one γ-ray per cascade is detected and thus the total efficiency to detect a cascade becomes $\epsilon_c = 1 - \prod_i(1 - \epsilon_i) \approx \sum_i \epsilon_i$, being ϵ_i the efficiency to detect a single gamma of energy E_i. Second, the efficiency to detect one γ-ray is proportional to its energy. Under these conditions the efficiency for detecting a cascade will be proportional to the known cascade energy (E_c) and independent of the actual cascade path: $\epsilon_c = k \sum_i E_i = k \cdot E_c$. However, the second condition is not true and needs from mathematical manipulation of the detector response. The measured counts for each deposited energy must be weighted with an energy (pulse height) dependent weighting function (WF); this is known as the *Pulse Height Weighting Technique(PHWT)*.

To calculate the WF, the response distribution of each detector must be well known. Since its experimental determination is impossible due to the lack of monoenergetic γ-ray sources in the whole energy range of interest (0-12 MeV), Monte Carlo simulations, performed in this work using the *Geant4* toolkit [21], are the best solution. In these simulations the details on the experimental setup and the 8-target sample were implemented in order to consider, among other effects, the attenuation of photons in the ^{242}Pu targets. From these accurate simulations we extract response functions for each detector to a large number of γ-ray energies, $R_{ij} = \frac{counts_{ij}}{N_i}$, where j represents the bin number for a certain deposited energy and i is the γ-ray energy. The response functions are appropiately broaden to consider the detectors' energy resolution before being used as the input to obtain the WF. The energy dependence of this resolution is extracted from the fit of the experimental broadenings with simulated responses to calibration sources at different γ-ray energies. The WF's are assumed to be 4th-8th-degree polynomials, $W_j = \sum_k a_k \cdot (E_j)^k$, and their parameters, a_k, are obtained with the *Minuit* function minimization code. The obtained WF's satisfy (within 2%), in the whole energy range from 100 keV-10 MeV, the proportionality condition

$$\sum_j W_j \cdot R_{ij} = E_i\,;\ \ where\ \sum_j R_{ij} = \epsilon_i.$$

4 Preliminary analysis and first results

The digitalized movies, in the so-called raw data format, are transfered and stored in the CERN Advanced STORage manager (*CASTOR*). This raw data files are processed with a dedicated pulse shape analysis routine [22]. It has been extensively validated to ensure the correct reconstruction of the amplitude, area, time, and all other relevant features of each recorded signal. The output of this program provides files in *ROOT* format, including the correlation of the complete set of variables that characterize the signals. Last, the data reduction is performed with the TTOFsort routine (developed by F. Gunsing), that implements the energy calibration of the detectors, applies the weighting functions and performs a coicidence rejection to satisfy the condition of detecting just one γ-ray, among many other options. This program provides histograms of detected counts as a function of the time-of-flight, deposited energy, etc...

The measured counts as a function of the time-of-flight (TOF) must then be converted into equivalent neutron energy (E_n). Provided that we are in the non-relativistic regime, the relation between these two quantities is given by $E_n(eV) = (72.29\frac{L(m)}{TOF(\mu s)})^2$, where L is the flightpath (185m for EAR1). The left panel of Figure 2 shows the counting rates for the ^{242}Pu sample and the different background contributions, as a function of the neutron energy, between thermal and the aimed 500 keV, normalized to the average pulse intensity of $7{\cdot}10^{12}$ protons. The so-called *dummy* sample is a replica of the

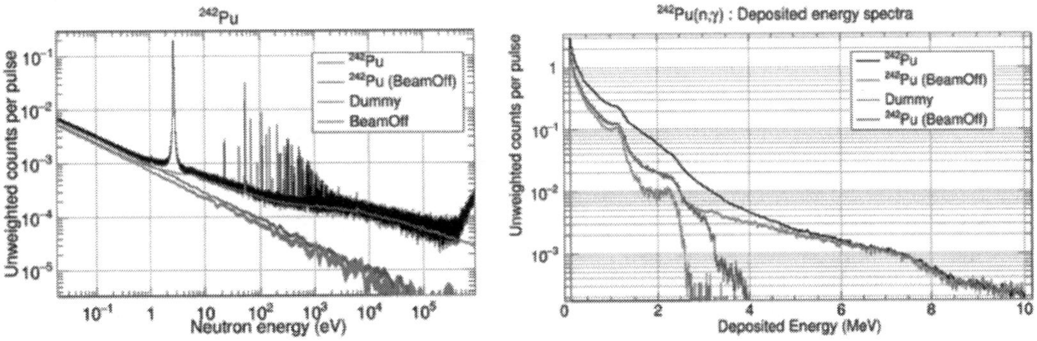

Figure 2. Left: Sketch of the individual targets and the assembled stack with 8 targets. Right: Picture of the ^{242}Pu sample assembly.

^{242}Pu assembly without ^{242}Pu deposits inside and is the main neutron related background component. In addition to the contribution of the backings in the sample, we also have to consider the room backgrounds. with and without the ^{242}Pu targets in place. The ratio of signals from capture on ^{242}Pu to background can be enhanced with an addecuate choice of the condtions in amplitude (or deposited energy E_{dep}). The corresponding deposited energy distributions are presented in the right panel of Figure 2, where the preliminary thresholds in E_{dep} have been set to 150 keV and 10 MeV. This figure shows that the excess of counts coming from ^{242}Pu are below the neutron separation energy of the compound nucleus, S_n=5.035 MeV. The increase of the beamoff background when the ^{242}Pu is in place, is thought to come from the spontaneous fission γ-rays.

In order to extract the neutron capture yield, Y_{exp}, the different background contributions shown in Figure 2 are subtracted from the total counts following a two step process. First the beamoff contributions are substracted: $A =^{242} Pu -^{242} Pu\_beamoff$ and $B = Dummy - Beamoff$. Then $C = A - B$, represents the counts coming from reactions on ^{242}Pu. However, not every reaction on ^{242}Pu is induced by neutrons. Indeed, for targets with a high atomic number Z, an additional background contribution comes from in-beam γ-rays induced reactions (compton $\propto Z^2$ and pair production $\propto Z$). A high Z material with negligible neutron capture cross section, ^{nat}Pb in our case, was measured to estimate the level of the in-beam γ-ray induced background, assumming that the γ-ray induced background, B_γ, dependence on Z and n(atoms/barn) is given by $B_\gamma \propto Z^{3/2} \cdot n$. After scaling the counts in Pb to the expected background in our measurement, the γ-ray background is found to be negligible at low energies and as high as 35% of ^{242}Pu after substracting the *dummy* in the hundreds of keV region. In this energy range the reduction of uncertainties will be mainly limited by the sytematic uncertainties related to the dummy sample since it represents up to 80% of the total counts.

According the to PHWT (see Section 3), from the weighted counts after background substraction, C_w, we extract the experimental capture yield as

$$Y_{exp}(E_n) = f_c \cdot f_{sat} \cdot \frac{C_w(E_n)}{\Phi(E_n) \cdot BIF \cdot (S_n + E_n)}$$

where Φ represents the accurately measured neutron flux of n_TOF-EAR1 [23], *BIF* stands for Beam Interception Fraction, being the unity since the sample covers the full beam; f_c is a factor that corrects for the counts lost below the amplitude threshold, the multiple detection of photons and the presence of conversion electrons; and $f_{norm} = \frac{1}{Y_{sat}}$, is the normalization factor obtained from the measured saturated value of the capture yield of ^{197}Au at $E_n = 4.9eV$, Y_{sat}. In Figure 3, we show

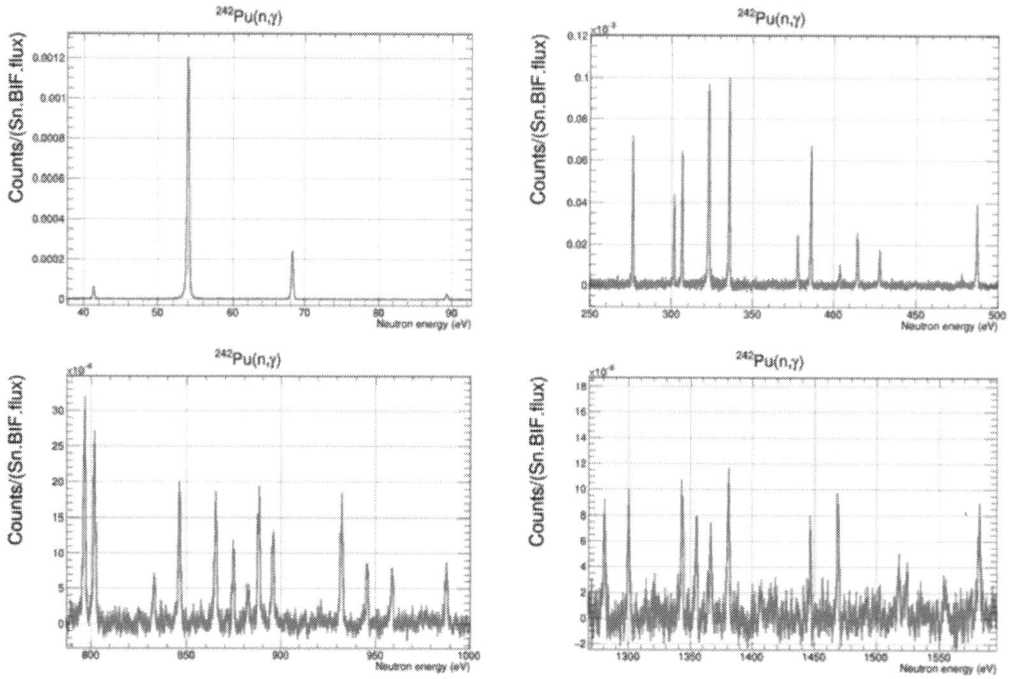

Figure 3. Capture counts per incident neutron (not in Yield units) obtained in the RRR from the tenths of eV to ~1.5keV

the preliminary results obtained after dividing the capture counts by the neutron flux, the BIF and S_n, and thus proportional to the capture yield. Four different energy ranges are shown to illustrate first the high level of statistics in the RRR below 1 keV and the promising results above this energy that hopefully will allow a resonance analysis up to 2keV. Beyond 2 keV we cannot clearly resolve every resonance due to the increasing level density and the limited statistics and we will thus employ the Unresolved Resonance Region (URR) formalism. The capability to extract a reliable cross section in the URR up to 500 keV with this experimental technique and background conditions will be validated with a carefoul analysis of the ^{197}Au(n,γ) measurement, which presents a standard cross section above 150 keV.

5 Summary and conclusions

The use of MOX fuels in innovative nuclear systems requires a better knowledge of the neutron radiative capture cross section on ^{242}Pu. Following the demands of the Nuclear Energy Agency the new measurement aims to reduce the current uncertainties in the 0.5 to 500 keV region down to 7-12% depending on the energy range. Furthermore, the new experimental data in the RRR, up to at least 1 keV, will allow to reduce the current 10% deviations in the average resonance parameters between the different libraries. The measurement has succesfuly taken place at n_TOF-EAR1, that features a very high energy resolution, using 95 mg of 99% pure ^{242}Pu electrodeposited on 8 thin targets assembled together and covering the full neutron beam. The first results after the preliminay quality checks and background substraction are promising and we expect to fulfill the proposed goals. A final capture yield will be ready soon and will be followed by a resonance analysis. In the last stage, we aim to collaborate with the evaluators in CEA Cadarache (JEFF) or ORNL (ENDF) in order to include the new experimental data in the upcoming release of the ^{242}Pu(n,γ) cross section evaluation.

Acknowledgments

This measurement has received funding from the EC FP7 Programme under the projects NEUTANDALUS (Grant No. 334315) and CHANDA (Grant No. 605203), and the Spanish Ministry of Economy and Competitiveness projects FPA2013-45083-P and FPA2014-53290-C2-2-P.

References

[1] N. Colonna et al, Energy Environ. Sci.**3**, 1910-1917(2010)

[2] *International Atomic Energy Agency, Status and advances in Mox fuel technology* IAEA Technical Reports Series **415** (2003)

[3] F. Poortmans et al., Nucl. Phys A **207**, 342-352(1973)

[4] R.W.Hockenbury et al., SP **425**, 584-586 (1975)

[5] K. Wisshak and F. Kaeppeler, Nucl. Sc. and Eng. **66**, 363 (1978)

[6] K. Wisshak and F. Kaeppeler, Nucl. Sc. and Eng. **69**, 39 (1979)

[7] *NEA High Request Priority List* http://www.nea.fr/dbdata/hprl/

[8] Salvatores and R. Jacqmin, *Uncertainty and target accuracy assessment for innovative system using recent covariance data evaluations* (NEA/WPEC-26, 2008)

[9] G. Noguere, E. Dupont, J. Tommasi and D. Bernard. Technical note CEA Cadarache, NT-SPRC/LEPH-05/204 (2005)

[10] J. Tommasi, E. Dupont and P. Marimbeau., Nucl. Sci. Eng.**154**, 119-133 (2006)

[11] E. Rich, G. Noguere, C. De Saint Jean and O. Serot. Proceedings of the Int. Conf. on Nuclear Data for Science and Technology, Nice, France, 2007

[12] C. Guerrero, E. Mendoza et al. *Radiative capture on* ^{242}*Pu for MOX fuel reactors*, CERN-INTC-2013-027 (INTC-P-387) (2013)

[13] C. Guerrero et al., Eur. Phys. J. A **49**, 27(2013)

[14] C. Weiss et al., Nucl Instrum. and Meth. A **799**, 90–98 (2015)

[15] C.Guerrero et al., Nucl. Instrum. Methods A **608**, 424 (2009)

[16] P. Mastinu et al., *New C6D6 detectors: reduced neutron sensitivity and improved safety*, CERN-n_TOF-PUB-2013-002 (2013)

[17] S. Marrone et al., Nucl. Instrum. Methods A **517**, 389 (2004).

[18] *CHANDA:solving CHAllenges in Nuclear DAta*. Project funded by FP7-EURATOM-FISSION, European Commission.

[19] R.L. Macklin, J.H. Gibbons, Phys. Rev. **159**, 1007 (1967)

[20] Abondano et al., Nucl. Instrum. and Meth. A **521**, 454–467(2004)

[21] S. Agostinelli et al., Nucl. Instrum. Methods A **506**, 250 (2003)

[22] P. Zugec et al., *Pulse Shape Analysis routines for n_TOF-Phase3*. Nucl. Instrum. Methods A. To be published (2015)

[23] M. Barbagallo et al., (The n_TOF Collaboration), EPJ A **49**, 156 (2013)

EPJ Web of Conferences **111**, 03001 (2016)
DOI: 10.1051/epjconf/201611103001

New ^{56}Fe Evaluation for the CIELO project

G. P. A. Nobre[1,a], M. Herman[1], D. Brown[1], R. Capote[2], A. Trkov[2], L. Leal[3], A. Plompen[4], Y. Danon[5], Jing QIAN[6], Zhigang GE[6], Tingjin LIU[6], Hnalin LU[7], and Xichao RUAN[7]

[1] *Brookhaven National Laboratory, Upton, NY 11973-5000, USA*
[2] *International Atomic Energy Agency, Vienna-A-1400, PO Box 100, Austria*
[3] *Oak Ridge National Laboratory, Oak Ridge, TN 37831, USA*
[4] *EC-JRC-IRMM, Geel, Belgium*
[5] *Rensselaer Polytechnic Institute, Troy, NY, USA*
[6] *China Nuclear Data Center (CNDC), P.O.Box 275-41, Beijing 102413, P.R.China*
[7] *China Institute of Atomic Energy (CIAE), P.O.Box 275-41, Beijing 102413, P.R.China*

Abstract. The Collaborative International Evaluated Library Organisation (CIELO) aims to provide revised and updated evaluations for ^{239}Pu, ^{238,235}U, ^{56}Fe, ^{16}O, and ^{1}H through international collaboration. This work, which is part of the CIELO project, presents the initial results for the evaluation of the ^{56}Fe isotope, with neutron-incident energy ranging from 0 to 20 MeV. The ^{56}Fe(n,p) cross sections were fitted to reproduce the ones from IRDFF dosimetry file. Our preliminary file provides good cross-section agreements for the main angle-integrated reactions, as well as a reasonable overall agreement for angular distributions and double-differential spectra, when compared to previous evaluations.

1 Introduction

The Collaborative International Evaluated Library Organisation (CIELO) is a pilot project from Subgroup 40 of the Working Party on International Nuclear Data Evaluation Co-operation (WPEC - SG40). Its main goal is to test the scheme of a broad international collaboration to produce new evaluated nuclear reaction data files. For this purpose evaluators from across the world will re-evaluate all the six main nuclides for reactor applications: ^{239}Pu, ^{238,235}U, ^{56}Fe, ^{16}O, and ^{1}H. This present work describes efforts to produce an updated file for ^{56}Fe, with special focus on the fast region.

2 Experimental data

There is a great amount of experimental data for ^{56}Fe and also for ^{Nat}Fe, of which 91.72% consists of the mass-56 isotope. To perform this evaluation we referred to careful reviews and previous evaluations, such as Refs. [1–3]. We also took into account recent experimental results [4].

3 Reaction calculations

For the reaction calculations we used the EMPIRE code [5] and for fitting its parameters we employed the KALMAN code, which is part of the EMPIRE package.

[a] e-mail: gnobre@bnl.gov

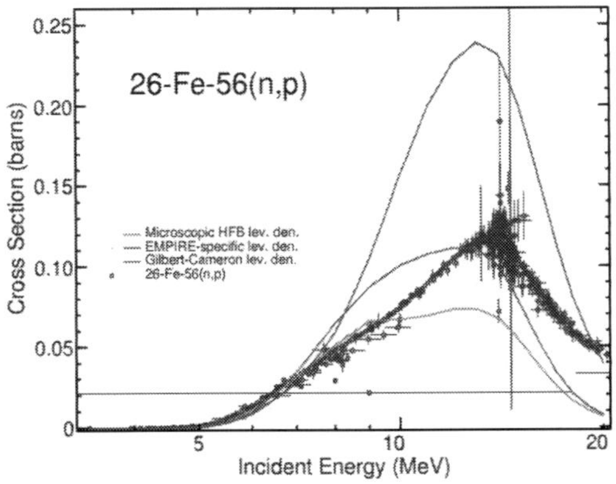

Figure 1. Different results for ^{56}Fe(n,p) cross section obtained by EMPIRE default calculations using different level-density models: microscopic Hartree-Fock-Bogoliubov (green curve); EMPIRE-specific level densities adjusted to RIPL-3 experimental D_{obs} and to discrete levels (red cruve); Gilbert-Cameron level-density model (gray curve).

3.1 Models adopted

In this section we will describe the models which were adopted in the calculations leading to the results that will be presented in the following sections.

The coupled channels approach was adopted to describe the direct reaction part, with levels not explicitly coupled being handled through Distorted-Wave Born Approximation (DWBA). The optical model potential used was the ^{56}Fe-specific dispersive potential by Soukhovitskii and Capote [6]. To account for the correlations between incident and exit channels in elastic scattering (width fluctuation correction) the model by Hofmann, Richert, Tepel and Weidenmueller (HRTW) [7] was employed up to incident energy of 8 MeV. The γ strength function adopted was the version 1 of the Modified Lorentzian (MLO1) from Ref. [8]. The pre-equilibrium contribution to neutron emission was treated in terms of the TUL formulation [9] of the multi-step direct mechanism for incident energies above 3 MeV complemented with the NVWY [10] multi-step compound in the whole energy range. The classical exciton model (PCROSS option), with Iwamoto-Harada extension for clusters, was used for the remaining ejectiles. The multiplier on the free path of excitons was set to 2.5. The adopted optical model, while performing superbly at higher incident energies, overshoots experimental data below 3 MeV. To compensate for this deficiency an energy-dependent factor scaling all cross sections was applied up to 3 MeV. We did not attempt a more rigorous approach of refitting parameters of the optical potential since in the final evaluation the region in question will be represented by experimental data.

3.2 Selection of level-density model

The ^{56}Fe(n,p) reaction has been carefully evaluated and extremely well defined for the IRDFF dosimetry file [1]. Calculations show strong sensitivity of the ^{56}Fe(n,p) cross sections to level densities (LD), indicating that this reaction may provide a very stringent test of the employed LD model. In Fig. 1 we show preliminary results obtained by default EMPIRE calculation using three different LD models.

We initially adopted the microscopic HFB LD model for further fitting as it reproduces the observed data in the region below ~ 9 MeV (see Fig. 1) and at higher energies we could manually alter pre-calculated RIPL-3 tables to impose agreement with the IRDFF file. It was observed that (n,p) cross sections depend strongly on both ^{56}Fe (incoming channel) and ^{56}Mn (outgoing channel) LD. After preliminary fittings of ^{56}Mn LD parameters we reached a reasonably good agreement, as seen in Fig. 2(a) (red curve). It is observed that a very minor change in LD, from red to blue curves in Fig. 2(b), already leads to a noticeable change in calculated cross sections (Fig. 2(a)). Moreover, the green curve in Fig. 2(b) represents a fit of the experimental LD measured by Schiller *et al.* [4], which is not dramatically different from the red curve. However, the corresponding cross section (green

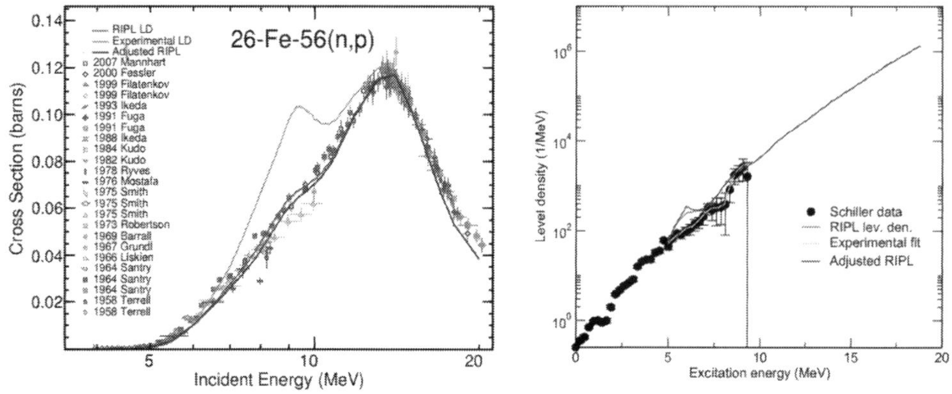

(a) ^{56}Fe(n,p) cross section. Data taken from EXFOR [4]. (b) ^{56}Fe level densities. Data taken from Ref. [4].

Figure 2. ^{56}Fe(n,p) cross section (left) and the corresponding level density used in calculation (right).

curve in Fig. 2(a)) is in significant disagreement with experimental data. This is likely to be due to different assumptions about spin and parity distributions during processing of the measurements.

Two LD models were selected for further testing: the HFB microscopic model and the Gilbert-Cameron (GC) one. In the case of the former we fitted the pairing-like and pseudo-a parameters for ^{56}Mn, in order to reproduce the IRDFF (n,p) cross section. We also introduced minor changes to the ^{56}Fe LD. In the case of GC model we fitted only the a level density parameter, aiming to reproduce IRDFF (n,p) and experimental α production [4] cross sections. All other LD parameters from GC model were calculated internally to ensure consistency and physically reasonable matching between LD and discrete levels. The energy-dependent tuning of the equilibrium decay width for the α channel was introduced for incident energies between 9 and 18 MeV to account for model deficiencies.

In Fig 3 we present some of the results obtained by our calculations and fits, for both microscopic HFB and Gilbert-Cameron LD models. Fig. 3(a) shows that calculations from both models are in good agreement with IRDFF, even though GC model does not reproduce the region around 6-7 MeV as well the microscopic HFB model. This is understandable in view of the stronger constraints on the GC model in contrast with the arbitrary modifications applied to the HFB model. For the total inelastic (Fig. 3(b)) we can observe a small difference between the results from the two LD models, both in agreement with recent experiments. Fig 3(c) show that GC model leads to slight improvement of the agreement with experimental data of the ^{56}Fe(n,2n) reaction. Additionally, Figs. 3(d) and 3(e) (which correspond to double-differential spectra at 8.17 MeV and α production cross section, respectively) display a significantly better agreement of GC model calculations when compared with the HFB ones.

3.3 Angular distributions

For angular distributions we assembled a full file according to the following prescription. *Integrated cross sections:* Up to 846 keV, which is the threshold for the first inelastic state, the evaluated file consists of fitted resonance parameters. The total cross section, for neutron-incident energy between 846 keV and 4 MeV, is given by JEFF-3.2 [3], which corresponds to smoothed Berthold *et al.* experimental data [4]. Angle-integrated inelastic cross sections for the excited levels are given by experimental data from Negret *et al.* [11]. All other channels, except the elastic one, are taken from EMPIRE calculations with GC LD , as the elastic is defined as the difference between the total cross section and

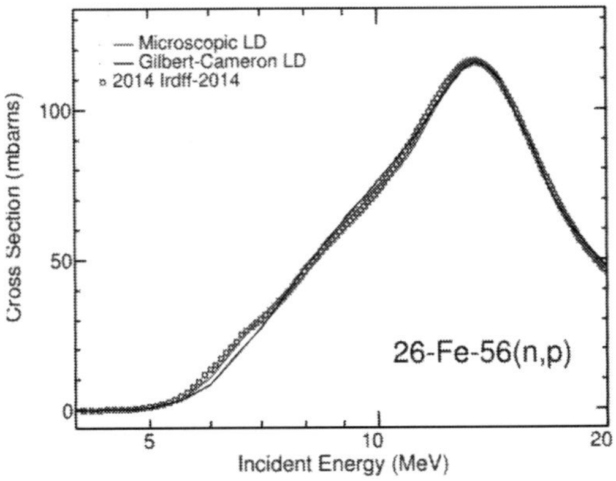

(a) ^{56}Fe(n,p) cross section. Data taken from IRDFF [1].

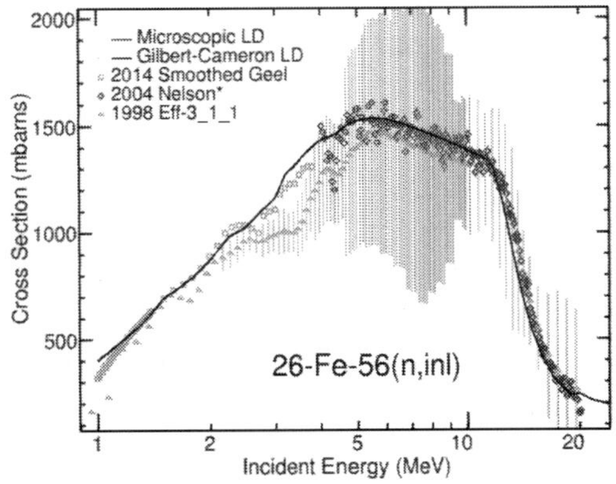

(b) ^{56}Fe inelastic cross section. Data taken from EXFOR [4].

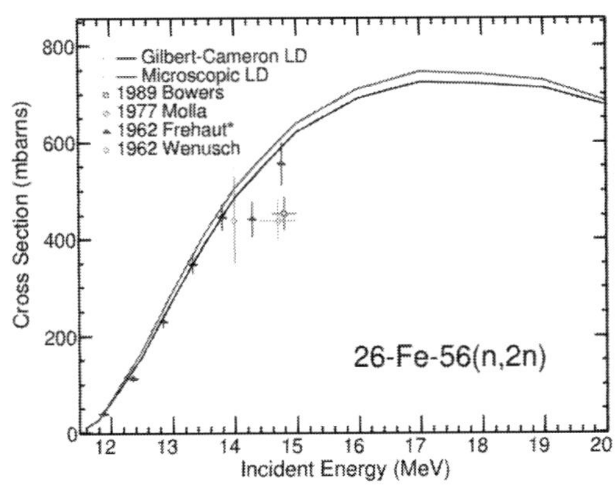

(c) ^{56}Fe(n,2n) cross section. Data taken from EXFOR [4].

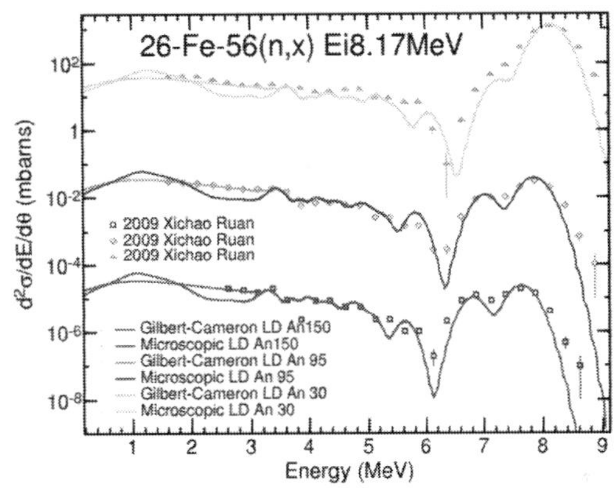

(d) ^{56}Fe double-differential spectra at 8.17 MeV for 30, 95 and 150 degrees. A successive factor of 10^3 was applied for the last two angles to facilitate the visualization. Data taken from EXFOR [4].

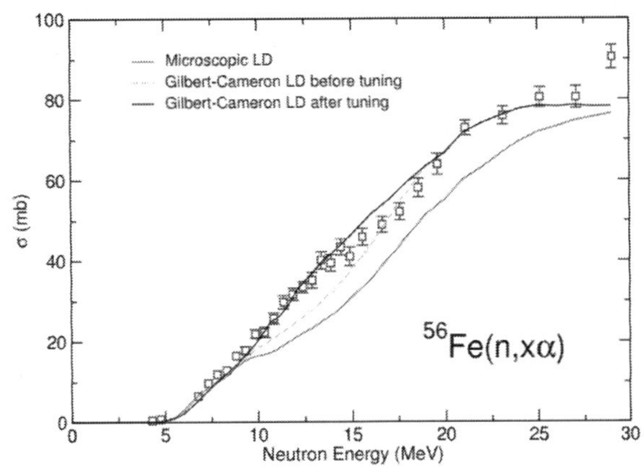

(e) ^{56}Fe(n,xα) cross section. Data taken from Ref. [4].

Figure 3. Comparison of results for both LD models for different neutron-induced reactions on ^{56}Fe.

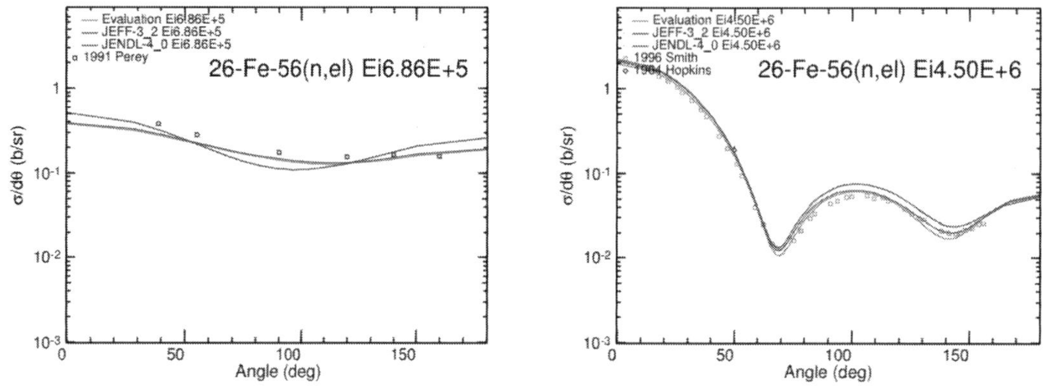

Figure 4. (color online) Elastic angular distributions for neutron-incident energy of 686keV (left panel) and 4.50 MeV (right panel) as examples.

the sum of all other channels. *Angular distributions:* The angular distributions for incident energies lying within the resonance region and up to 846 keV are taken from JENDL-4.0 [2]. For all other values of incident energy, differential cross sections are obtained directly from EMPIRE calculations.

Due to the the large availability of experimental data in the literature we were able to generate more than 2800 elastic angular distribution plots. In Fig. 4 we show as example two of them: one for the lower incident-energy regime and one for a higher energy.

4 Conclusions

In this work we presented the latest developments on a new reaction evaluation for ^{56}Fe, as part of the CIELO project. Calculations show great sensibility of cross sections on the level-density model adopted. Comparisons indicate that employing Gilbert Cameron LD model provides better results. ^{56}Fe(n,p) cross sections were succesfully fitted to the IRDFF dosimetry file while other reaction channels are in good agreement with experimental data. An assembled file was produced with associated elastic angular distributions calculated, reaching reasonably good agreement with experimental data.

References

[1] R. Capote *et al.*, Journal of ASTM International (JAI) **9**, JAI104119 (2012).
[2] K. Shibata *et al.*, J. Nucl. Sci. Tech. **48(1)**, 1-30 (2011).
[3] `https://www.oecd-nea.org/dbforms/data/eva/evatapes/jeff_32/`
[4] N. Otuka et al., Nucl. Data Sheets **120**, 272 (2014).
[5] M. Herman *et al.*, Nucl. Data Sheets **108**, 2655 (2007).
[6] E. Soukhovitskii and R. Capote, Unpublished (RIPL reference index: 614).
[7] H. M. Hofmann *et al.*, Ann. Phys. **90**, 403 (1975); H. M. Hofmann *et al.*, Z. Phys. **297**, 153 (1980).
[8] V. A. Plujko *et al.*, Journal Nucl. Science and Tech. **39**, 811 (2002).
[9] T. Tamura, T. Udagawa, and H. Lenske, Phys. Rev. **C26**, 379 (1982).
[10] H. Nishioka *et al.*, Ann. Phys. **172**, 67 (1986).
[11] A. Negret, C. Borcea, and A. J. M. Plompen, Phys. Rev. C **88**, 027601 (2013).

EPJ Web of Conferences **111**, 03002 (2016)
DOI: 10.1051/epjconf/201611103002

Evaluation of Neutron Capture Cross Sections and Covariances on ^{99}Tc and ^{129}I in the keV Energy Region

Nobuyuki Iwamoto[1,a]

[1] *Nuclear Data Center, Nuclear Science and Engineering Center, Japan Atomic Energy Agency, 2-4 Shirakata, Tokai, Ibaraki 319-1195, JAPAN*

Abstract. Neutron capture cross sections and covariances on radioactive ^{99}Tc and ^{129}I have been required for developing environmental load-reducing technology. Their evaluation was performed by using nuclear reaction calculation code CCONE and Baysian code KALMAN with data assumed on the basis of measured data. The obtained total and capture cross sections are in good agreement with the measured data. The resulting uncertainties of capture cross section were 12-18% and 20-29% for ^{99}Tc and ^{129}I, respectively, in the keV energy region.

1 Introduction

Neutron capture cross sections and covariances on long-lived fission products (LLFPs) as well as major and minor actinides have been required for development of environmental load-reducing technology. Neutron cross section data for the LLFPs of ^{99}Tc (half-life 0.21 Myr) and ^{129}I (15.7 Myr) are included in major evaluated libraries (e.g., JENDL-4.0 [1], ENDF/B-VII.1 [2] and JEFF-3.2 [3]). Their covariance data were, however, not included in JENDL-4.0 and JEFF-3.2, although they are already taken into account in ENDF/B-VII.1. The cumulative yields of radioactive ^{99}Tc and ^{129}I are 6.1-5.2% and 0.5-3.0%, respectively, in ^{235}U fission at thermal to fast neutron energies [4]. They have long lasting radiotoxicity due to long half-lives. Thus, it is well known that they are the candidates for nuclear transmutation by neutron capture reactions in, e.g., accelerator driven system. In the present work the evaluation of neutron capture cross sections and covariances for ^{99}Tc and ^{129}I was performed in the 1-10^3 keV energy region (hereafter referred to as keV energy region).

2 Evaluation Method

The evaluation of neutron cross sections was done by using nuclear reaction calculation code CCONE (version 0.8) [5]. The detailed methodology of cross section evaluation for ^{99}Tc can be found in Ref. [6]. For ^{129}I the coupled-channels optical model based on the rigid rotator model was adopted with the potential form of Kunieda et al. [7] for target nucleus. The coupled levels were taken from the Evaluated Nuclear Structure Data File (ENSDF) [8]. The deformation parameters fixed by fitting to measured data were relatively small (0.05 and 0.03 for ^{127,129}I, respectively), compared to neighboring even-even nuclides (0.12-0.18). Similar results were recently obtained by Shibata [9].

[a] e-mail: iwamoto.nobuyuki@jaea.go.jp

Article available at http://www.epj-conferences.org or http://dx.doi.org/10.1051/epjconf/201611103002

The Hauser-Feshbach statistical model with width fluctuation correction was used with discrete level information retrieved from Reference Input Parameter Library (RIPL-3) [10]. The nuclear level density was adopted by the composite Gilbert-Cameron formula with Fermi-gas model [11]. The level density parameters were determined if s-wave resonance spacings were available from measurements. Otherwise, the systematic values of Ref. [11] were used. Generalized Lorentzian form was employed for gamma strength function of E1 radiation [12]. The M1 and E2 transitions for gamma-decays were also taken into account by the Lorentzian form [12].

Sensitivities of the parameters, related to the optical model potential (OMP) and reaction models, to the cross sections were also calculated by the CCONE code, and then, the sensitivity matrix **C** was generated from these sensitivities for covariance evaluation. The uncertainties for the potential depths of OMP were given to be 5%. The uncertainties for potential radius and diffuseness were 3% and 10%, respectively. The relatively large uncertainties (10-50%) were assigned for the deformation parameters, since their information was not directly obtained for odd isotopes. The uncertainties of level density parameter were chosen to be 3%, which was small relative to those of other parameters for the reaction models. Thus, the initial parameter covariance matrix was made from these parameter uncertainties without considering the correlation among the parameters.

The covariance evaluation was done by using Baysian code KALMAN [13], which is based on the generalized least-square approach. In this code the parameter covariance matrix was used as prior covariance matrix **X**. Experimental covariance matrix **V** was made by assuming data with errors at energy points. The data and energy were taken from the present cross section data. The data errors (σ) consisted of statistical (σ_{sta}) and systematic (σ_{sys}) components like $\sigma^2 = \sigma^2_{sta} + \sigma^2_{sys} = 2\,\sigma^2_{sta}$. The correlation among the data was prescribed as follows:

$$Corr\,(V_{i,j}) = \begin{cases} 1.0 & \text{if } i = j, \\ 0.5 & \text{if } i \neq j, \end{cases} \tag{1}$$

instead of directly using statistical and systematic errors derived from measurements. One reason to make this assumption is that error given in experimental data is relatively small, but the dispersion of experimental data is much larger than the errors. The posterior covariance was calculated by

$$\mathbf{P} = (\mathbf{X}^{-1} + \mathbf{C}^T\mathbf{V}^{-1}\mathbf{C})^{-1} \tag{2}$$

$$= \mathbf{X} - \mathbf{X}\mathbf{C}^T(\mathbf{C}\mathbf{X}\mathbf{C}^T + \mathbf{V})^{-1}\mathbf{C}\mathbf{X}, \tag{3}$$

from the prior and experimental covariance matrices. Finally, the cross section covariance matrix was obtained by the usual equation $\mathbf{M} = \mathbf{C}\mathbf{P}\mathbf{C}^T$.

3 Results

3.1 ^{99}Tc

The evaluated results of total cross section and covariance are shown as a function of incident neutron energy in Figure 1. The gray shaded area represents the prior error which was given to cover the experimental data from uncertainties of the OMP parameters. The assumed errors at 5 energy points were shown by green bars. The red shaded area shows the posterior error obtained by the KALMAN calculation. The prior and posterior errors are also shown by blue dashed and solid lines at the right axis. The result of total cross section reasonably explains the experimental data of Foster Jr and Glasgow [14]. The posterior error was obtained to be 3-6% in the keV energy region.

Figure 2 shows the evaluated result of capture cross section. The evaluated errors are almost consistent with most of the measured data, except for Chou et al. [15] who derived the data slightly

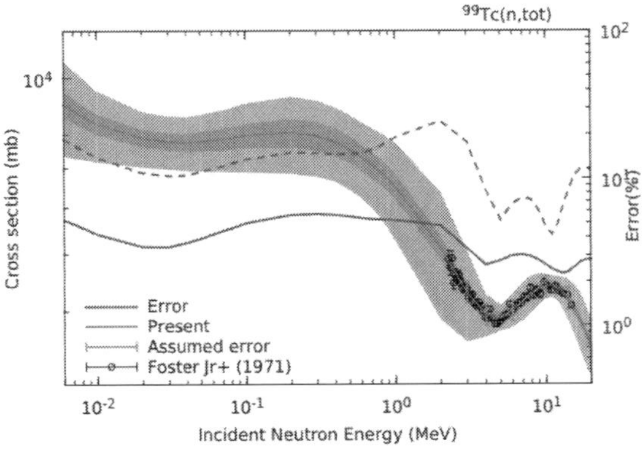

Figure 1. Evaluated results of total cross section and its uncertainty of ^{99}Tc.

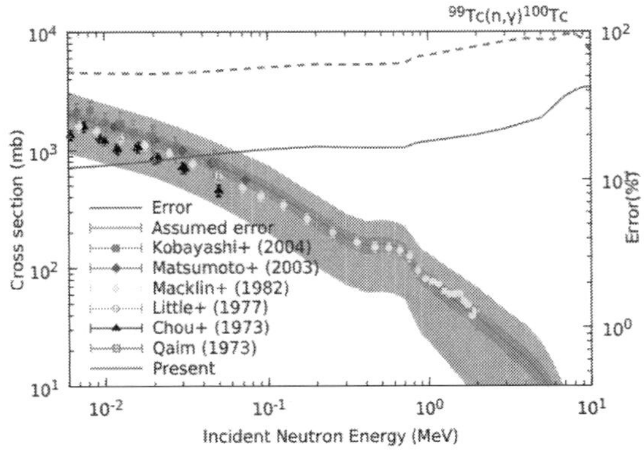

Figure 2. Evaluated results of capture cross section and its uncertainty of ^{99}Tc.

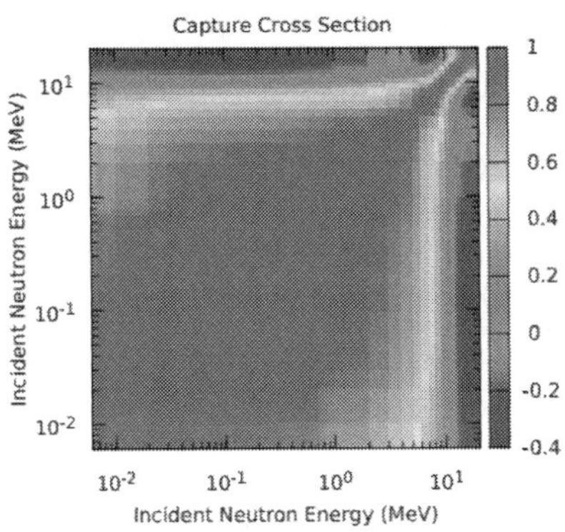

Figure 3. Correlation matrix for capture cross section of ^{99}Tc.

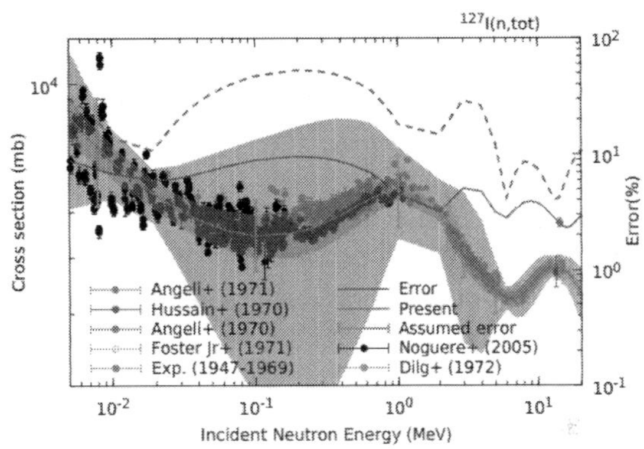

Figure 4. Evaluated results of total cross section and its uncertainty of ^{127}I.

smaller. The prior error was firstly assumed to be 50% or higher, but the posterior error was well constrained by the assumed errors and other information coming from measured capture gamma-ray spectra [16]. The evaluated result of uncertainty was 12-18% in the keV energy region.

Figure 3 shows the resulting correlation matrix of capture cross section. The strong and positive correlation is found below 10 MeV. This is because the positive change of parameters such as level density parameter and normalization factor of gamma strength functions for the compound nucleus, makes the cross section uniformly increase in this region as seen in the figure on parameter sensitivities for ^{129}I below.

3.2 ^{129}I

In contrast to ^{99}Tc, the limited number of experiments was made for radioactive ^{129}I. Therefore, the neutron cross sections of ^{129}I were evaluated, together with those of stable ^{127}I, because many experimental efforts have been done for ^{127}I. The present evaluation of total cross sections was based on data recently measured by Noguere et al. [17, 18]. The result for ^{127}I is illustrated in Figure 4. The adopted OMP parameters have mass-number dependence, but the potential differences coming

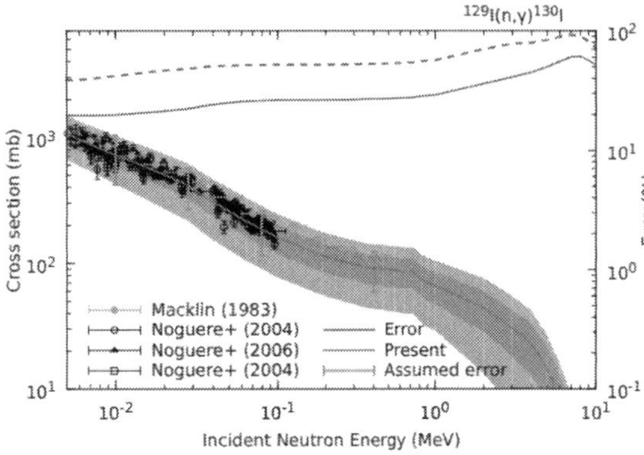

Figure 5. Evaluated results of capture cross section and its uncertainty of ^{129}I.

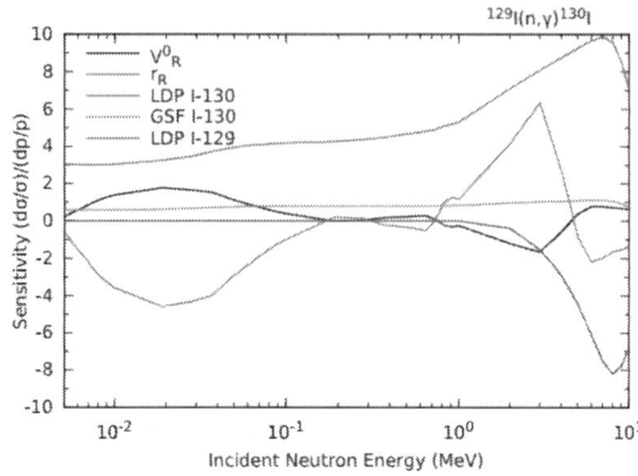

Figure 6. Sensitivities of model parameters to capture cross section for ^{129}I.

from the mass-number difference are not so large that the total cross sections show almost similar characteristics. The present OMP parameters reasonably explain the measured data of both ^{127,129}I. For covariance evaluation the representative points were assumed to encompass all measured data in the energy range of 5 keV to 20 MeV for ^{127}I, whereas measured data for ^{129}I were provided only below the neutron energies of 60 keV [17]. Thus, the overall uncertainty evaluation comes from the assumed data of ^{127}I. The uncertainty of 5-10% for ^{129}I in the keV energy region was derived by the present method of covariance evaluation, although individual errors of measured data are typically much smaller. This result can be found from Figure 4. Figure 5 shows the evaluated capture cross section on ^{129}I. The data of Macklin [19] are marginally consistent with those of Noguere et al. [17, 20]. The present evaluation takes into account both the measured data and gives the errors of 20-29% in the keV energy region.

The sensitivities of the parameters for the OMP and reaction models to the capture cross section on ^{129}I are represented in Figure 6, which were calculated by the CCONE code. The capture cross section is very sensitive to the level density parameter of compound nucleus ^{130}I (LDP I-130). The negative sensitivity for level density parameter of target ^{129}I (LDP I-129) is found above 1MeV. This is due to the contribution of inelastic scattering cross section. The gamma strength function of ^{130}I (GSF I-130) has a weak, but positive sensitivity to the cross section. The sensitivities for OMP parameters (e.g., constant leading term V_R^0 of potential depth and radius r_R in the real volume term) are changed to be positive and negative with changing neutron energy.

4 Summary

The evaluation of neutron capture cross sections and covariances on well-known LLFPs of ^{99}Tc and ^{129}I was performed by using CCONE-KALMAN code system, in order to recognize the present uncertainties of capture cross sections which are needed to develop environmental load-reducing technology. The measured data of stable ^{127}I were used for the better cross section and uncertainty evaluations of ^{129}I due to the lack of experimental information. The calculated data for ^{99}Tc and ^{129}I well reproduce the measured data of total and capture cross sections. The obtained uncertainties of capture cross section resulted in 12-18% for ^{99}Tc and 20-29% for ^{129}I in the keV energy region. The capture cross sections of the evaluated libraries (JENDL-4.0, ENDF/B-VII.1 and JEFF-3.2) are within the present uncertainties.

Acknowledgements

Present study includes the result of "Research and Development for accuracy improvement of neutron nuclear data on minor actinides" entrusted to the Japan Atomic Energy Agency by the Ministry of Education, Culture, Sports, Science and Technology of Japan (MEXT).

References

[1] K. Shibata, O. Iwamoto, T. Nakagawa, N. Iwamoto, A. Ichihara, S. Kunieda, S. Chiba, K. Furutaka, N. Otuka, T. Ohsawa, T. Murata, H. Matsunobu, A. Zukeran, S. Kamada, J. Katakura, J. Nucl. Sci. Technol. **48**, 1 (2010)

[2] M.B. Chadwick, M. Herman, P. Obložinský, M.E. Dunn, Y. Danon, A.C. Kahler, D.L. Smith, B. Pritychenko, G. Arbanas, R. Arcilla, R. Brewer, D.A. Brown, R. Capote, A.D. Carlson, Y.S. Cho, H. Derrien, K. Guber, G.M. Hale, S. Hoblit, S. Holloway, T.D. Johnson, T. Kawano, B.C. Kiedrowski, H. Kim, S. Kunieda, N.M. Larson, L. Leal, J.P. Lestone, R.C. Little, E.A. McCutchan, R.E. MacFarlane, M. MacInnes, C.M. Mattoon, R.D. McKnight, S.F. Mughabghab, G.P.A. Nobre, G. Palmiotti, A. Palumbo, M.T. Pigni, V.G. Pronyaev, R.O. Sayer, A.A. Sonzogni, N.C. Summers, P. Talou, I.J. Thompson, A. Trkov, R.L. Vogt, S.C. van der Marck, A. Wallner, M.C. White, D. Wiarda, P.G. Young, Nucl. Data Sheets, **112**, 2887 (2011)

[3] OECD Nuclear Energy Agency Data Bank.

[4] J. Katakura, JAEA-Data/Code 2011-025 (Japan Atomic Energy Agency) (2012)

[5] O. Iwamoto, J. Nucl. Sci. Technol. **44**, 687 (2007)

[6] N. Iwamoto, J. Nucl. Sci. Technol. **49**, 244 (2012)

[7] S. Kunieda, S. Chiba, K. Shibata, A. Ichihara, E.Sh. Sukhovitskii, J. Nucl. Sci. Technol. **44**, 838 (2007)

[8] Evaluated Nuclear Structure Data File. Available from http://www.nndc.bnl.gov/ensdf/

[9] K. Shibata, J. Nucl. Sci. Technol. **52**,1174 (2015)

[10] R. Capote, M. Herman, P. Oblozinsky, P.G. Young, S. Goriely, T. Belgya, A.V. Ignatyuk, A.J. Koning, S. Hilaire, V.A. Plujko, M. Avrigeanu, O. Bersillon, M.B. Chadwick, T. Fukahori, Z. Ge, Y. Han, S. Kailas, J. Kopecky, V.M. Maslov, G. Reffo, M. Sin, E.Sh. Soukhovitskii, P. Talou, Nucl. Data Sheets, **110**, 3107 (2009)

[11] A. Mengoni, Y. Nakajima, J. Nucl. Sci. Technol. **31**,151 (1994)

[12] J. Kopecky, M. Uhl, Phys. Rev. **C41**, 1941 (1990)

[13] T. Kawano, K. Shibata, JAERI-Data/Code 97-037 (Japan Atomic Energy Agency) (1997) [in Japanease]

[14] D.G. Foster Jr., D.W. Glasgow, Phys. Rev. **C3**, 576 (1971)

[15] J.-C. Chou, H. Werle, J. Nucl. Energy, **27**, 811 (1973)

[16] T. Matsumoto, M. Igashira, T. Ohsaki, J. Nucl. Sci. Technol. **40**, 61 (2003)

[17] G. Noguere, A. Brusegan, A. Leprêtre, N. Herault, O. Bouland, G. Rudolf, *Int. Conf. on Nucl. Data for Sci. and Technol., Santa Fe 2004 Vol.2* (AIP, New York, 2005) 1462

[18] G. Noguere, A. Brusegan, A. Leprêtre, N. Herault, O. Bouland, G. Rudolf, Private communication (2005)

[19] R.L. Macklin, Nucl. Sci. Eng. **85**, 350 (1983)

[20] G. Noguere, O. Bouland, A. Brusegan, P. Schillebeeckx, P. Siegler, A. Leprêtre, N. Herault, G. Rudolf, Phys. Rev. **C74**, 054602 (2006)

EPJ Web of Conferences **111**, 03003 (2016)
DOI: 10.1051/epjconf/201611103003

Monte Carlo simulation of transfer reactions using extended R-matrix theory picturing surrogate-type WFCF features

Olivier H. Bouland[a]

CEA, DEN, DER, SPRC, Physics Studies Laboratory, Cadarache, F-13108 Saint-Paul-lez-Durance, France

Abstract. This article supplies an overview of issues related to the interpretation of surrogate measurement results for neutron-incident cross section predictions; difficulties that are somehow masked by the historical conversion route based on Weisskopf-Ewing approximation. Our proposal is to handle the various difficulties by using a more rigorous approach relying on Monte Carlo simulation of transfer reactions with extended *R*-matrix theory. The multiple deficiencies of the historical surrogate treatment are recalled but only one is examined in some details here; meaning the calculation of in-out-going channel Width Fluctuation Correction Factors (WFCF) which behavior witness partly the failure of Niels Bohr's compound nucleus theoretical landmark. Relevant WFCF calculations according to neutron-induced surrogate- and cross section-types as a function of neutron-induced fluctuating energy range [*0 - 2.1 MeV*] are presented and commented in the case of the $^{240}Pu^*$ and $^{241}Pu^*$ compound nucleus isotopes.

1 Introduction

The aim of this paper is to supply an overview of issues related to the conversion of surrogate measurements to neutron-induced cross sections; difficulties that are somehow masked by persistent use of the so-called Weisskopf-Ewing approximation (WE). The difficulties will be enlightened by using the more rigorous approach developed in our AVXSF-LNG (AVerage CROSS Section Fission – Lynn and Next Generation) code [1] at CEA Cadarache in collaboration with LANL/T2. Our code is based on Monte Carlo simulation of transfer reactions (also referred as surrogates) and makes use of extended *R*-matrix theory.

Surrogate measurements are substitute technique to determine reaction cross sections for compound nuclei that are difficult to measure directly by neutron physics spectroscopy (NPS) or to predict with some degree of confidence from 'systematics' or theory. The surrogate method comes as an alternative to form the compound nucleus (CN), let say *C**, usually formed in NPS as [*n+target→ C**]. Alternatively another projectile-target combination, more accessible experimentally, can be involved such that [*projectile + Surrogate_Target → C*+ ejectile*]. By measuring the number of coincidences between the ejectile and the observable that identifies the desired exit reaction channel (*c'*), normalized to total number of measured surrogate events (i.e; how many times the selected CN is formed), the experimental probability, let's say $\beta^{CN}_{sp,c'}(E^*)$, related to CN formation at excitation

[a] Corresponding author: olivier.bouland@cea.fr

energy E^* by given surrogate reaction (labeled '*sp*' since the CN entrance channel is of single-particle type) and subsequent decay in reaction channel c', is obtained experimentally. Combination (see later in the text) of $\beta^{CN}_{sp,c'}(E^*)$ with calculated neutron-induced CN formation cross section is then routinely carried out using the WE frame and benchmarked, whenever possible, with neutron-incident cross sections that were directly measured.

2 Historical surrogate modeling

The starting point and appropriate formalism for describing CN reactions is Hauser-Feshbach (HF) statistical theory [2], meaning the pure Hauser-Feshbach equation (as stated by equation 1 of Ref. [1]) that carries 'Niels Bohr's hypothesis' of independence of formation and decay of the CN [3] in first order. A more realistic picture involves $W_{c,c'}$, the in-out-going channel WFCF. The formulation for the average partial reaction cross section, $\sigma_{c,c'}(E_c)$, according to entrance channel c and exit channel c' applied to neutron-induced reactions at given neutron energy E_n is then addressed as

$$\sigma_{n,c'}(E_n) = \sum_{J^\pi} \sigma_n^{CN}(E_n, J, \pi) \sum_{s'=|I'-i'|}^{|I'+i'|} \sum_{l'=|J-s'|}^{|J+s'|} \frac{T_{c'}^{J^\pi(l's')}(E_{c'})}{\sum_{c''} T_{c''}^{J^\pi(l''s'')}(E_{c''})} \times W_{n,c'}^{J^\pi} , \tag{1}$$

where $\sigma_n^{CN}(E_n, J, \pi)$ is the neutron-induced compound nucleus formation cross section for given (J,π) couple; the expression of which is,

$$\sigma_n^{CN}(E_n, J, \pi) = \pi \lambda\!\!\!^{-2} g_J \sum_{s=|I-\frac{1}{2}|}^{|I+\frac{1}{2}|} \sum_{l=|J-s|}^{|J+s|} T_n^{J^\pi(ls)}(E_n), \tag{2}$$

with g_J, the statistical spin factor or weight according to CN total angular momentum J as $g_J = (2J+1)/(2(2I+1))$ and where $T_n^{(J^\pi ls)}$ are the neutron entrance transmission coefficients.

Historical surrogate treatment assumes that customary in-out-going channel WFCF can be neglected (i.e ; $W_{c,c}\approx 1$) although, by matter of fact, we know that this correction is substantial [4] for any open channels for excitation energies up to $2\ MeV$ above neutron emission threshold (S_n) as far as actinide nuclides are concerned. According to fission decay, the amount of average cross section correction depends on both the number of fission channels involved and the magnitude of their average widths. The larger the sub-barrier effect is (case of fertile heavy isotopes), the larger the amount of fluctuations ($W_{n,f} \approx 35\%$ compared to 20% at 1 keV neutron-incident energy respectively for fertile and fissile isotopes [5]). Obviously $W_{n,f}$ tends to unity as the total number of open channels increases. The absence of WFCF in the cross section formulation is then the first step leading to the WE limit of HF theory. **Equation 1** then switches back to pure HF that can be re-written in a concise manner as,

$$\sigma_{n,c'}(E_n) = \sum_{J^\pi} \left[\sigma_n^{CN}(E_n, J, \pi) \times P_{c'}^{J^\pi}(E_{c'}) \right] , \tag{3}$$

where $P_{c'}^{J^\pi}(E_{c'})$, the individual CN decay probability into reaction channel c', is also commonly referred as branching ratio (BR) to channel c' in associated surrogate literature. Equation 3 can be factorized such that,

$$\sigma_{n,c'}(E_n) = \sigma_n^{CN}(E_n) \sum_{J^\pi} \left[\frac{\sigma_n^{CN}(E_n, J, \pi)}{\sigma_n^{CN}(E_n)} \times P_{c'}^{J^\pi}(E_{c'}) \right], \tag{4}$$

to make provision for $\sigma_n^{CN}(E_n,J,\pi)/\sigma_n^{CN}(E_n)$, the fraction of CN excited states formed with *(J, π)* couple that is commonly labeled as $F_{sp}^{CN}(E^*,J,\pi)$ in surrogate literature. This gives us the opportunity to unfold the experimental (coincidence) surrogate probability accordingly such that writing,

$$\beta_{sp,c'}^{CN}(E^*) = \sum_{J^\pi}\left[F_{sp}^{CN}(E^*,J,\pi)\times P_{c'}^{J^\pi}(E_{c'})\right], \tag{5}$$

Straightforward connection can be made under the so-called WE limit of the Hauser-Fechbach theory between neutron-induced cross section data and $\beta_{sp,c'}^{CN}$; the actually observed surrogate probability that is mostly measured as a function of a single variable meaning the CN excitation energy (E^*). This probability involves the additional crude simplification that BR quantities are independent of spin and parity, meaning substituting $P_{c'}^{J^\pi}(E_{c'})$ by $P_{c'}(E_{c'})$. In the framework of independence between formation and decay processes in compound nucleus system supplemented by the latter hypothesis, branching ratios are automatically pulled off the spin-parity summation in **Eqs. 3** and **5**. The replacement of $P_{c'}(E_{c'})$ in the **Eq. 3** by its equivalence given by **Eq. 5** returns the well-suited and well-known WE expression that is,

$$\sigma_{n,c'}^{WE}(E_n) = \left[\sum_{J^\pi}\sigma_n^{CN}(E_n,J,\pi)\right]\times\beta_{sp,c'}^{CN}(E^*), \tag{6}$$

$$\text{since } \left[\sum_{J^\pi}F_{sp}^{CN}(E^*,J,\pi)\right] = 1. \tag{7}$$

3 Refined transfer reaction data modeling based on extended *R*-matrix Monte Carlo simulations

At first sight, this WE strategy (**Eq. 6**) supplies a suitable and quick estimate of the desired neutron-induced cross section, inaccessible from NPS, without any need to extract individual CN decay reaction probabilities, $P_{c'}^{J^\pi}(E_{c'})$, from surrogate data using both **Eq. 5** and input simulations of the $F_{sp}^{CN}(E^*,J,\pi)$ direct-reaction CN population fractions. The retrieval procedure of Eq. 6 assumes, of course, that the neutron-induced total formation cross section, $\sum_{J^\pi}\sigma_n^{CN}(E_n,J,\pi)$ is ideally computed using convenient optical potential. Unfortunately this WE strategy (**Eq. 6**) has shown its limits over the last decade especially in terms of neutron-induced capture cross section predictions [6] from associated surrogate-reaction measurements although it worked surprisingly pretty well in terms of fission cross section retrieval [7]. Thinking about the conceptual differences between **Eqs. 4** and **5** and according to **Eq. 1**, we realize that

A/ the absence of WFCF prevents at least from dealing with the genuine outgoing channel width correlations; meaning in the case of observed fission, correlations between the fission and total CN decay widths. However, the absence of WFCF can also alter the surrogate data conversion to neutron cross section data since we expect conceptual differences in WFCF calculations between surrogate- and neutron-type measurements; this question being the purpose of this short paper,

B/ the use of the unitary property (**Eq. 7**) washes out *(J,π)* population fraction differences encountered between neutron-induced and surrogate reactions. This is equivalent to state that final results are not sensitive to the actual entrance spin-parity probability distribution although many prospective calculations, spread over four decades (already in Ref. [8]), have shown that distribution patterns are strongly spin-parity reaction dependent and this, at least for nucleus excitation energies up to $(S_n + 2)$ *MeV*.

In consequence of the WE strategy expected deficiencies, we have chosen an approach free of the WE assumption for simulating $\beta^{CN}_{sp,c'}$, the surrogate probability actually measured. This more rigorous method is implemented in our **AVXSF-LNG** code and relies in particular on a Monte Carlo simulation of the transfer reactions using extended *R*-matrix theory. An article in preparation [9] will supply more details. However a refined analytical formulation of the fission channel, allowed also by our code, can illustrate our purpose. This calculates $\beta^{CN}_{sp,f}$ as

$$\beta_{sp,f}(E^*) = \sum_{J^\pi}\left[F^{CN}_{sp}(E^*, J,\pi) \times \left[\sum_{\mu\in J^\pi} P^{\mu}_{f}(E_{c'})W^{\mu}_{II}\right] \times W^{J^\pi}_{sp,f}\right], \qquad (8)$$

where $P^{\mu}_{f}(E_{c'})$ and W^{μ}_{II} are respectively the individual CN fission decay probability over given Aage Bohr [1] transition channel μ and appropriate class-II compound nucleus state WFCF; the behavior and the effect of latter being detailed in [5].

Equation 8 enlightens the importance of the in-out-going channel WFCF although this latter might be inadequately calculated using its customary definition as remembered by the following,

$$\left\langle \frac{\Gamma^{J^\pi}_{c}\Gamma^{J^\pi}_{c'}}{\sum_{c''}\Gamma^{J^\pi}_{c''}} \right\rangle = \frac{\left\langle\Gamma^{J^\pi}_{c}\right\rangle\left\langle\Gamma^{J^\pi}_{c'}\right\rangle}{\sum_{c''}\left\langle\Gamma^{J^\pi}_{c''}\right\rangle} \times W_{c,c'}, \qquad (9)$$

where $\Gamma^{J^\pi}_{c}$ and $\Gamma^{J^\pi}_{c'}$ are respectively channel widths corresponding to the c entrance and c' exit channels.

4 Expected WFCF differences between surrogate- and neutron-induced-type measurements

Straight application of **Eq. 9** to the fissile $^{240}Pu^*$ and fertile $^{241}Pu^*$ compound nucleus isotopes in the case of neutron-induced reaction carried out analogously to the general single variable integral established by Dresner [10] (see also [1] for practical details) can be visualized respectively under **Figs. 1** and **2**. In both cases, we recognize the well-known enhancement of the elastic cross section (greater than +100% as symbolized by the solid-orange-opened circle curve on both graphic insets) whereas the inelastic channel is strongly depreciated (at least of -30%) when neutron inelastic threshold is crossed at $S_n + 7$ *keV*.

We can wonder whether this conventional WFCF prescription remains valid when surrogate reactions are involved and more precisely when a direct reaction is the first stage leading ultimately to CN formation. Considering the very different nature of the entrance channel in the latter case compared to neutron-induced reactions, we might consider the extreme case of no correlation between CN formation and decay widths as it is suggested in Ref.[9] according to *(t,pf)* surrogate measurements. **Eq. 9** then reduces to the following

$$\left\langle \frac{\Gamma^{J^\pi}_{sp}\Gamma^{J^\pi}_{c'}}{\sum_{c''}\Gamma^{J^\pi}_{c''}} \right\rangle \equiv \Gamma^{J^\pi}_{sp}\frac{\left\langle\Gamma^{J^\pi}_{c'}\right\rangle}{\sum_{c''}\left\langle\Gamma^{J^\pi}_{c''}\right\rangle} \times W_{sp,c'}, \qquad (10)$$

where $\Gamma_{sp}^{J^\pi}$ is the entrance channel width related to the single particle stripped from the incident light particle (e.g; a dineutron for a *(t,p)* entrance vector) and transferred into corresponding sp-orbit of target nucleus that carries two neutrons less than the formed CN.

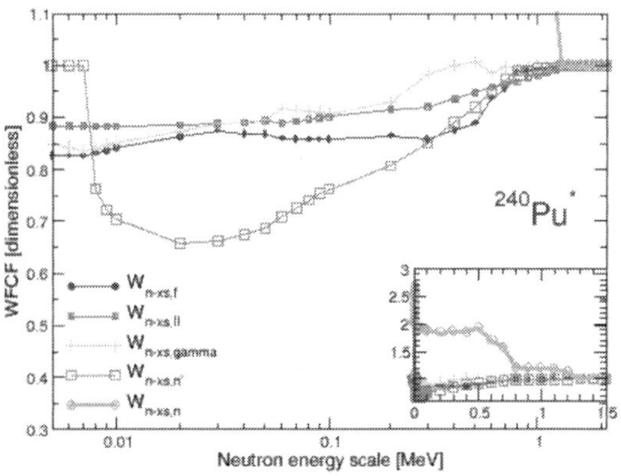

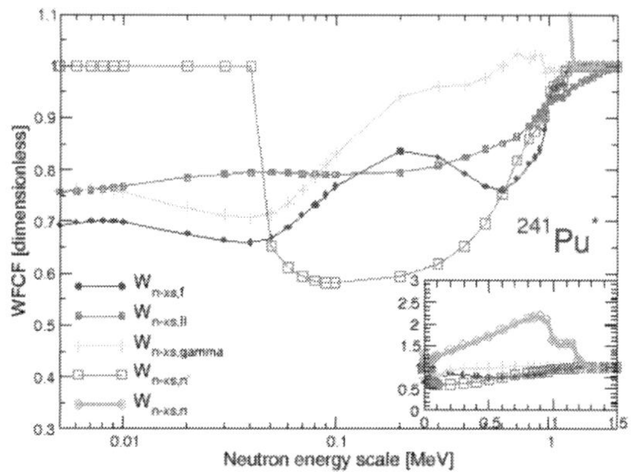

Figures 1 (left side) and 2 (right side): customary WFCF curves (**Eq. 9**) as a function of neutron incident energy respectively for the fissile $^{240}Pu^*$ and fertile $^{241}Pu^*$ CN isotopes. WFCF curves are related respectively to overall fission ($W_{n\text{-}xs,f}$), class-II CN state intrinsic properties ($W_{n\text{-}xs,II}$), gamma emission ($W_{n\text{-}xs,\gamma}$), neutron elastic scattering ($W_{n\text{-}xs,n}$) and inelastic emission ($W_{n\text{-}xs,n'}$). Customary WFCF behavior manifests by an amount of flux taken from each exit channel that is not also an entrance channel (fission, gamma and inelastic neutron emission) and redistributed to the entrance-exit channel (meaning the neutron channel according to a neutron elastic reaction).

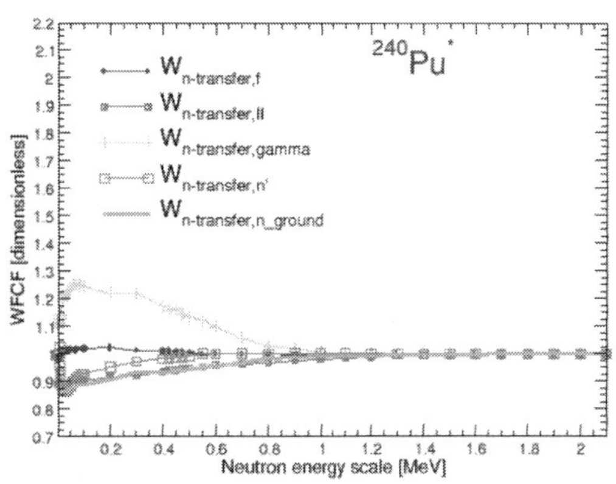

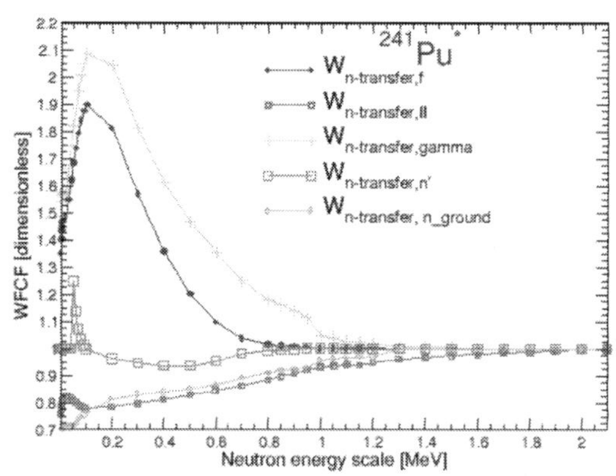

Figures 3 (left side) and 4 (right side). Analogous WFCF numerical Dresner calculations as **Figs. 1** and **2** except that in-out-going channel width correlation is assumed to be negligible in transfer reaction treatment as addressed by **Eq. 10**. For genuine comparisons to the WFCF cross section-type calculations, same J^π CN state populations have been supplied according to a neutron entrance channel configuration.

Figures 3 and **4** supply a quick overview of how modified are WFCF as assumed in the surrogate extreme limit (**Eq. 10**) of none in-out-going channel width correlation but with preservation of channel width correlations between the various CN exit channels. By analogy to the customary situation, each WFCF tends to unity when the total number of channels becomes very large (in practice above *(S_n +1.5) MeV* of excitation energy for these plutonium isotopes). However the absence of elastic reaction prevents any elastic enhancement (since no exit channel is similar by nature to the direct reaction entrance channel) but it happens that both radiative and fission decays now endorse the role of the enhanced channel. In this picture, the WFCF coefficients related to $^{241}Pu^*$ carry the maximum of strength (up to a factor 2.1 for the γ-decay probability – **Fig. 4** - $W_{n\text{-transfer},\gamma}$) because of the $^{241}Pu^*$ neutron-incident sub-threshold fission nature that maintains at high level partial cross section fluctuations. By reciprocity, neutron emission channels are depreciated accordingly to the total amount of reaction rate redistributed. This statement can be materialized as

$$\left(WFCF_{sp,f} \times P_f + WFCF_{sp,\gamma} \times P_\gamma\right) \equiv \left| \sum_i \left(WFCF_{sp,n_i} \times P_{n_i}\right) \right|$$

where n_i corresponds to given CN neutron emission channel i including the channel corresponding to the residual nucleus left in its ground state.

5 Summary

This paper clearly recalls the four key ingredients, $F_{sp}^{CN}(E^*, J, \pi)$, $P_f^{\mu}(E_{c'})$, $W_{sp,f}^{J^\pi}$ and W_{II}^{μ}, which full knowledge is crucial for any valuable comparison between predicted and observed surrogate probabilities over low excitation energy range ($4.0 < E^*\,[MeV] < S_n+2.0$) as it was suggested by recent measured γ-decay surrogate-type probabilities. This paper acknowledges in particular the dedicated behavior of surrogate WFCF as assumed in the extreme limit (**Eq. 10**) of none in-out-going channel width correlation but with preservation of channel width correlations between the various CN exit channels. This special behavior is worth to be noted since we are not usually accustomed to deal with it. In particular we realize that both radiative and fission decays can now endorse the role of the enhanced channel. The enhancement factor observed on both fission and γ-decay probabilities is especially strong (up to +100%) for fertile heavy isotopes in the energy range above S_n although the actual consequence of wrong prediction on small fission decay probability (of *0.1* magnitude) for fertile nuclides remains limited in neutron reactor applications. On the opposite, this might have great importance for neutron-induced capture cross section retrievals. As conclusion, we own nowadays some capability for more precise surrogate-type probability calculations preventing the fragile use of the historical Weisskopf-Ewing assumption that especially does not include the calculation of WFCF.

The author is indebted to Dr Olivier Serot for discussions and careful reading of the paper material.

References

1. O. Bouland, J.E. Lynn, P. Talou, Phys. Rev. C **88,** 054612 (2013).
2. W. Hauser, H. Feshbach, Phys. Rev. **87**, 366 (1952).
3. N. Bohr, Nature **137**, 344 (1936).
4. S. Hilaire, Ch. Lagrange, A.J. Koning, Annals of Physics **306**, 209-231 (2003).
5. O. Bouland, J.E. Lynn and P. Talou, Nuclear Data Sheets, **118** 211-215 (2014).
6. G. Boutoux, B. Jurado, V. Méot, O. Roig *et al.,* Phys. Lett. B **712**, 319-325 (2012).
7. G. Kessedjian, B. Jurado, M. Aiche, G. Barreau *et al.,* Phys. Lett. B **692**, 5. 297-301 (2010).
8. B.B. Back, O. Hansen, H.C. Britt, J.D. Garrett, Phys. Rev. C **9**, 5, 1924 (1974).
9. O. Bouland, J.E. Lynn, P. Talou, *Revisiting fission-probability data using R-matrix Monte-Carlo simulations*, to be submitted to Phys. Rev. C (2016).
10. L. Dresner, *Proc. Int. Conf. on Neut. Inter. with the nucleus*, Columbia Univ., CU-**175**, 71 (1957).

EPJ Web of Conferences **111**, 03004 (2016)
DOI: 10.1051/epjconf/201611103004

A Lane consistent optical model potential for nucleon scattering on actinide nuclei with extended coupling

José Manuel Quesada[1,a], Roberto Capote[2,b], Efrem S. Soukhovitskiĩ[3,c], and Satoshi Chiba[4,d]

[1]*Departamento de Física Atómica, Molecular y Nuclear, Universidad de Sevilla , Apartado 1065, 41080, Sevilla, España*
[2]*NAPC–Nuclear Data Section, International Atomic Energy Agency, Vienna International Centre, A–1400, Vienna, Austria*
[3]*Joint Institute for Energy and Nuclear Research 220109, Minsk-Sosny, Belarus*
[4]*Research Laboratory for Nuclear Reactors, Tokyo Institute of Technology 2-12-1-N1-9 Ookayama, Meguro-ku, Tokyo 152–8550, Japan*

Abstract. An extension for odd-A actinides of a previously derived dispersive coupled-channel optical model potential (OMP) for ^{238}U and ^{232}Th nuclei is presented. It is used to fit simultaneously all the available experimental databases including neutron strength functions for nucleon scattering on ^{232}Th , ^{233,235,238}U and ^{239}Pu nuclei. Quasi-elastic (p,n) scattering data on ^{232}Th and ^{238}U to the isobaric analogue states of the target nucleus are also used to constrain the isovector part of the optical potential. For even-even (odd) actinides almost all low-lying collective levels below 1 MeV (0.5 MeV) of excitation energy are coupled. OMP parameters show a smooth energy dependence and energy independent geometry.

1 Introduction

The need of higher accuracy data for fast fission reactors ([1, 2]) requires improving the description of neutron induded reactions at energies from a few keV up to 6–7 MeV. While the ground-state band of even-even actinides below 500 keV is well described by a rigid rotor model, above 500 keV several vibrational bands are observed that need to be considered. The situation for odd actinides is even more complex, as no pairing gap exists, therefore low-lying excited states are dominated by single-particle (1QP) states. Rotational bands built on 1QP states are observed at very low excitation energies (e.g. K=1/2+,7/2- and 5/2+ bands dominate the low lying excitation spectra for ^{235}U and ^{239}Pu nuclei) and could be excited in neutron induced reactions. Therefore low-lying rotational bands, built on vibrational band-heads for even-even and on single-particle bandheads for odd nuclei, need to be taken into account to describe neutron inelastic scattering on actinides. Authors were not able to find published optical model potentials for neutron scattering on odd actinides, even if many publications described applications to even-even actinides [3–9]. The authors have derived a dispersive and isospin

[a]e-mail: quesada@us.es
[b]e-mail: Roberto.CapoteNoy@iaea.org
[c]e-mail: esukhov@tut.by
[d]e-mail: chiba.s.ae@m.titech.ac.jp

dependent optical model potential for actinides [10–12] assuming a rigid rotor coupling of levels in the rotational ground state band. The coupling of vibrational bands improved the description of neutron scattering on even-even actinides in the energy region from 500 keV up to about 3 MeV, which is critical for fast neutron fission reactors, as was discussed at the WONDER 2012 workshop [13]. The main purpose of this contribution is to extend the previously derived isospin-dependent dispersive optical model potential for even-even actinides [13] to account for extended multiple-band couplings in odd-A actinides, while keeping the quality of the previously achieved description of scattering data. Preliminary findings for odd-nuclei were presented at ND2013 [14]. The work presented here for including the coupling of rotational bands built on single particle bandheads in odd nuclei is expected to reduce the observed discrepancies in neutron inelastic scattering on major fissile actinides ^{233}U, ^{235}U and ^{239}Pu [2].

2 The formalism

The latest developments of the dispersive coupled-channel optical model potential with full coupling for even-even and odd actinides are reviewed in this section. The basic ingredients of our methodology were presented previously [13], therefore we briefly outline the extension of the coupling model. A full description with all technical details will be presented in a coming publication.

2.1 The target hamiltonian and nuclear wave-functions

The existence of the pairing gap in even-even actinides allows a clear separation between single particle and collective excitations (rotational and vibrational). Therefore the low lying excited states can be described as members of rotational bands built on vibrational excitations. However, the nuclear structure description of odd actinides is particularly difficult due the coupling between the nuclear rotation with the unpaired nucleon [15–17], which does not permit the separation of the single-particle, rotational and vibrational degrees of freedom.

In the model, in even-even actinides, the nucleus comprises the even-even core where only paired nucleons are present. Additionally, for odd-mass nuclei, a single unpaired nucleon moves in the nuclear mean field created by the even-even core. We further assume that the nuclear ground state may be statically deformed, but it is axially symmetric. Dynamical deformations (including non-axial) are assumed to be small, and $E_{rot}(E_p) \ll E_{vib}$, i.e. the adiabatic assumption holds for the separation of the rotational (single-particle) and vibrational motion ([13]). Under those assumptions, the nuclear Hamiltonian can be written as,

$$H = H_{rot} + H_{vib} + H_p + H_{int} \tag{1}$$

where H_{rot} is the rotational energy operator; H_{vib} is the vibrational energy operator; H_p is the energy operator of the unpaired nucleon (single-particle operator); and H_{int} is the interaction energy operator of the unpaired nucleon with the nuclear even-even core field.

For even-even actinides the nuclear Hamiltonian $H \equiv H_{rot} + H_{vib}$ describes the collective motion only, neglecting the interaction of vibrational and rotational states. Adiabatic approximation for the collective motion means that the even-even nuclear wave function can be factorized into a rotational and a vibrational part $\Psi = \Phi_{rot}(\Theta)\left|n_{\lambda_{ph}}\right\rangle$, where the rotational wave function $\Phi_{rot}(\Theta)$ depends on the Euler angles Θ, and the vibrational wave function depends on the number n of excited vibrational phonons of multipolarity λ_{ph} considered. The phonon parity is given by $\pi_{ph} = (-1)^{\lambda_{ph}}$.

Assuming that the nucleus behaves as a non-axial rotor and denoting K as the total angular momentum projection on the symmetry axis and I ($I \geq K$) as the total angular momentum, then the nuclear wave function for even-even nuclei Ψ can be written in the form ([18, 19])

$$\Psi(IM\tau\Theta) = \sum_{K=0}^{I} A_K^\tau \left[\frac{2I+1}{16\pi^2(1+\delta_{K0})}\right]^{1/2} \left[D^I_{MK}(\theta',\varphi') + (-1)^{I+\lambda_{ph}} D^I_{M-K}(\theta',\varphi')\right] \left|n^K_{\lambda_{ph}}\right\rangle \tag{2}$$

being $(-1)^{\lambda_{ph}}$ the parity of the phonon. The coefficients A_K^τ are the K-mixing coefficients that depend on all deformation parameters.

For odd actinides we neglect vibrational excitations. Following refs.[19, 20], we assume the extreme extracore single-particle model for the Hamiltonian H_p describing the unpaired nucleon. The nucleon single-particle wavefunctions, satisfying $H_p\chi_\Omega = E_p\chi_\Omega$, being E_p the single-particle energy, are expanded in the spherical single-particle basis functions as $\chi_\Omega = \sum_{nj} c^\Omega_{nj} |nj\rangle$.

The nuclear wave function of odd-A actinides can be expressed in the extreme single-particle model:

$$\Psi(IM\tau\Theta) = \sum_{K=1/2}^{I} \sum_{\Omega} A^\tau_{K\Omega} \left[\frac{2I+1}{16\pi^2}\right]^{1/2} \left[D^I_{MK}(\theta',\varphi')\chi_\Omega + (-1)^{I-1/2} D^I_{M-K}(\theta',\varphi')\pi_\chi \chi_{-\Omega}\right] \tag{3}$$

being π_χ the parity of the intrinsic wave function.

It is worth mentioning that the odd-nucleus wave function given by eq.(3) has exactly the same collective angular operator structure (i.e. Wigner functions) that the even-nucleus wave function given by eq.(2). Such analogy allows using the same matrix element derived for the even-even nuclear wave function to obtain the odd-A matrix element as will be shown below.

2.2 The reduced matrix elements of the couplings

According to ref.[13], a key ingredient in our approach is the assumption of static large axial deformations on which small dynamic deformations $(\delta\beta_2, \gamma, \beta_3, \eta)$, with axial and non-axial components, are built

$$\begin{aligned} R(\theta',\varphi') \;=\; & R_0\left\{1 + \sum_{\lambda=2,4,6,8} \beta_{\lambda 0} Y_{\lambda 0}(\theta')\right\} + R_0\left[\delta\beta_2 - \beta_{20}\frac{\gamma^2}{2}\right] Y_{20}(\theta') + \\ & R_0\frac{\beta_{20}\gamma}{\sqrt{2}}\left[Y_{22}(\theta',\varphi') + Y_{2-2}(\theta',\varphi')\right] + R_0\beta_3 \cos\eta Y_{30}(\theta') + \\ & R_0\beta_3\frac{\sin\eta}{\sqrt{2}}\left[Y_{32}(\theta',\varphi') + Y_{3-2}(\theta',\varphi')\right] \end{aligned} \tag{4}$$

Generalizing [13], the coupling matrix elements for even-even nuclei are given by

$$\begin{aligned} < IK\|\left[D^\lambda_{;\mu} + (-1)^\mu D^\lambda_{;-\mu}\right]\|I'K' >= & \frac{\sqrt{2I'+1}}{\sqrt{(1+\delta_{K0})(1+\delta_{K'0})}} \times \frac{[1+(-1)^{\lambda+\mu+\lambda_{ph}+\lambda'_{ph}}]}{2} \times \\ & [< I'\lambda K'\mu|IK > + (-1)^{I'+\lambda'_{ph}} < I'\lambda - K'\mu|IK > + \\ & (-1)^{I+\lambda_{ph}} < I'\lambda K'\mu|I-K > + (-1)^{I+I'+\lambda_{ph}+\lambda'_{ph}} < I'\lambda - K'\mu|I-K >] \end{aligned} \tag{5}$$

where $\lambda + \mu + \lambda_{ph} + \lambda'_{ph} = \lambda + \lambda_{ph} + \lambda'_{ph} = even$ (as $\mu = even$).

Generalizing [14] the reduced matrix elements for odd-A nuclei are given by

$$
\begin{aligned}
< IK \| \left[D^{\lambda}_{;\mu} + (-1)^{\mu} D^{\lambda}_{;-\mu} \right] \| I'K' >= \sqrt{2I'+1} \frac{[1 + (-1)^{\lambda} \pi_{\chi} \pi_{\chi'}]}{2} \sum_{nj} c^{K}_{nj} c^{K'}_{nj} \\
[< I' \lambda K' \mu | IK > + (-1)^{I'-1/2} \pi_{\chi'} < I' \lambda - K' \mu | IK > + \\
(-1)^{I-1/2} \pi_{\chi} < I' \lambda K' \mu | I - K > + (-1)^{I+I'-1} \pi_{\chi} \pi_{\chi'} < I' \lambda - K' \mu | I - K >]
\end{aligned}
\tag{6}
$$

where the following selection rules are implicit:

- for intraband transitions $\chi_K = \chi_{K'}$, therefore $\pi_{\chi}\pi_{\chi'} = +1$, $\lambda = even$.
- for inter-band transitions if $\pi_{\chi} = \pi_{\chi'}$ then $\lambda = even$; if $\pi_{\chi} \neq \pi_{\chi'}$ then $\lambda = odd$. In practice, these selection rules forbid inter-band transitions in the cases under study as each inter-band matrix element will be proportional to the product of any two of the small dynamical deformations quantities.

2.3 Results and conclusions

A dispersive and Lane consistent OMP with multiple-band coupling and energy independent geometry has been developed for nucleon scattering on odd actinides: $^{233,235,238}U$, ^{239}Pu, and ^{232}Th. It's has been optimized by a least-squares fit of a regional set of OMP parameters from (n,n),(p,p) & (p,n) reactions. It included couplings based on rigid rotor with dynamical corrections for rotational bands built on vibrational (even-even) or single-particle band-heads (odd nuclei). Fitted static deformations are in reasonable agreement with FRDM deformations theoretically derived by Moller and Nix [21]

References

[1] M. Salvatores *et al.*, "Uncertainty and target accuracy assesment for innovative systems using recent covariance data evaluations", *International Evaluation Cooperation*, Volume **26**, NEA no. 6410, Organization for Economic Cooperation and Development (OECD), Nuclear Energy Agency, Paris, France (2008).

[2] A.J. Plompen, T. Kawano, and R. Capote Noy, "Inelastic scattering and capture cross-section data of major actinides in the fast neutron region", Technical Report **INDC(NDS)-0597**, International Atomic Energy Agency, Vienna, Austria (2012).

[3] D.W. Chan *et al.*, *Phys. Rev.* **C26** (1982) 841.

[4] D.W. Chan *et al.*, *Phys. Rev.* **C26** (1982) 861.

[5] E. Sheldon. L.E. Beghian, D.W. Chan *et al.*, *J.Phys.G:Nucl. Phys.* **12**, 443 (1986).

[6] T. Kawano, N. Fujikawa, and Y. Kanda, "Evaluation of ^{238}U inelastic scattering cross section", Technical Report **JAERI-M-94-019**, **INDC(JPN)-169**, Japan Atomic Energy Research Institute, Ibaraki-ken, Japan (1994).

[7] E. Sh. Sukhovitskiĭ and Yu.V. Porodzinskiĭ, *Proc. of 9-th Int.Symposium on Capture Gamma-Ray Spectroscopy and Related Topics, Budapest , Hungary, Vol.1, p. 335 (1996).*

[8] V.M. Maslov, Yu.V. Porodzinskij, N.A. Tetereva, M. Baba, and A. Hasegawa, *Nucl. Phys.* **A764**, 212–245 (2006).

[9] M.J. Lopez Jimenez, B. Morillon, and P. Romain, *Ann. Nucl. Energy* **32**, 195–213 (2005).

[10] E.Sh. Soukhovitskiĭ, R. Capote, J.M. Quesada, and S. Chiba, *Phys. Rev.* **C72**, 024604 (2005).

[11] R. Capote, E. Sh. Soukhovitskiĭ, J. M. Quesada, and S. Chiba, *Phys. Rev.* **C72**, 064610 (2005).

[12] R. Capote, S. Chiba, E.Sh. Soukhovitskiĩ, J. M. Quesada, and E. Bauge, *J. of Nucl. Sc. Tech.* **45**, 333–340 (2008).

[13] J.M Quesada, E.S. Soukhovitskiĩ, R. Capote, and S. Chiba, *EPJ Web of Conferences* **42**, 02005 (2013).

[14] J.M. Quesada, R. Capote, E.Sh. Soukhovitskkiĩ, and S. Chiba, *Nucl. Data Sheets* **118**, 270–272 (2014).

[15] A.S. Davydov, *Nucl. Phys.* **24**, 682 (1961).

[16] A.S. Davydov and R.A. Sardaryan, *Nucl. Phys.* **37**, 106–118 (1962).

[17] J.P. Davidson, *Rev. Mod. Phys.* **37**,105–158 (1965).

[18] A. Bohr, *Mat. Fys. Medd. Dan. Vid. Selsk.* **26**, No. 14 (1952).

[19] M. A. Preston and R. K. Bhaduri. *Structure of the Nucleus,* Addison-Wesley Publishing Company, Inc. (1975).

[20] K. Hetch and G. R. Satchler, *Nucl. Phys.* **32**, 286–318 (1962).

[21] T. Belgya, O. Bersillon, R. Capote, T. Fukahori, G. Zhigang, S. Goriely, M. Herman, A. V. Ignatyuk, S. Kailas, A. Koning, P. Obložinský, V. Plujko, and P. Young, Handbook for calculations of nuclear reaction data, Reference Input Parameter Library-2. Technical Report **IAEA-TECDOC-1506**, International Atomic Energy Agency, Vienna, Austria (2006).

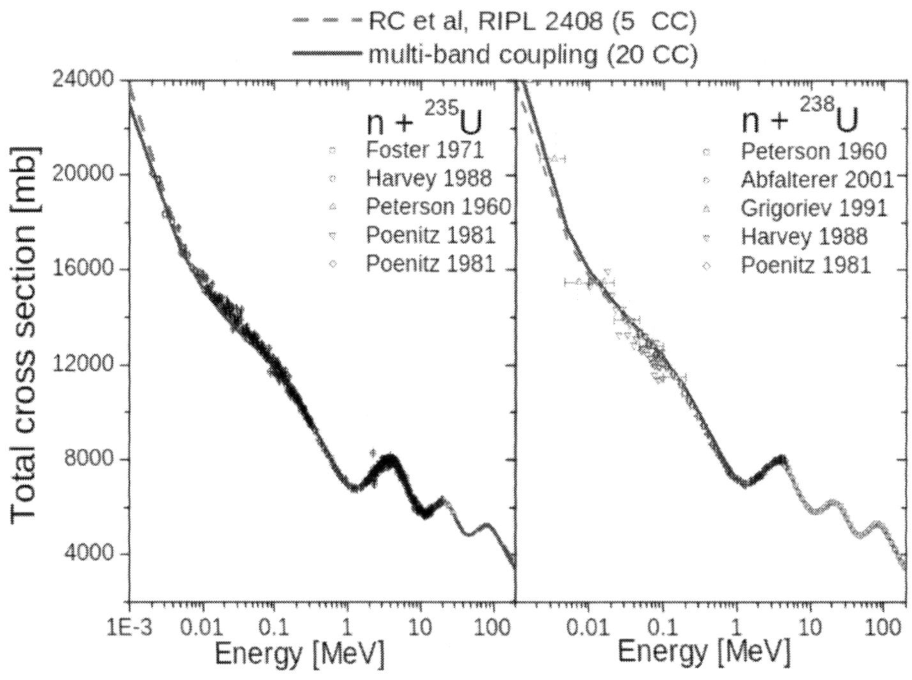

Figure 1. Calculated total cross sections using the DCCOMP with multiple-band coupling for neutron induced reactions on ^{235,238}targets.

Table 1. Chi-square values of the optimum fit.

Nuclide	χ^2	
	Present	[12]
^{232}Th	2.45	2.55
^{233}U	0.09	0.61
^{235}U	1.38	2.05
^{238}U	1.52	2.10
^{239}Pu	1.18	2.64

EPJ Web of Conferences **111**, 04001 (2016)
DOI: 10.1051/epjconf/201611104001

Study on neutron scattering in light water

Juan Pablo Scotta[1a], Ignacio Marquez Damian[2], Gilles Noguere[1], and David Bernard[1]

[1]SPRC/LEPh, CEA Cadarache, F-13108 Saint-Paul-Lez-Durance, France
[2]Neutron Physics Department, Centro Atomico Bariloche, Argentina

Abstract. It is presented a method to produce covariance matrices of the light water total cross section from thermal scattering laws of the JEFF-3.1.1 nuclear data library and CAB model. The generalized least square method was used to fit the LEAPR module parameters of the processing tool NJOY with light water experimental transmission measurements at 293.6K with CONRAD code. The marginalization technique was applied to account for systematic uncertainties.

1. Introduction

The probability of an incident neutron with energy and solid angle (E;Ω) of scattering a light water molecule and being emitted at a different energy and different solid angle (E';Ω') can be expressed with the double differential scattering cross section:

$$\frac{d^2\sigma}{d\Omega dE}(E \rightarrow E', \Omega \rightarrow \Omega') = \frac{\sigma_b}{4\pi kT}\sqrt{\frac{E'}{E}}e^{-\frac{\beta}{2}}S(\alpha,\beta) \tag{1}$$

where σ_b is the characteristic bound cross section of the material, T is the temperature and k is the Boltzmann constant.

The double differential cross section depends on the thermal scattering law function $S(\alpha,\beta)$, where α and β are the dimensionless momentum and energy transfer respectively:

$$\alpha = \frac{E'+E-2\sqrt{E'E}\cos\theta}{AkT}\ ;\ \beta = \frac{E'-E}{kT} \tag{2}$$

where cos θ is the cosine of the scattering angle in the laboratory system and A is the ratio between the mass of the scattering molecule and the neutron mass.

[a] Corresponding author: juan-pablo.scotta@cea.fr

In order to evaluate the function S(α,β) some approximations are often used: the incoherent and the Gaussian approximation. The processing of the thermal scattering law is done with LEAPR module of NJOY code [1], and is computed with these approximations.

The main parameter that defines the dynamics of the function S(α,β) is the frequency spectrum of the material, which defines its excitation states. When using LEAPR module, the frequency spectrum is a resultant of three different components: a solid type spectrum, a translational spectrum and discrete oscillators.

2. Models for H into H2O

It will be presented two different models that describe the frequency distribution of H into H_2O.

The first one is IKE model, developed by M. Mattes and J. Keinert [2]. It is based on experimental frequency spectrum measured by B. C. Haywood and D. I. Page [3]. The rotational mode of hydrogen bound to the water molecule is represented by a solid type spectrum with a weight w_R. Two discrete oscillators, with energies E_1 and E_2, define the intra-molecular modes of vibration (bending and stretching modes), with corresponding weight w_1 and w_2. Finally, an effective mass describes the molecular clustering effect of the water molecule with a weight w_t. The sum of all the weights must be equal to unity.

The nuclear data library JEFF-3.1.1 adopted this model in its database. Table 1 resumes the LEAPR parameters of this model at 293.6K.

The second model, CAB model, was calculated by I. Marquez Damian [4] and is based on molecular dynamic simulations with GROMACS code. It was obtained an equivalent model to IKE model but the translational mode was replaced by the diffusional model of Egelstaff-Schofield [5] to better describe low energy dynamics. The key parameter in this model is the diffusion constant c. Table 1 resumes the LEAPR parameters of CAB model at 293.6K.

Figure 1 compares the frequency spectrum for IKE model (used in the JEFF-3.1.1 nuclear data library) and CAB model at 293.6K.

The total cross section is obtained integrating equation (1) in all emitted neutron energies and solid angles. It was computed for JEFF-3.1.1 library and CAB model at 293.6 K. Figure 2 compares them with experimental data [6- 8].

Table 1. IKE and CAB model parameters of H into H2O at 293.6K for the LEAPR module of NJOY.

Frequency spectrum parameters	JEFF - 3.1.1	CAB Model
Translational weight ω_t	0.021739	0.007918
Solid-type spectrum weight ω_R	0.489130	0.522080
Scissoring mode oscillator energy $\mathbf{E}_2$ (meV)	205	205
Scissoring mode oscillator weight ω_2	0.16304	0.15667
Stretching modes oscillator energies $\mathbf{E}_{1,3}$ (meV)	436	430
Stretching modes oscillator weights $\omega_{1,3}$	0.3261	0.31333
Dimensionless diffusion constant **c**	0	3.969

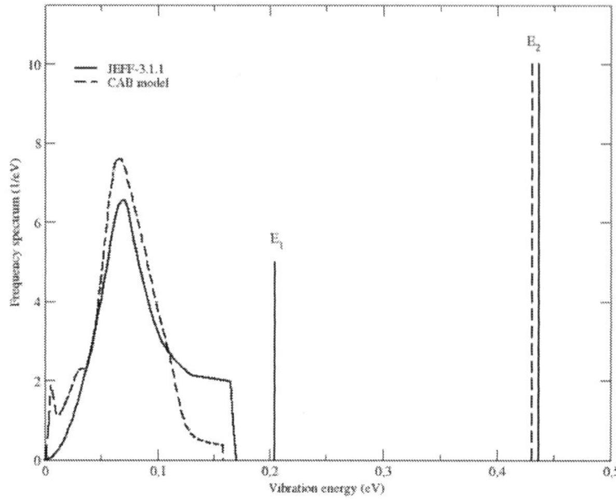

Figure. 1. Frequency spectrum for JEFF-3.1.1 and CAB model at 293.6 K.

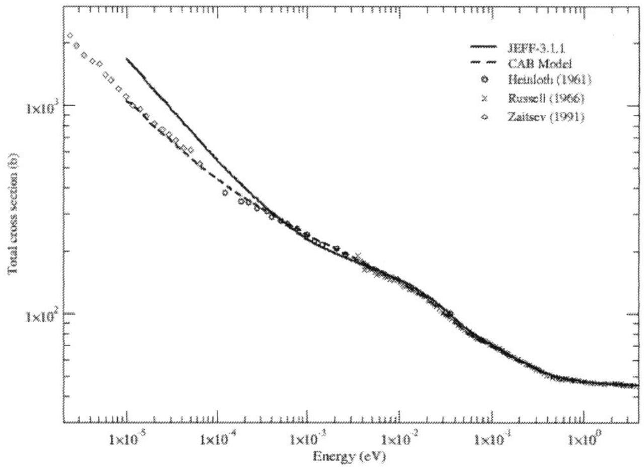

Figure. 2. Total cross section for JEFF-3.1.1 and CAB model, at 293.6 K.

3. Covariance matrices of differential and total cross section

3.1 Governing equations

In this work, the full covariance matrix between the model parameters is defined as follow:

$$\Sigma = \begin{pmatrix} \Sigma_{11} & \Sigma_{12} \\ \Sigma_{21} & \Sigma_{22} \end{pmatrix} \tag{3}$$

where each element of the matrix is obtained by:

$$\Sigma_{11} = M_x + \left(G_x^T G_x\right)^{-1} G_x^T G_\theta M_\theta G_\theta{}^T G_x \left(G_x^T G_x\right)^{-1} \tag{4}$$

$$\Sigma_{12} = -(G_x^T G_x)^{-1} G_x^T G_\theta M_\theta \tag{5}$$

$$\Sigma_{22} = M_\theta \tag{6}$$

in which M_x stands for the covariance matrix between the best-fit values of the model parameter and M_θ represents the covariance matrix between the nuisance parameters that account systematic uncertainties. The nuisance parameters included in the present study were the normalization and the background. In our context, the model parameter vector x and the nuisance parameter vector θ are:

$$\vec{x} = (c, \omega_t, \omega_R, \Delta, E_1, E_2, \omega_1, \omega_2);\ \vec{\theta} = (N, B) \tag{7}$$

The matrices G_x and G_θ correspond to the derivatives of the calculated quantity z to the model and the nuisance parameters respectively:

$$G_x = \begin{pmatrix} \frac{\partial z_1}{\partial c} & \frac{\partial z_1}{\partial \omega_t} & \cdots & \frac{\partial z_1}{\partial \omega_2} \\ \frac{\partial z_2}{\partial c} & \frac{\partial z_2}{\partial \omega_t} & \cdots & \frac{\partial z_2}{\partial \omega_2} \\ \vdots & \vdots & \cdots & \vdots \\ \frac{\partial z_k}{\partial c} & \frac{\partial z_k}{\partial \omega_t} & \cdots & \frac{\partial z_k}{\partial \omega_2} \end{pmatrix};\ G_\theta = \begin{pmatrix} \frac{\partial z_1}{\partial N} & \frac{\partial z_1}{\partial B} \\ \frac{\partial z_2}{\partial N} & \frac{\partial z_2}{\partial B} \\ \vdots & \vdots \\ \frac{\partial z_k}{\partial N} & \frac{\partial z_k}{\partial B} \end{pmatrix} \tag{8}$$

To calculate the covariance matrices of the total cross sections CONRAD code [9] was used. The Generalized Least Square method (GLS) was used in association with the marginalization technique in which the zero variance penalty condition accounts for systematic uncertainties [10].

The "best estimates values" of the LEAPR module parameters were obtained from water transmission experimental measures.

3.2 Sensitivity analysis of the model parameters

A sensitivity analysis of the LEAPR module parameters to the H_2O transmission at 293.6 K was done. Their sensitivity as a function of the incident neutron energy is plotted in figure 3 for the H_2O transmission calculated with CAB model.

The parameters ω_R and Δ are one order of magnitude more sensitive to the transmission than the rest. This highlights the relevance of the continuous part of the frequency spectrum, described by these two parameters. Their anticorrelation suggests that only one of them should be taken into account in the marginalization technique.

The parameters c and ω_t follow an equivalent trend because they are physically correlated. The parameters not included in figure 3 (E_1, E_2, ω_1 and ω_2) have a negligible sensitivity. Their uncertainties are not taken into account in the present study.

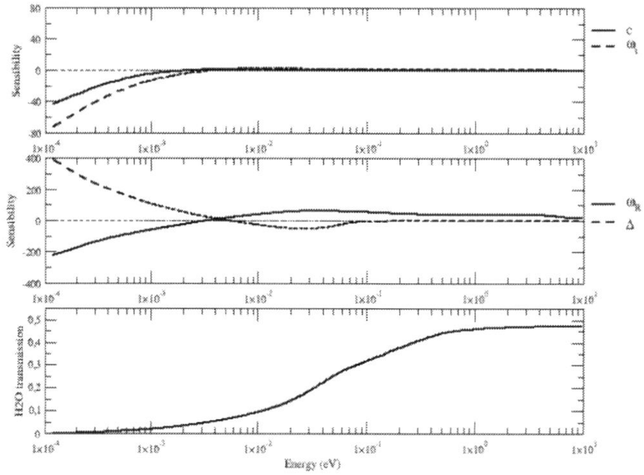

Figure. 3. Sensitivity (in %/%) of the LEAPR model parameters to the H2O transmission as a function of the energy calculated with CAB model.

3.3 Results

According to our sensitivity analysis, the uncertainty on the total cross section is mainly driven by the parameters Δ and ω_t. The relative uncertainties and the correlation matrices of the total cross section for JEFF-3.1.1 and CAB model at 293.6 K are shown in figure 4.

In the cold neutron energy range, below 10^{-3} eV, the uncertainty associated to JEFF-3.1.1 is larger than the uncertainty associated to CAB. This is explained by the better agreement of the CAB model with the low energy total cross section measurements.

Around 5 meV, the uncertainty found for CAB model is low. Additional work is needed for improving this underestimated magnitude.

Above the thermal energy, the relative uncertainties remain lower than 2%. The main contribution is related to the ^{1}H free atom cross section whose uncertainty is lower than 1% [11].

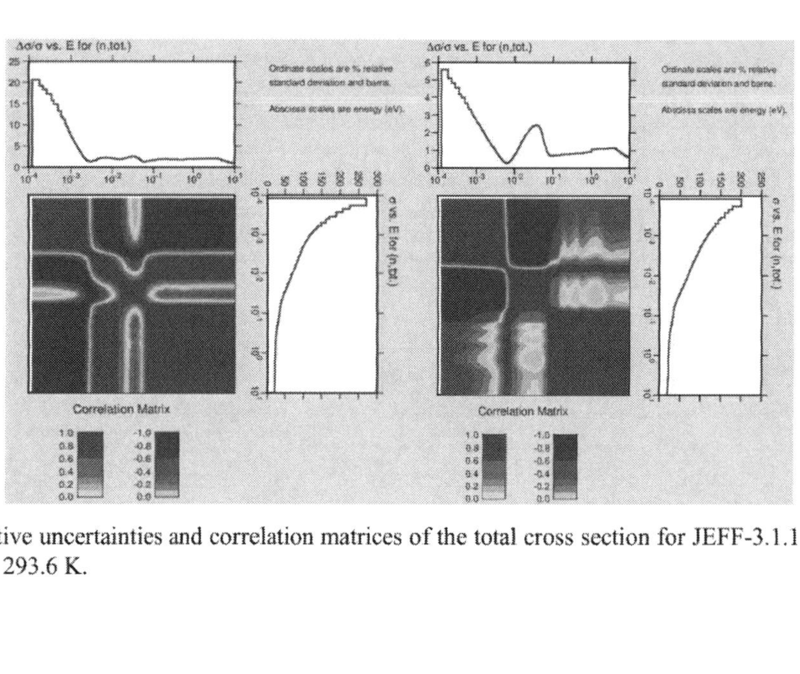

Figure. 4. Relative uncertainties and correlation matrices of the total cross section for JEFF-3.1.1 (left) and CAB model (right) at 293.6 K.

4. Conclusions

The covariance matrices of the light water total cross sections from thermal scattering laws of JEFF-3.1.1 nuclear data library and CAB model were obtained by using CONRAD code. In the cold neutron energy range, differences in the relative uncertainties between both models are important. The uncertainties associated to JEFF-3.1.1 are larger because the agreement with the low energy total cross section is poor.

Present work is being done to propagate uncertainties due to the thermal scattering laws to integral calculations. Preliminary calculations performed on MOX fuel provide uncertainties on the multiplication factor ranging from 100 pcm with the CAB model to 200 pcm with JEFF-3.1.1.

The next step would be to determine uncertainties on the water potential parameters used in the molecular dynamic simulation code GROMACS, in order to obtained directly uncertainties on the frequency spectrum introduced in LEAPR module.

References

1. R.E. MacFarlane and A.C. Kahler, Nucl. Data Sheets **111**, 2729 (2010).
2. M. Mattes and J. Keinert, "Thermal Neutron Scattering Data for the moderator Materials H2O, D2O and ZrHx in ENDF-6 Format and as ACE Library for MCNP(x) Codes", International Atomic Energy Agency, INDC(NDS)-0470, 2005.
3. D.I. Page, B.C. Haywood, "The Harwell scattering law program: frequency distributions of moderators", Harwell Report AERE-R 5778, 1968.
4. J. I. Marquez Damian et al, Ann. Nucl. Ener. **65**, 280 (2014).
5. P.A. Egelstaff and P. Schofield, Nucl. Sci. Eng. **12**, 260 (1962).
6. J.L. Russell Jr., J.M. Neill and J.R. Brown, "Total cross section measurements of H2O", General Atomic Report GA-7581, 1966.
7. K.N. Zaitsev, V.N. Petrov, S.P. Kuztnesov, O.A. Langer, I.V. Meshkov, A.D. Perekrestenko, Sov. At. Energy **70**, 238 (1991).
8. K. Heinloth, Z. Phys. **163**, 218 (1961).
9. C. De Saint Jean et al., "CONRAD, a new software for nuclear data analysis", Nuclear Energy Agency, JEFFDOC-1142, 2006.
10. G. Noguere, P. Archier and C. De Saint Jean, Nucl. Sci. Eng. **172**, 164 (2012).
11. S. Mughabghab, "Atlas of Neutron Resonance", Elsevier, 2006.

EPJ Web of Conferences **111**, 05001 (2016)
DOI: 10.1051/epjconf/201611105001

Investigating Prompt Fission Neutron Emission from ^{235}U(n,f) in the Resolved Resonance Region

Alf Göök[1,a], Franz-Josef Hambsch[1,b], and Stephan Oberstedt[1]

[1] *European Commission, Joint Research Centre, Institute for Reference Materials and Measurements,Retieseweg 111, B-2440, Geel, Belgium*

Abstract. Investigations of prompt emission in fission is of importance in understanding the fission process in general and the sharing of excitation energy among the fission fragments in particular. Experimental activities at IRMM on prompt neutron emission from fission in response to OECD/NEA nuclear data requests is presented in this contribution. Main focus lies on currently on-going investigations of prompt neutron emission from the reaction ^{235}U(n,f) in the region of the resolved resonances. For this reaction strong fluctuations of fission fragment mass distributions and mean total kinetic energy have been observed [Nucl. Phys. A 491, 56 (1989)] as a function of incident neutron energy in the resonance region. In addition fluctuations of prompt neutron multiplicities were also observed [Phys. Rev. C 13, 195 (1976)]. The goal of the present study is to verify the current knowledge of prompt neutron multiplicity fluctuations and to study correlations with fission fragment properties.

1 Introduction

Prompt neutron emission in fission is of particular importance in understanding the fission process. Knowledge of the properties of prompt fission neutrons, their multiplicities and energy distributions could give answers to questions related not only to the neutron emission itself, but also to questions relevant to the formation of the fission fragments, the sharing of excitation energy among them and the time scale of the process. For nuclear modeling and improved evaluation of nuclear data the knowledge about fluctuations in the prompt neutron multiplicity as a function of incident neutron energy is requested for the major actinides ^{235}U and ^{239}Pu [1, 2]. Fluctuations in fission fragment mass and total kinetic energy (TKE) in both isotopes have been observed in resonance neutron induced fission [3, 4]. Independently, fluctuations in the number of emitted neutrons have also been observed [5]. In view of the fact that both neutron number and fission fragment properties have been found to vary it is necessary to study the correlations of prompt neutron multiplicity and fission fragments properties in the resonance region. Experimental investigations of the correlations of prompt fission neutron multiplicity with fragment properties in resonance neutron induced fission on ^{235}U and ^{239}Pu, are taking place at the GELINA facility of the JRC-IRMM.

[a] e-mail: alf.gook@ec.europa.eu
[b] e-mail: franz-josef.hambsch@ec.europa.eu

Article available at http://www.epj-conferences.org or http://dx.doi.org/10.1051/epjconf/201611105001

2 Experimental Method

The experimental technique used to study prompt neutron emission in fission at the GELINA facility is based on techniques pioneered by Bowman [6]. The method involves extracting fission fragment masses by measurement of either velocity (2v) or energy (2E) of the two fragments, and coincident measurement of fission neutron time-of-flight at known angles with respect to the fission axis. This gives the basic information needed for reconstructing the neutron emission in the rest frame of the fully accelerated fragment, cf. Fig. 1. In an experiment on prompt fission neutrons from ^{252}Cf(sf), Budtz-Jørgensen et al. [7] exploited the combination of a twin Frisch-grid ionization chamber for fission fragment properties (2E technique) and a liquid scintillator for neutron detection. This serves as the basis for our experimental setup (which is described in Sect. 3) to study prompt neutron emission. The ionization chamber has a large solid angle, which not only facilitates the fragment-neutron coincidence rate, but also introduces a less biased selection of coincident events.

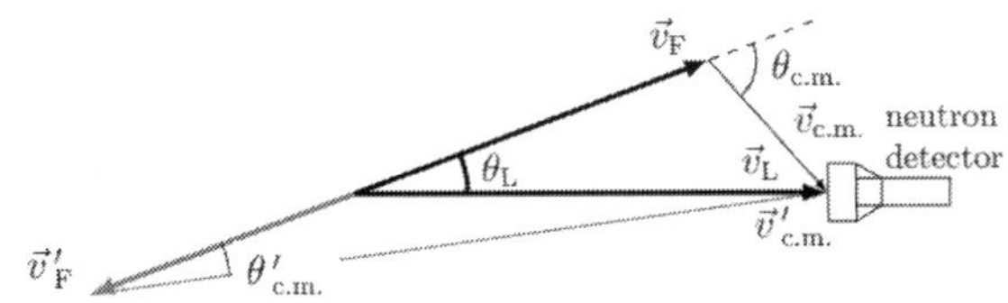

$$v^2_{c.m.} = v^2_L + v^2_F - 2v_L v_F \cos\theta_L$$
$$\cos\theta_{c.m.} = (v_L\cos\theta_L - v_F)/v_{c.m.}$$

Figure 1. Vector diagram of the kinematics of neutron emission from a fully accelerated fragment and the transformation into the fragment rest frame (c.m.), black lines represent the fragment detected in the same hemisphere as the neutron, while red lines represent the complimentary fragment. Vectors and angles drawn with thick lines are directly measured in the experiment. The labels are; fragment velocity $\vec{v}_F$, complimentary fragment velocity $\vec{v}\,'_F$, neutron velocity in the laboratory frame $\vec{v}_L$ and neutron velocity in the center-of-mass frame $\vec{v}_{c.m.}$.

2.1 Prompt Fission Neutrons from ^{252}Cf(SF)

An experimental investigation of prompt fission neutron multiplicity correlations with fission fragment properties in the spontaneous fission of ^{252}Cf was undertaken to verify setup and analysis procedures relevant for studying neutron emission in resonance-neutron induced fission. Beside the purely experimental interest, recent advances in the theoretical description of prompt emission in fission [8–10] require detailed correlation data for verifying the models. The experiment is essentially a reproduction of the experiment performed by Budtz-Jørgensen et al. [7] with the traditional analog acquisition system now replaced by a fully digital one and subsequent digital signal processing. Experimental details and main results have been published elsewhere [11], for the sake of clarity, a few selected results will be presented in the following. On the left hand side of Fig. 2 the mean number of neutrons emitted per fission is shown as a function of the TKE of the fission fragments, the dependence is nearly linear (except for at low TKE). A least square fit to the data from the present experiment gives an inverse slope of (12.6 ± 0.2) MeV/n. Due to the correlation between TKE and fragment mass distributions this quantity must not be interpreted as the energy needed for a pair of fragments to emit one more neutron, as noted already in Ref. [18]. The discrepant data from Refs. [12, 13] can be explained by failure in taking into account the recoil energy imparted to the fission fragment by the detected neutron [19]. A detailed discussion of this can be found in Ref. [11]. The mass dependence of the prompt fission neutron multiplicity per fragment from this experiment is compared to literature data [7, 14–17] in the right hand side of Fig. 2. The data from this study agree quite well with experimental data available in the literature.

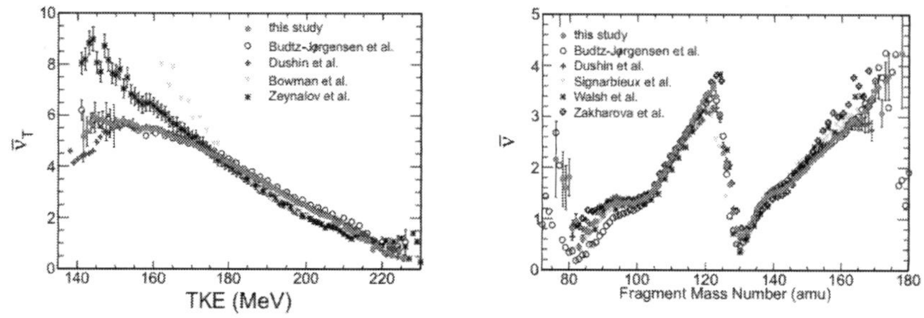

Figure 2. Left: Average neutron multiplicity per fission as a function of TKE, in comparison with experimental data from literature [7, 12–14]. Right: Average prompt fission neutron multiplicity as a function of fission fragment mass, in comparison with data from literature [7, 14–17].

3 Setup for the ^{235}U(n,f) Experiment

The experimental setup for investigating correlations of prompt neutrons with fission fragments in resonance-neutron induced fission on ^{235}U and ^{239}Pu is illustrated in Fig. 3. The fission target is placed inside the ionization chamber about 9.2 m away from the neutron production target of the GELINA facility. An array of neutron detectors (SCINTIA) is employed in order to facilitate the fission-neutron coincidence count rate. The SCINTIA array consists of 7 NE213 equivalent liquid scintillators (Scionix LS-301) and 5 para-therphenyl detectors. When employing an array of neutron detectors, each detector forms an axis of symmetry around which the orientation of the fission axis needs to be known. Hence the traditional ionization chamber is no longer sufficient to reconstruct the kinematics in the fragment rest frame. To remedy the situation, the ionization chambers anode plate is replaced by a position sensitive readout structure, which allows determination of all three space components of the fission fragments direction of travel. A 22 channel fully digital data acquisition is used. For each fission event the digital waveforms of all channels, sampled at 400 MS/s with 14-bit resolution, are stored on disk for off-line treatment. A register counting the number of oscillations of the sampling clock (800 MHz) is reset by a pulse generated just before the electron bunch from the GELINA hits the neutron production target. The value of the register as the fission trigger arrives is stored together with the waveform data, giving the incident neutron time-of-flight for each event. In Fig. 4 a time-of-flight spectrum of the ongoing ^{235}U(n,f) experiment is displayed.

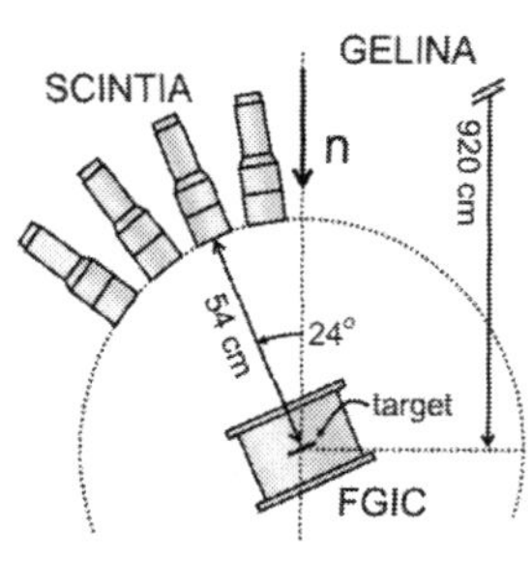

Figure 3. Schematic view of the experimental setup for investigating resonance neutron induced fission.

4 Preliminary results on $^{235}U(n,f)$

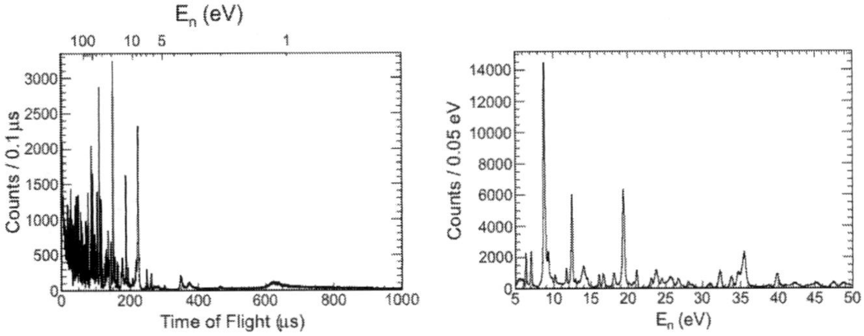

Figure 4. Left: Incident neutron time of flight spectrum from the ongoing experiment on $^{235}U(n,f)$. Right: The same spectrum converted to incident neutron energy in the range [5,50] eV (right).

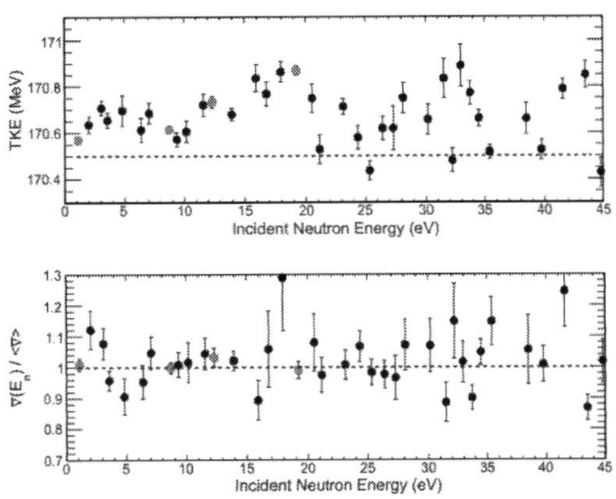

Figure 5. Top: Average TKE as a function of incident neutron energy for given resonance or resonance group in the range up to 45 eV, the dashed line refers to the thermal value. Bottom: Neutron multiplicity for the same resonances or resonance groups relative to the weighted average neutron multiplicity. The red points highlight strong resonances in the fission cross section of ^{235}U.

The top panel of Fig.5 shows the average TKE as a function of incident neutron energy in the range up to 45 eV. Each point corresponds to either a single resonance or resonance group. The red points highlight the strongest resonances in the fission cross section of ^{235}U. A clear fluctuating behavior exceeding the experimental uncertainty is observed. This was already pointed out in Ref. [3]. However, the statistical significance is much improved in the present work leading to a clearer picture of the energy dependent changes compared to the earlier work. The lower panel of Fig. 6 shows the neutron multiplicity for a given resonance or resonance group relative to the weighted average neutron multiplicity. Again, the strongest resonances are highlighted in red. At the current moment of data collection, an anti-correlation, i.e., high TKE corresponding to lower neutron multiplicity, is observed when all points in the given incident neutron energy interval are plotted. This is expected from energy conservation. Of course, also possible changes in the mass distribution need to be accounted for. Nevertheless, there seem to be no sizeable fluctuations in the neutron multiplicity in the strong resonances.

5 Conclusions

An investigation of prompt fission neutrons and fission fragments in the spontaneous fission of ^{252}Cf has been performed as a preparatory step to investigate prompt fission neutron emission in resonance-neutron induced fission. A literature comparison of the results show that the experimental setup and analysis is under control.

Prompt fission neutron and fission fragment correlations in resonance neutron induced fission on ^{235}U are taking place at the GELINA facility of the IRMM. We are confident that the new data will shed light in the understanding of the observed neutron multiplicity fluctuations in the resonance region and the behavior of the neutron multiplicity as a function of mass and incident neutron energy.

Acknowledgements

The authors are grateful to the co-workers of the SN3S unit of the IRMM.

References

[1] C. Lubitz, R. Roussin, NEA/WPEC **18** (1999)
[2] C. De Saint Jean, R. McKnight, NEA/WPEC **34** (2014)
[3] F.J. Hambsch, H.H. Knitter, C. Budtz-Jørgensen, J.P. Theobald, Nuclear Physics A **491**, 56 (1989)
[4] F.J. Hambsch, I. Ruskov, L. Dematté, in *Proc. Scientific Workshop on Nuclear Fission Dynamics and the Emission of Prompt Neutrons and Gamma Rays (Theory-1), EUR 24802* (Sinaia, Romania, 2011), p. 41
[5] R.E. Howe, T.W. Phillips, C.D. Bowman, Phys. Rev. C **13**, 195 (1976)
[6] H.R. Bowman, S.G. Thompson, J.C.D. Milton, W.J. Swiatecki, Phys. Rev. **126**, 2120 (1962)
[7] C. Budtz-Jørgensen, H.H. Knitter, Nucl. Phys. **A490**, 307 (1988)
[8] S. Lemaire, P. Talou, T. Kawano, M.B. Chadwick, D.G. Madland, Phys. Rev. C **72**, 024601 (2005)
[9] B. Becker, P. Talou, T. Kawano, Y. Danon, I. Stetcu, Phys. Rev. C **87**, 014617 (2013)
[10] O. Litaize, O. Serot, Phys. Rev. C **82**, 054616 (2010)
[11] A. Göök, F.J. Hambsch, M. Vidali, Phys. Rev. C **90**, 064611 (2014)
[12] H.R. Bowman, J.C.D. Milton, S.G. Thompson, W.J. Swiatecki, Phys. Rev. **129**, 2133 (1963)
[13] S. Zeynalov, F.J. Hambsch, S. Oberstedt, Jour. Korean Phys. Soc. **59**, 1396 (2011)
[14] V. Dushin, F.J. Hambsch, V. Jakovlev, V. Kalinin, I. Kraev, A. Laptev, D. Nikolaev, B. Petrov, G. Petrov, V. Petrova et al., Nucl. Inst. and Meth. A **516**, 539 (2004)
[15] R. Walsh, J. Boldeman, Nucl. Phys. A **276**, 189 (1977)
[16] V. Zakharova, D. Ryazanov, Sov. J. Nucl. Phys. **30**, 19 (1979)
[17] C. Signarbieux, J. Poitou, M. Ribarg, J. Matuszek, Phys. Lett. **39B**, 503 (1972)
[18] H. Nifenecker, C. Signarbieux, R. Babinet, J. Poitou, in *Proc. 3rd Symp. on Physics and Chemistry of Fission* (IAEA, Vienna, 1974), Vol. 2, p. 51
[19] A. Gavron, Nucl. Inst. and Meth. **115**, 99 (1974)

EPJ Web of Conferences 111, 05002 (2016)
DOI: 10.1051/epjconf/201611105002

Even-odd effects in the prompt fission emission of even Z actinides

Anabella Tudora[1,a], Franz-Josef Hambsch[2], Georgiana Giubega[1] and Iuliana Visan[1]

[1] University of Bucharest, Faculty of Physics, 405 Atomistilor Str., POB MG-11, Magurele 77125, Romania
[2] EC-JRC-Institute for Reference Materials and Measurements (IRMM), Retieseweg 111, Geel, 2440 Belgium

Abstract. The investigation of even-odd effects in the prompt emission of even Z actinides showed a sawtooth shape of ν(Z) with staggering in the asymmetric fission region. Average prompt emission quantities as a function of A, e.g. ν(A), of even Z fragmentations are higher than those of odd Z fragmentations and they exhibit oscillations with a periodicity of about 5 mass units in the asymmetric fission region. This periodicity is not due to the Z even-odd effect in fragment distributions. The even-odd effect in <ν>(TKE) is increasing with increasing TKE and it decreases with increasing mass of the fissioning nucleus. The global even-odd effect in total average prompt emission quantities is decreasing with increasing mass of the fissioning nucleus. In the case of an even-odd fissioning nucleus, ^{234}U(n,f), the global even-odd effect in prompt emission quantities exhibits a very slow variation with the incident neutron energy.

1 Introduction

Among the characteristics of fission fragment distributions the Z even-odd effect has repeatedly come into attention as an interesting feature because it contains information about the saddle to scission process, playing an important role in the fission modelling. The neutron even-odd effect received less attention since only the charge number Z and the post neutron mass number are known experimentally with sufficient accuracy. Also it should be kept in mind that due to the even-odd staggering in neutron separation energy, the prompt neutron evaporation brings an even-odd effect of its own. Contrary to the even-odd effects in fission fragment distributions that were extensively studied, the even-odd effects in the prompt neutron and γ-ray emission received less attention. Our recent works focused on this subject. The first studies referred to the even-odd effects in the prompt emission of even-even actinides fissioning spontaneously, ^{252}Cf(SF), $^{236-244}$Pu(SF) and induced by thermal neutrons ^{233,235}U(n_{th},f), ^{239}Pu(n_{th},f). The results are reported in Refs. [1, 2]. The study is continued with an even-odd fissioning nucleus, ^{234}U(n,f) at incident neutron energies (En) ranging from 0.2 MeV to 5 MeV. Because the fragment distributions play an important role in the calculation of different average prompt emission quantities as a function of A, of Z, of TKE and total average ones, it is also interesting to see the impact on the prompt emission of the following basic features of the Z even-odd effect in experimental fragment distributions (mentioned in the majority of publications, e.g. [3, 4] and references therein): i) the effect globally decreases for heavier fissioning systems (some authors mentioning that decreases with the fissility parameter), ii) the effect grows with the fragmentation

[a] Corresponding author: anabellatudora@hotmail.com

asymmetry, iii) the amplitude of the even-odd structure decreases with increasing excitation energy of the fissioning nucleus, iv) the Z even-odd effect grows with the fragment kinetic energy.

The Point-by-Point (PbP) model of prompt emission provides as primary results the multi-parametric matrices of different quantities characterizing both the fragments and the prompt emission, generically labelled q(A,Z,TKE) (e.g. ν(A,Z,TKE), Eγ(A,Z,TKE), a(A,Z,TKE)). Different average quantities q(Z), q(A), q(TKE) and <q> are obtained by averaging the PbP matrices q(A,Z,TKE) over the multiple distributions Y(A,Z,TKE) in different ways (for details see Refs. [1, 2, 5, 6] and references therein). These distributions are taken as Y(A,Z,TKE) = Y(A,TKE) p(Z,A) in which Y(A,TKE) are experimental distributions (usually reconstructed from the single ones Y(A), TKE(A) and σ_{TKE}(A)) and the isobaric charge distributions p(Z,A) are provided by the Zp model of Wahl [7-9]. The global even-odd effect in Y(Z) projections, defined according to Gönnenwein [3]:

$$\delta_{Y(Z)} = \left(\sum_{Ze} Y(Ze) - \sum_{Zo} Y(Zo) \right) \Big/ \sum_{Z} Y(Z) \quad (1)$$

as well as the even-odd effects in the charge polarization ΔZ(A)=Zp(A)-Z_{UCD}(A) and the root-mean-square rms(A) of the Gaussian functions fitting the isobaric charge distribution p(Z,A) are investigated. The charge polarisation plays an important role in the construction of the fragmentation range of the PbP treatment. I.e. at each A (covering a large range from symmetric fission up to a very asymmetric split) 3 up to 5 charge numbers Z being taken as the nearest integers above and below the most probable charges Zp(A) (that are the values on which the fitting Gaussians are centred). TKE covers a large range (usually from 100 MeV to 200 MeV with a step of 5 MeV).

We have investigated the global Z and N even-odd effects in different total average quantities characterizing the prompt emission, defined as [1, 2]:

$$\delta_{<q>} = \left(<q>_{even\,Z/N} - <q>_{odd\,Z/N} \right) / <q> \quad (2)$$

The behaviour of average q(Z), of average q(A) of even Z and odd Z fragmentations and of average neutron separation energy <Sn> as a function of N were investigated, too. The variation of the Z even-odd effect in different average quantities as a function of TKE is emphasized by the function [1, 2]:

$$\delta_{<q>}(TKE) = \left(<q>_{even\,Z}(TKE) - <q>_{odd\,Z}(TKE) \right) / <q>(TKE) \quad (3)$$

Note, cold fission is coming close to neutron-less fission. Because this work is referring to prompt neutron emission, cold fission fragmentations are not taken into account. To avoid any limit situation the fragmentations with TXE less than 9 MeV were excluded (see Refs. [1, 2] and references therein).

2 Even-odd effects in the prompt emission of even Z actinides

2.1 Even-even nuclei fissioning spontaneously or induced by thermal neutrons

As it was reported in [1, 2] the global even-odd effect in the Y(Z) projections is decreasing with increasing mass of the fissioning system: from about 22 % for ^{235,233}U(n_{th},f) to about 10-14% for even-even Pu isotopes fissioning spontaneously and by thermal neutrons and up to about 4 % for ^{252}Cf(SF). ΔZ(A) and rms(A) exhibit oscillations with a periodicity of about 5 mass units in the asymmetric fission region. The oscillation amplitudes are also decreasing with increasing mass of the fissioning system, they being the most pronounced for ^{233,235}U(n_{th},f) and almost vanished in the case of ^{252}Cf(SF).

These features of even-odd effects in fragment distributions and charge polarizations are reflected in the prompt emission. In all cases ν(Z) exhibit sawtooth shapes and a visible staggering in the asymmetric fission region. In the case of ^{252}Cf(SF) the ν(Z) staggering is visibly less pronounced compared to U and Pu isotopes. The average quantities as a function of A (e.g. TXE(A), ν(A), Eγ(A)) of even Z fragmentations are higher than those of odd Z fragmentations and both exhibit oscillations with a periodicity of about 5 mass units. This periodicity is not due to the Z even-odd effects in fragment distributions. Only the magnitude of oscillation amplitudes reflects the size of the effect. Average quantities as a function of TKE of even Z fragmentations are also higher than those of odd Z ones. The function of Eq.(3) is plotted in **Fig.1**. In the case of ^{235,233}U(n_{th},f), ^{239}Pu(n_{th},f) and ^{252}Cf(SF) (upper part) the increase of the even-odd effect in prompt neutron multiplicity with TKE and the decrease of this effect for heavier fissioning systems is visible. In the case of $^{236-244}$Pu(SF) (lower part) the even-odd effect is also increasing with TKE but the decrease of the effect for heavier fissioning nuclei is not visible.

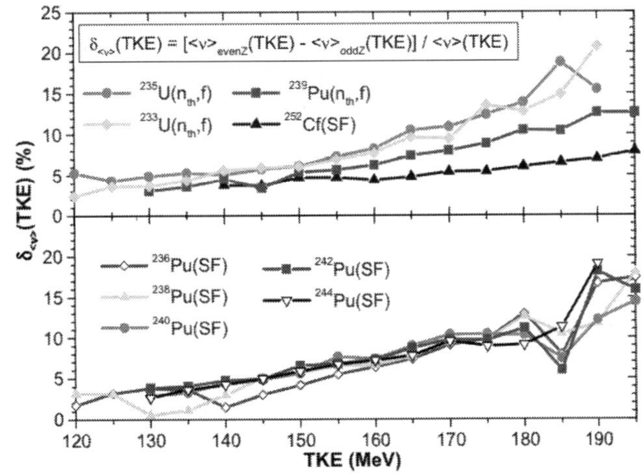

Figure 1. The even-odd effect in prompt neutron multiplicity as a function of TKE of ^{233,235}U(n$_{th}$,f), ^{239}Pu(n$_{th}$,f) and ^{252}Cf(SF) in the upper part and of $^{236, 238, 240, 242,244}$Pu(SF) in the lower part.

The global even-odd effect (given by Eq.(2)) in <TXE>, <ν> and other quantities is decreasing with increasing mass of the fissioning system, e.g. δ<ν> is of about 9% for ^{233,235}U(n$_{th}$,f), of about 8% for ^{239}Pu(n$_{th}$,f) and 6% for ^{252}Cf(SF). In the case of Pu(SF) δ<ν> is almost constant (of about 7-8%). In all cases the global even-odd effect in the average prompt neutron energy <E> and average prompt γ-ray energy <Eγ> is 2.5 – 3 times lower than the effect in prompt neutron multiplicity. Details about the even-odd effect in the prompt emission of the thermal neutron induced fission of ^{233,235}Uand ^{239}Pu and ^{252}Cf(SF) are given in Ref. [1] and of $^{236-244}$Pu(SF) in Ref.[2].

2.2 Even-odd fissioning nuclei: ^{234}U(n,f) at En ranging from 0.2 MeV to 5 MeV

The experimental Y(A,TKE) distributions of ^{234}U(n,f) measured at IRMM at 14 incident energies ranging from 0.2 MeV to 5 MeV and the Zp model with parameters from systematics, mainly developed by Wahl [8, 9], allow the investigation of even-odd effects. The even-odd factors F_Z and F_N entering the Zp model play the most important role. Based on existing systematics of these factors [10] several parameterisations (showing different decreases of F_Z and F_N with increasing En) can be deduced. They led to different Y(A,Z,TKE) distributions and charge polarizations. The obtained global even-odd effects in Y(Z) at the 14 incident energies (according to Eq.(1)) are plotted with different symbols in the left part of **Fig.2**. The uncertainties in $\delta_{Y(Z)}$ resulting from the propagation of uncertainties in the experimental distributions are included in the size of symbols. As it can be seen $\delta_{Y(Z)}$ exhibits a clear decrease with En in all cases. The linear fits to all $\delta_{Y(Z)}$ are plotted with dashed lines in the same colour as the respective symbol. The obtained $\delta_{Y(Z)}$ values, varying from about 26% (the highest result) to about 4 - 5% at 5 MeV are in overall agreement with the compilation of Ref. [11] giving an even-odd effect of (23 ± 8.9)% for the fission of ^{234}U induced by neutrons with fission spectrum energies. Our Y(Z) results (plotted with red circles and blue squares in the right part of **Fig.2**) describe well the experimental data of the electromagnetic induced fission of ^{235}U obtained in the SOFIA experiment at GSI [12] (black stars connected with a thick line) in the asymmetric fission region. The differences appearing at symmetric fission are due to the different ways to induce the fission of ^{235}U.

The two facts illustrated in Fig.2 (the $\delta_{Y(Z)}$ results in overall agreement with the compilation of Ref. [11] and the good description of the Y(Z) data of ^{235}U(em.f)) can be considered as a validation of the Y(A,Z,TKE) distributions based on experimental Y(A,TKE) data and the Zp model with parameters from systematics.

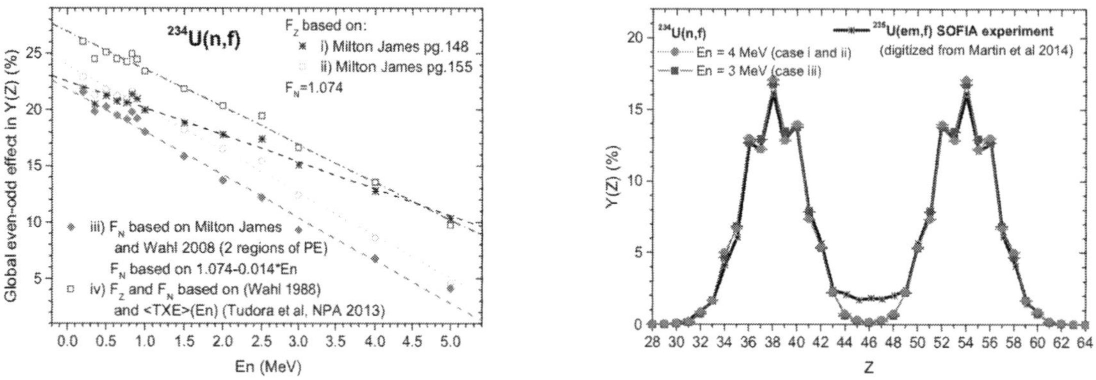

Figure 2: Left part: the global even-odd effect in Y(Z) as a function of En obtained with different F_Z and F_N parameterisations based on existing systematics. Right part: comparison of present Y(Z) results of ^{234}U(n,f) (red circles and blue squares) with the experimental data of ^{235}U(em.f) (SOFIA experiment at GSI) given with black stars connected with a thick line.

The impact of different prescriptions of Zp model parameters on the even-odd effects in prompt emission was also investigated. The differences between the probabilities of even Z and odd Z fragmentations able to emit prompt neutrons given by different Zp model parameter prescriptions are low, of about 2 - 3%. This fact leads to insignificant differences between the global even-odd effects in different prompt emission quantities (e.g. $\langle\nu\rangle$, <TXE>, $\langle E\gamma\rangle$) obtained with different prescriptions of Zp model parameters (the differences being less than the uncertainties in the global even-odd effects).

The global even-odd effects in different prompt emission quantities show a very slowly decreasing trend with increasing En. Practically they can be considered as constant with En. E.g. $\delta\langle\nu\rangle$ is varying between 8% - 9%, being at the same level of magnitude as in the case of neighbouring even-even nuclei ^{235,233}U(nth,f). This slow variation with En is due to a) the N even-odd staggering in <Sn>, the prompt neutron emission being favoured for odd N fragments and b) the even-odd character of the fissioning nucleus leading to pairs with even N and odd N complementary fragments. The favoured prompt emission for odd N fragments is often acting contrary to the Z even-odd effect in fragment distributions (higher probabilities of even Z pairs than of odd Z pairs).

The even-odd effects in different average quantities as a function of Z, of A, of TKE have the same characteristics as those of even-even fissioning nuclei. E.g. $\nu(Z)$ exhibit a sawtooth shape and staggering in the asymmetric fission region. The interesting behaviour observed experimentally in the case of $\nu(A)$, consisting in the multiplicity increase with En mainly for heavy fragments, is exhibited by $\nu(Z)$, too. Different average quantities as a function of A (e.g. $\nu(A)$, TXE(A)) of even Z fragmentations are higher than those of odd Z fragmentations. Their oscillation periodicity of about 5 mass units is not due to the Z even-odd effects in fragment distributions.

The increase of the Z even-odd effect in different quantities with increasing TKE, expressed by the function of Eq.(3), is illustrated in **Fig.3** for the prompt neutron multiplicity. The very low even-odd effect at low TKE values can be explained by the behaviour of the experimental Y(A,TKE) distributions as a function of A. At low TKE values the yields Y(A.TKE) are high in the mass region near symmetric fission where the even-odd effect in Z is very low. In terms of multi-modal fission here the symmetric super-long mode SL is dominant. At medium TKE values the yields are high in the mass region of asymmetric fission (with maximums placed at A_H of about 140-145) where the Z even-odd effect is pronounced. Here the fission mode S2 is the dominant one. At high TKE values the yields are high in the mass region of asymmetric fission, too (with maximums placed at A_H of about 132-135) where the even-odd effect is pronounced. At high TKE the fission mode S1 is dominant.

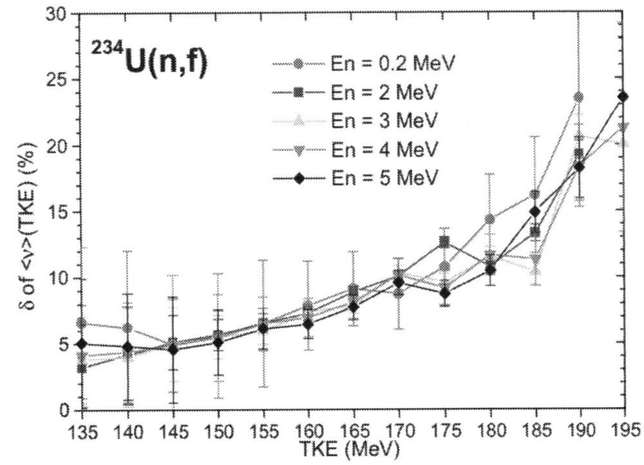

Figure 3. Even-odd effect in prompt neutron multiplicity as a function of TKE (expressed by Eq.(3)) at a few En ranging from 0.2 MeV to 5 MeV.

3 Conclusions

In the case of ^{234}U(n,f) even if different prescriptions of the Zp model parameters led to differences in the even-odd effects in Y(Z) and ΔZ(A) insignificant differences in the global even-odd effects in prompt emission were obtained, due to the low differences in the probabilities of even Z and odd Z fragmentations able to emit prompt neutrons provided by different Zp model parameter prescriptions.

Even if the global even-odd effect in Y(Z) is decreasing with increasing En, the global even-odd effects in different total average prompt emission quantities shows a very slow variation with En.

ν(Z) exhibits a sawtooth shape with staggering in the asymmetric fission region. It increases with En mainly for heavy fragments as in the case of ν(A). The Z even-odd effect in <ν>(TKE) is increasing with TKE. The very low even-odd effect at low TKE values is explained by the behaviour of the Y(A,TKE) yields as a function of A for low, medium and high TKE values in correlation with the presence of the even-odd effects in the asymmetric fission region.

References

1. A.Tudora, F.-J.Hambsch, G.Giubega, I.Visan, Nucl.Phys.A **929**, 260-292 (2014)
2. A.Tudora, F.-J.Hambsch, G.Giubega, I.Visan, Nucl.Phys.A **933**, 165-188 (2015)
3. F.Gonnenwein in *The Nuclear Fission Process* (CRC Press Boca Raton NY USA) ch. 8 (1991)
4. N.Caamano, F.Rejmund, K.-H.Schmidt, J.Phys.G Nucl.Part.Phys. 38 035101 (2011)
5. A.Tudora, F.-J.Hambsch, I.Visan, G.Giubega, Nucl.Phys.A **940**, 242-263 (2015)
6. A.Tudora, F.-J.Hambsch, S.Oberstedt, Nucl.Phys.A **917**, 43-70 (2013)
7. A.C.Wahl, At.Data Nucl.Data Tables **39**, 1-156 (1988).
8. A.C.Wahl in *Fission Product Yield Data for the Transmutation of Minor Actinide Nuclear Waste* (IAEA STI/PUB/1286) chapter 4.2 (2008)
9. A.C.Wahl in *Compilation and evaluation of fission yield nuclear data* (IAEA TECDOC-1168) Sections 4.1, 4.2, 7.4 (2000)
10. R.W.Mills, M.F.James in *Compilation and evaluation of fission yield nuclear data* (IAEA TECDOC-1168) Section 7.2 (2000)
11. T.R.England, B.F.Rider *Evaluation and compilation of fission product yields* LA-U-94-3106 ENDF-349 (1993)
12. J.F.Martin for the SOFIA collaboration, P(ND)-2 International workshop on perspectives of nuclear data for the next decade (organised by CEA and NEA) Bruyères-le-Châtel, France (2014)

EPJ Web of Conferences **111**, 05003 (2016)
DOI: 10.1051/epjconf/201611105003

Energy Correlation of Prompt Fission Neutrons

Zs. Elter[1,a] and I. Pázsit[1,b]

[1] *Chalmers University of Technology, Division of Nuclear Engineering, Göteborg, Sweden, SE-41296*

Abstract. In all cases where neutron fluctuations in a branching process (such as in multiplicity measurements) are treated in an energy dependent description, the energy correlations of the branching itself (energy correlations of the fission neutrons) need to be known. To date, these are not known from experiments. Such correlations can be theoretically and numerically derived by modelling the details of the fission process. It was suggested earlier that the fact that the prompt neutrons are emitted from the moving fission targets, will influence their energy and angular distributions in the lab system, which possibly induces correlations. In this paper the influence of the neutron emission process from the moving targets on the energy correlations is investigated analytically and via numerical simulations. It is shown that the correlations are generated by the random energy and direction distributions of the fission fragments. Analytical formulas are derived for the two-point energy distributions, and quantitative results are obtained by Monte-Carlo simulations. The results lend insight into the character of the two-point distributions, and give quantitative estimates of the energy correlations, which are generally small.
Key Words: fission neutrons; energy correlations; angular correlations.

1 Introduction

The statistical properties of the detection counts of neutrons, generated in subcritical multiplying media by an external source, have long been used in reactor physics and nuclear safeguards to determine parameters of the system under investigation. In reactor cores, the subcritical reactivity can be extracted from the variance to mean or from the temporal correlations of the detections; and in safeguards, the fissile mass can be determined from coincidence an multiplicity measurements [1]. The possibility of extracting such parameters is due to the fact that the branching, i.e. the neutron multiplication, generates temporal correlations between the neutrons born in the same chain.

As long as only the number of the detector counts is used, as has been the case until very recently, it is sufficient to know the number distribution of the neutrons born in a (spontaneous or induced) fission event as input data to the corresponding master equation. More precisely, since only the first few statistical moments of the detector counts are used, it is sufficient to know the moments of the fission number distribution to the same order. However, as was pointed out in [2], if the detection statistics is extended to include energy dependence, then it is not sufficient to know the number distribution and the average energy spectrum ("one-point energy distribution") of the fission neutrons. Rather, to calculate the statistics of two or three neutrons detected in coincidence in an energy dependent

[a] e-mail: zsolt@nephy.chalmers.se
[b] e-mail: imre@chalmers.se

case, one needs to know the two- and three-point energy distribution of the source (fission) neutrons, respectively, and so on.

Unfortunately, there are no experimental data available on the energy and angular correlations of fission neutrons. Instead, possible sources of neutron correlations are being sought in the modeling and simulation of the fission process. One possible mechanism for such correlations for spallation and fission neutrons was suggested in [3] already in 2000. It was argued that if the energies of the fission neutrons are independent of each other in any individual fission event, but the (independent and identical) energy distributions of the neutrons depend on the actual number of the neutrons (i.e the energy spectra are different for different neutron multiplicities on fission), then neutron energies become correlated when considering all possible fission events with different neutron numbers.

Recently it has also been recognised by the nuclear safeguards community that another source of the correlations between the angles and energies of the neutrons, as measured in the laboratory system, can arise from the neutron emission process. Namely, from the fact that most of the neutrons (except the so-called "pre-scission" neutrons) are emitted from the moving fission fragments as evaporation neutrons. Although the energies and emission angles of the evaporation neutrons are independent from each other in the center of mass system, their energies and directions in the lab system will be jointly affected by the fact that they are emitted from the same fission fragment, or from the two separate fission fragments, whose energies and directions are strongly correlated to each other by conservation of energy and momentum. These are now also being incorporated into numerical simulations of the intra-nuclear cascade codes [4, 5].

Although a large bulk of investigations were made in the safeguards community on how the emission from a moving target biases the angular directions of the neutrons in the laboratory system, the two-point (in angle) covariance function of these neutrons was neither calculated, nor measured. In addition, the effect of the emission from the moving fission fragments on the energy correlations in the lab system has not received attention yet. The purpose of this paper is to explore the properties of the energy and angular correlations of fission neutrons in the lab system through calculating the covariances, and even the two-point distributions of the neutron energies first qualitatively through analytical considerations, and then quantitatively through numerical simulations.

2 Analytic demonstration

In the analytical considerations, it is shown that both the energy and the angular correlations are generated in the lab system through the *random properties* of the fission fragment parameters.

The simplification will be used here that instead of calculating energy correlations, correlations and covariances in the velocities will be treated throughout in this section. In addition a one-dimensional model will be employed, where all velocity vectors are aligned along a line, hence they can be treated as (signed) scalars. The reason for this is that the scalar velocities of the fission fragment and the emitted neutrons become additive, leading to simple convolution-type expressions for the distribution of the neutron velocities in the lab system. The same is not true for the energies, where more complicated relationships exist between the energies in the CM and the lab system.

For the sake of simplicity, only the case of two neutrons emitted from the same moving fission fragment and that the two emitted neutrons move towards the positive x-axis will be considered. Denote the two neutron velocities in the CM system as v_1 and v_2, and assume that they are independent and identically distributed with a probability density $f_v(v)$. The independence implies that one has

$$f_2(v_1, v_2) = f_v(v_1) f_v(v_2) \quad and \quad \langle v_1 \, v_2 \rangle = \langle v_1 \rangle \langle v_2 \rangle \tag{1}$$

The velocity of the fission fragment in the lab system is denoted by u, with a probability distribution $f_u(u)$. Here the terminology of Papoulis [6] is adopted with f_u and f_v denoting two different functions. It is also assumed that the velocity of the fission fragments is independent from the velocities of the neutrons, i.e.

$$f_2(v,u) = f_v(v)f_u(u) \quad \textit{and} \quad \langle v_i\, u\rangle = \langle v_i\rangle \langle u\rangle\,; \quad \text{for} \quad i = 1,2 \tag{2}$$

The velocities of the neutrons in the lab system will be denoted as V_i, $i = 1,2$. In the present simple model one will have

$$V_i = v_i + u, \quad i = 1,2 \tag{3}$$

There are of course further assumptions behind (3), such as that the emission of the first neutron does not alter the velocity of the fission fragment. This is equivalent to neglecting the neutron mass in comparison to the mass of the fission fragments. But for the qualitative considerations these do not play a role.

The one-point and two-point distributions of the V_i are denoted as $p_V(V)$ and $p_2(V_1,V_2)$. The task of the analytical treatment is to calculate the distributions $p_V(V)$ and $p_2(V_1,V_2)$ from the distributions $f_v(v)$ and $f_u(u)$ and the functional relationship (3) between the v_i, u and the V_i.

Let us calculate first the covariance

$$\sigma^2_{V_1V_2} = \langle V_1V_2\rangle - \langle V_1\rangle\langle V_2\rangle = \langle (v_1+u)(v_2+u)\rangle - \langle v_1+u\rangle\langle v_2+u\rangle = \sigma_u^2 \tag{4}$$

in terms of the variances of the neutron and the fission fragments, since it does not require the knowledge of the full probability distributions. Eq. (4) shows that the non-zero covariance between the neutron velocities is brought about by the random character, i.e. the non-zero variance of the fission fragment velocity.

It is more challenging to calculate the two-point distribution $p_2(V_1,V_2)$, or rather the velocity covariance function in the lab system, i.e.

$$\mathrm{Cov}(V_1,V_2) \equiv p_2(V_1,V_2) - p_V(V_1)p_V(V_2) \tag{5}$$

which gives a direct measure of the correlations between the neutron velocities in the lab system.

In the simple 1-D model the covariance function (5) can be easily given in analytical form. Due to the independence of the fission fragment velocity in the lab system and the neutron velocities in the CM system, the one-point and two-point densities of the neutron velocities in the lab system are given by the following expressions:

$$p_V(V_i) = \int_0^{V_i} f_v(V_i-u)f_u(u)\,du \quad i = 1,2 \tag{6}$$

and

$$p_2(V_1,V_2) = \int_0^{\max(V_1,V_2)} f_v(V_1-u)f_v(V_2-u)f_u(u)du \tag{7}$$

From here one obtains the result

$$\mathrm{Cov}(V_1,V_2) =$$

$$\int_0^{\max(V_1,V_2)} f_v(V_1-u)f_v(V_2-u)f_u(u)du - \int_0^{V_1} f_v(V_1-u)f_u(u)\,du \int_0^{V_2} f_v(V_2-u)f_u(u)\,du \tag{8}$$

It is straightforward to generalise expression (8) for the case of energy or angular correlations in a 3-D case by assuming an isotropic distribution of emitted neutrons in the CM system and even for the case when a random number of neutrons is emitted per fission. The corresponding formulae become though rapidly exceedingly involved since one has to handle the distributions of complicated non-linear functions of random variables with known distributions (velocities and directions of the fission fragments and the emitted neutrons in the CM system). Quantitative results were therefore obtained by numerical methods.

3 Numerical simulations

As indicated above, it is much more expedient to calculate the covariances and the two-point distributions by Monte-Carlo based numerical simulations in case of 3-D events and realistic fission fragment energy distributions. The details of such a simulation implemented in MATLAB is described below.

The statistical properties of the input data of the model follow largely from the choice of the fissile material. This was chosen to be uranium 235. This choice specifies, among others, the number distribution of the fission neutrons. This data is though not used in the present simulations, since always two neutrons are followed up after the fission process. One needs to specify also the energy spectrum (probability density of the neutron energies), and the energy distribution of the fission fragments.

The total kinetic energy is distributed deterministically between the two fragments according to conservation of the energy and the momentum. However, in an ensemble of many fission events, the masses of the fragments are also random variables, which adds the randomness of the fission fragment energies.

Hence the simulation starts with sampling the energies of the fission fragments. Since the kinematics of the transformation from the CM system to the lab system is given in terms of velocities, the energies are converted to speeds, then with the sampling of random angles, they are converted into velocity vectors. After the transition from CM to lab system is performed, the speeds are converted to energies again.

The speed of the fission fragments in the laboratory frame was thus calculated in four steps:

1. The fission yield table of the U235 was sampled to acquire a random fission fragment mass m_{ff} (for this the JEFF 3.1 data base was used).

2. The value of the total kinetic energy TKE of both fragments is considered as a normal random variable with a mean 164.4 MeV, and standard deviation 4.5 MeV.

3. The kinetic energy of the investigated fragment (the one sampled in the first step) is calculated as $KE_{ff} = \frac{235-m_{ff}}{235} TKE$.

4. Finally the random fission fragment speed is $u = \sqrt{\frac{2KE_{ff}}{m_{ff}}}$

Hence, ultimately the probability density function of the fission fragment kinetic energy is a function of the probability distributions of the mass and the total kinetic energy.

The energies of the neutrons in the CM frame are sampled from the Maxwell spectrum. The neutron energies are independent from the energies of the fission fragments. As it was mentioned previously, the emission of the neutrons is not supposed to change the energy of the fission fragment. Such a "recoil" effect could actually be modeled in the simulations, but was not considered to be important in the present investigations, due to its very minor quantitative effect.

Having obtained the scalar velocities of the fission fragments in the lab system and the neutrons in the CM system, the neutron velocities are obtained in the lab system as

$$V_i = \sqrt{v_i^2 + u^2 + 2\,u\,v_i\,\cos(\alpha_i)\,\sin(\beta_i)} \tag{9}$$

where α_i and β_i are the azimuth and polar angles respectively between the directions of the fission fragment velocity in the lab system and that of neutron i in the CM system. The angles are supposed to be uniformly distributed.

3.1 Results

During the simulations, not only the covariance defined in Eq. (4) was determined, but also the statistical correlation was calculated to gain a normalised estimator. The statistical correlation is defined for two random variables as the ratio of the covariance and the standard deviations:

$$cor(X, Y) = \frac{\sigma^2_{X,Y}}{\sigma_X \sigma_Y} \cdot 100[\%] \tag{10}$$

The definition Eq. (10) gives a scalar quantity between -100 and 100.

The simulations were done for 1-D, 2-D and 3-D events. The 1D+ case is related to a 1D simulation when the neutrons are emitted only towards the positive axis. In the 2D case, the neutron and the fission fragment velocities are not only scalar values, but vectors. The neutron velocity in the lab system is the vectorial sum of the fission fragment speed in the lab and the neutron velocity in the center-of-mass system. The 3D simulation was the generalization of the 2D case.

The results for the statistical correlation for the 1D, 2D and 3D cases are summarized in Table 1. Though the correlation was found positive for both cases, it is seen that the correlation is increasing with the complexity of the geometry.

Table 1: Statistical correlation for 1D, 2D and 3D cases.

Case	Statistical correlation [%]	σ
1D+	8.83	0.029
1D	0.16	0.031
2D	1.11	0.033
3D	2.23	0.033

Fig. 1a shows the bivariate density function $p_2(E_1, E_2)$ of the two neutron energies in the different systems, whereas Fig. 1b displays the energy covariance map

$$\text{Cov}(E_1, E_2) = p_2(E_1, E_2) - p_E(E_1)p_E(E_2). \tag{11}$$

Fig. 1b shows the structure of the energy correlations. As expected, these are symmetric around the $E_1 = E_2$ line. There is a peak of positive correlations at low energies around $E_1 = E_2$, as well as two ridges of negative (exclusive) correlations for low and constant E_2 along the E_1a axis and a mirror image for low E_1 along the E_2 axis. These negative correlations mean that the joint occurrence of the energy pairs described above and shown by the elongated negative ridges is rather unlikely.

4 Conclusions

It was shown that both energy and angular correlations between fission neutrons in the lab system, emitted from the moving fission fragments, are generated due to the random energy and angular distributions of the fission fragments. Preliminary, simplified simulations show that the energy correlations generated this way are small. Further simulations in more realistic models for both energy and angular correlations will be performed in the continuation. As a first step, neutron emissions from both fission fragments can be taken into account. Furthermore, a random number of neutrons for all multiplicities can be accounted for, with also taking into account the fact that the neutron energy distributions depend on the number of neutrons emitted.

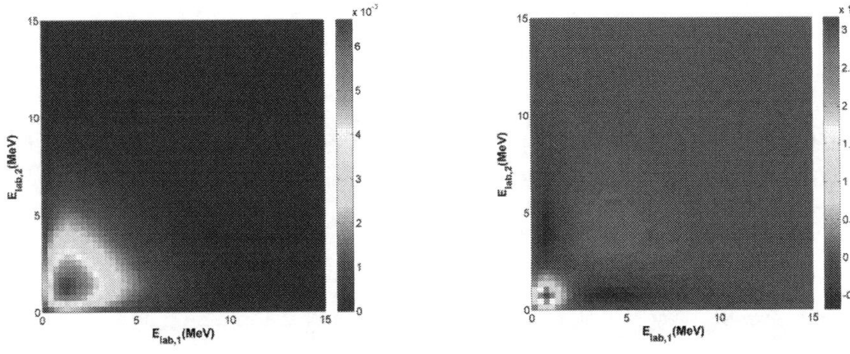

(a) Joint energy distribution

(b) Covariance map

Figure 1: Joint distributions for the realistic 3D case.

Acknowledgements

This work was financially supported by the Swedish Research Council, Grant No B0774801.

References

[1] I. Pázsit, L. Pál, *Neutron Fluctuations: a Treatise on the Physics of Branching Processes*, 1st edn. (Elsevier, New York, 2008)

[2] I. Pázsit, Physica Scripta **59**, 344 (1999)

[3] I. Pázsit, N. Sjöstrand, V. Fhager, Nuclear Instruments and Methods in Physics Research Section A: Accelerators, Spectrometers, Detectors and Associated Equipment **452**, 256 (2000)

[4] R. Vogt, J. Randrup, D.A. Brown, M.A. Descalle, W.E. Ormand, Physical Review C **85**, 024608 (2012)

[5] C. Hagmann, J. Randrup, R. Vogt, Nuclear Science, IEEE Transactions on **60**, 545 (2013)

[6] A. Papoulis, *Probability, Random Variables and Stochastic Processes*, 3rd edn. (McGraw-Hill, Inc, New York, 1991)

EPJ Web of Conferences **111**, 05004 (2016)
DOI: 10.1051/epjconf/201611105004

Evaluating the ^{239}Pu Prompt Fission Neutron Spectrum Induced by Thermal to 30 MeV Neutrons

D. Neudecker[1,a], P. Talou[1], T. Kawano[1], A.C. Kahler[1], M.E. Rising[1], and M.C. White[1]

[1] *Los Alamos National Laboratory, P.O. Box 1663, Los Alamos, NM 87545, USA*

Abstract. We present a new evaluation of the ^{239}Pu prompt fission neutron spectrum (PFNS) induced by thermal to 30 MeV neutrons. Compared to the ENDF/B-VII.1 evaluation, this one includes recently published experimental data as well as an improved and extended model description to predict PFNS. For instance, the pre-equilibrium neutron emission component to the PFNS is considered and the incident energy dependence of model parameters is parametrized more realistically. Experimental and model parameter uncertainties and covariances are estimated in detail. Also, evaluated covariances are provided between all PFNS at different incident neutron energies. Selected evaluation results and first benchmark calculations using this evaluation are briefly discussed.

1 Introduction

A prompt fission neutron spectrum (PFNS) provides the energy distribution of neutrons emitted promptly after fission and before the onset of delayed β-decay. The PFNS of ^{239}Pu is a quantity of high interest for nuclear data applications including reactor physics and global security. For instance, a reliably evaluated ^{239}Pu PFNS is key to accurately simulate the reactivity of chain reactions in a nuclear reactor. Therefore, significant efforts have been undertaken to provide improved evaluated nuclear data within the framework of an IAEA coordinated research project on PFNS of actinides [1] as well as for the CIELO project [2] and the next release of the ENDF/B-library.

Here, we present a new evaluation of the ^{239}Pu PFNS induced by neutrons with incident energies from thermal to 30 MeV. This evaluation is currently tested as part of a CIELO prototype file for incident energies above 5 MeV and was also distributed to the IAEA CRP [1]. Compared to the ENDF/B-VII.1 evaluation [3], improved experiment and model inputs were included in the evaluation as described in Section 2. Changes compared to ENDF/B-VII.1 are discussed in Section 3, along with first benchmark results. Section 4 offers a brief summary and an outlook for future work.

2 Evaluation methodology and input

Evaluated data ψ and covariances Cov^{ψ} are provided for incident neutron energies E_{inc} = {thermal, 0.5, 1, 2, 3, 4, 5, 5.5, 6, 6.5, 7, 8, 9, 10, 11, 11.5, 12, 13, 14, 15, 17.5, 20, 30} MeV and for each

[a] e-mail: dneudecker@lanl.gov

Table 1. The first author, main references, incident energies, outgoing energy range and range of relative uncertainties are given for the experimental database included in the evaluation.

Author	References	E_{inc}	E_{out} (MeV)	Rel. Unc. (%)
Starostov combined	[4–7]	thermal	0.025–11.3	2–73
Lajtai et al.	[8]	thermal	0.025–4.0	9–47
Knitter	[9]	0.215 MeV	0.28–13.87	4–347
Lestone et al.	[10]	1.5 MeV	1.5–11.5	2–112
Chatillon et al.	[11, 12]	1-15 MeV	0.3–8.25	5–120

E_{inc} for the same outgoing energy grid E from 10^{-5} to 30 MeV. They were obtained by the frequently employed generalized least squares algorithm:

$$\psi = N + \mathrm{Cov}^{\psi} S^{t} (\mathrm{Cov}^{\phi})^{-1} (\phi - SN) \text{ and } \mathrm{Cov}^{\psi} = \mathrm{Cov}^{N} - \mathrm{Cov}^{N} S^{t} \left(S\,\mathrm{Cov}^{N} S^{t} + \mathrm{Cov}^{\phi} \right)^{-1} S\,\mathrm{Cov}^{N}. \quad (1)$$

The variables N and ϕ denote model predicted PFNS values and experimental data and Cov^{N} and Cov^{ϕ} their respective covariances. The design matrix S and its transpose S^{t} are calculated by linear interpolation taking into account E_{inc} and the outgoing neutron energy E of N and ϕ.

Experimental input

The experimental data ϕ included in the evaluation through Eq. (1) are listed in Table 1. The data of Lestone et al. [10] and Chatillon et al. [11] were recently published and are thus not included in the ENDF/B-VII.1 evaluation. The latter data were corrected following the work of Granier [12]. We did not include Chatillon et al. data for $E_{\mathrm{inc}} > 15$ MeV as the measurement does not capture the full pre-equilibrium component of the PFNS contrary to our model values. If we include the data for $E_{\mathrm{inc}} > 15$ MeV, this physics discrepancy between model and experiment will lead to biased evaluated data. The data of [4–7] were statistically combined into one data set, which we term here "Starostov combined".

The experimental covariances Cov^{ϕ} appearing in Eq. (1) were estimated in great detail for each data set and covariances between different data sets were estimated as well [13]. The total uncertainties given in Table 1 are for most experiments larger than those found in their respective EXFOR entries. The uncertainties had to be increased after recent Monte Carlo transport studies [12–14] of the experimental set-ups of [4–9, 11] indicated possible biases of the experimental data due to e.g., multiple scattering. These effects were not considered for the evaluation of ENDF/B-VII.1. The experimental data and covariances for this evaluation will be published at https://www-nds.iaea.org/pfns/.

Model input

The model values N and covariances Cov^{N} were calculated using the code CoH [15]. It gives N in the laboratory frame with outgoing neutron energy E by

$$N(E) = \Big\{ \nu_1 N_1(E) P_{f1} + [\chi_1(E) + \nu_2 N_2(E)] P_{f2} + [\chi_1(E) + \chi_2(E) + \nu_3 N_3(E)] P_{f3} \quad (2)$$
$$+ [\chi_1(E) + \chi_2(E) + \chi_3(E) + \nu_4 N_4(E)] P_{f4} \Big\} \Big/ \Big\{ \nu_1 P_{f1} + [1 + \nu_2] P_{f2} + [2 + \nu_3] P_{f3} + [3 + \nu_4] P_{f4} \Big\}.$$

The spectra $\chi_i(E)$ describe the emission of $i = \{1, 2, 3\}$ neutron(s) prior to fission. These pre-fission neutrons are emitted in pre-equilibrium and compound processes, and result in a pre-fission neutron spectrum which is added to the PFNS. Only neutrons are considered that lead to a residual nucleus with enough energy to fission. The pre-equilibrium component was not considered in ENDF/B-VII.1. The variables $N_i(E)$ are the spectra of neutrons emitted promptly after fission in first-chance ($i = 1$), second-chance ($i = 2$), etc., processes and are calculated by an extended version [16] of the Los Alamos model [17] (LAM), while the exciton model was used for the pre-equilibrium component. The variables P_{fi} yield the probability that fission of ith chance occurs. The fission barrier parameters used to calculate P_{fi} were fitted to reproduce P_{fi} from ENDF/B-VII.0. N_i are weighted with ν_i, the average number of neutrons emitted.

The total kinetic energy carried away by the fission fragments and the energy release are important parameters of the LAM and are known—from measurements and theoretical considerations—to depend on E_{inc}. This E_{inc}-dependence was considered following Refs. [18, 19], while the total kinetic energy was assumed to be constant for the ENDF/B-VII.1 evaluation. Also, uncertainties of more model parameters following [16] were included for this evaluation compared to ENDF/B-VII.1.

3 Results and discussion

Evaluated data ψ and covariances $\mathbf{Cov}^{\psi}$ are obtained by Eq. (1) and the experimental and model inputs of Section 2. The experimental data set Starostov combined [4–7] and Lestone et al. [10] have the highest precision of our experimental database. The evaluated results presented here (termed "Current") at E_{inc} = thermal and 1.5 MeV follow both data sets closely as shown in the upper panels of Fig. 1. The ENDF/B-VII.1 evaluation [3] also follows the data of Lestone et al. closely at E_{inc} = 1.5 MeV. For E_{inc} = thermal, ENDF/B-VII.1 has a trend to lower PFNS values for $E_{\text{out}} \leq 0.8$ MeV compared to the combined Starostov data as well as the current evaluation.

The shapes of ENDF/B-VII.1 and the current evaluation differ distinctly for E_{inc} = 6 and 14 MeV in the lower panels of Fig. 1 due to the improvements made in the model description. In ENDF/B-VII.1, all neutrons evaporated from the initial and subsequent compound nuclei are counted, while in the current evaluation only those neutrons associated with the fission process are considered. This difference results in the bump at 300 keV in the current evaluation for E_{inc} = 6 MeV. This bump stems from the superposition of first and second chance fission contributions to the PFNS, where the latter drops off sharply at a threshold defined by the energy needed to overcome the fission barrier. At E_{inc} = 14 MeV, the difference in the shape (e.g., the peak around E = 8 MeV) arises from the pre-equilibrium component, which was not considered for ENDF/B-VII.1. For $E_{\text{inc}} > 1.5$ MeV, only the data Chatillon et al. [11] were deemed accurate and precise enough to be included in the evaluation. However, they do not conclusively support one evaluation over the other. The Chi-Nu project at LANSCE [20], which aims at delivering ^{239}Pu PFNS precision data for E_{inc} = 0.5–20 MeV, might provide more conclusive answers in the near future.

First benchmarks results are presented in Table 2 for the criticality k_{eff} of the Jezebel (PMF1), Flattop-Pu (PMF6) and a suite of thermal solution (PST) assemblies [21]. They were frist calculated using ENDF/B-VII.1 data, and then with a modified ENDF/B-VII.1 file, which was updated with the current evaluation, the findings of the WPEC SG34 [22] on the neutron multiplicity and CIELO ^{16}O data [2]. Jezebel and Flattop benchmarks are higher than 1 for ENDF/B-VII.1 and are now lower. The k_{eff} values for the suite of seven PSTs is slightly closer to 1 than the respective ENDF/B-VII.1 value, mostly due to new ^{16}O data. While these results look encouraging, one has to consider that a critical assembly benchmark is not a *sufficient* validation constraint of evaluated data, as many reactions contribute to calculations of assemblies and biases in several of them might compensate each other.

Evaluated covariances Cov^{ψ} are also provided as part of the same evaluation and for each incident energy as well as between different incident energies. In Ref. [23], it is shown that neglecting those cross-correlations between different incident energies results in an under-prediction of the k_{eff} uncertainty caused by the ^{239}Pu PFNS of Jezebel and Flattop by up to 50%.

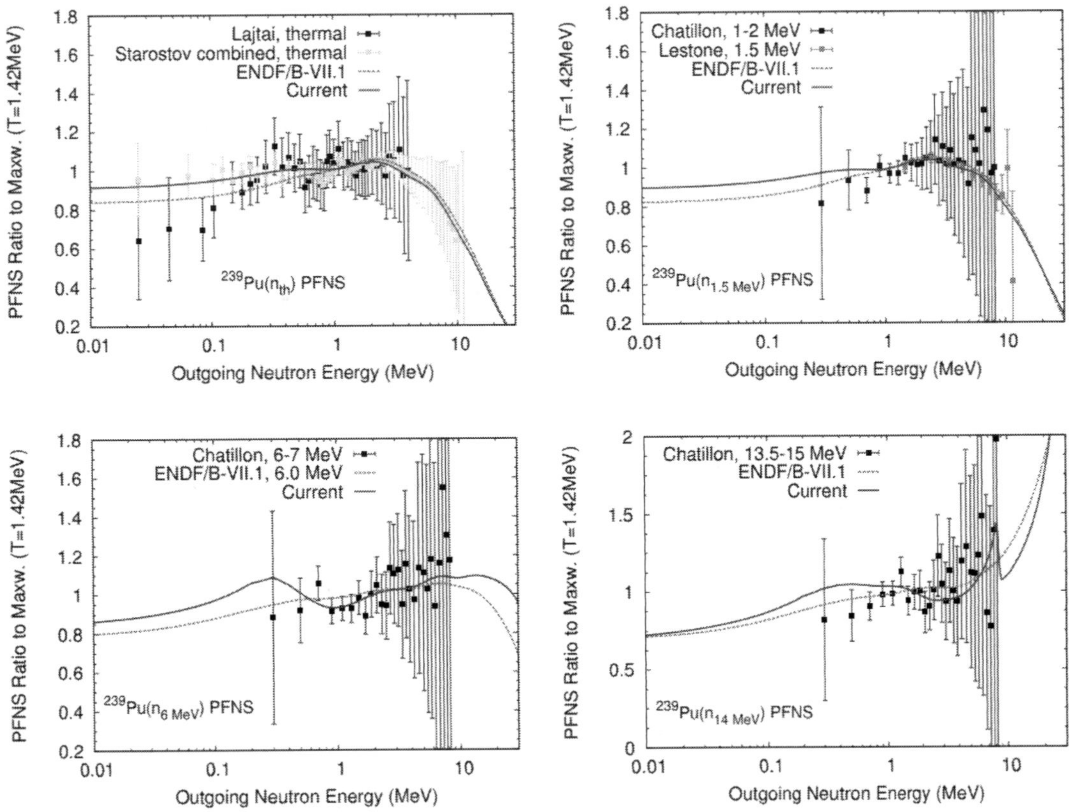

Figure 1. The evaluated data for E_{inc} = thermal, 1.5, 6 and 14 MeV ("Current") are compared to the ENDF/B-VII.1 evaluation and experimental data of the same E_{inc} (scaled to the current evaluation). The current evaluated data at E_{inc} = 1.5 MeV were obtained by averaging evaluated data at 1 and 2 MeV in order to compare to the data of Lestone et al.

Table 2. Benchmark calculations of k_{eff} for Jezebel (PMF1), Flattop-Pu (PMF6) and an average of a suite of seven thermal solution assemblies (PST) is shown using ENDF/B-VII.1 and the current evaluation. The suite of PSTs comprises PST1.4, PST4.1, PST12.10, PST12.13, PST18.6, PST34.4 and PST 34.15. The naming convention follows the ICSBEP handbook [21].

Evaluation	PMF1	PMF6	PST averages
ENDF/B-VII.1	1.00061	1.00111	1.00388
Current	0.99908	0.99946	1.00295

4 Summary and outlook

We presented an evaluation of the ^{239}Pu prompt fission neutron spectrum (PFNS) for incident neutron energies from thermal to 30 MeV, which was submitted to the IAEA Coordinated Research Project on PFNS [1] and is partially implemented in a CIELO test file [2]. Compared to the ENDF/B-VII.1 evaluation [3], (1) recently published experimental data [10, 11] were implemented, (2) detailed experimental uncertainties were estimated including uncertainties associated with previously unknown possible biases [13, 14], (3) the original Los Alamos model [17] was extended [16], (4) the pre-equilibrium component to the PFNS was considered and (5) the incident energy dependence of important model parameters was included in the model description following [18, 19]. The evaluated results shown in Fig. 1 follow high precision experimental data and differ in shape—especially above the second chance fission threshold—from the ENDF/B-VII.1 evaluation. Initial criticality benchmarks show small but acceptable changes compared to ENDF/B-VII.1.

Benchmarking is ongoing, partially within the CIELO project, to study also the effect of compensating errors in other reactions and isotopes. Finally, an incident energy dependent evaluation of the ^{235}U PFNS along the same lines is in progress.

Acknowledgements

We would like to acknowledge the valuable input of R. Capote, M.B. Chadwick, J.J. Egan, Th. Granier, R.C. Haight, H.Y. Lee, J.P. Lestone, W. Mannhart, J.M. O'Donnell, N. Otsuka, V. Pronyaev, A. Sardet, T.N. Taddeucci, D.L. Smith, P. Staples and J. Taieb. This work at LANL was sponsored by the NNSA, US DoE under Contract No. DE-AC52-06NA25396.

References

[1] R. Capote et al., Nuclear Data Sheets, submitted (2016)
[2] M.B. Chadwick et al., Nuclear Data Sheets **118**, 1 (2014)
[3] M.B. Chadwick et al., Nuclear Data Sheets **112**, 2887 (2011)
[4] V.N. Nefedov et al., IAEA Vienna Report INDC(CCP)-0457 (2014), EXFOR-No. 40871
[5] A.A. Boytsov et al., IAEA Vienna Report INDC(CCP)-0459 (2014), EXFOR-No. 40873
[6] B.I. Starostov et al., IAEA Vienna Report INDC(CCP)-0458 (2014), EXFOR-No. 40872
[7] B.I. Starostov et al., IAEA Vienna Report INDC(CCP)-293/L, 19 (1989), EXFOR-No. 40930
[8] A. Lajtai et al., Proceedings of the Conf. on Nuclear Data for Basic and Applied Sciences, Santa Fe 1985 **1**, 613 (1985), EXFOR-No. 30704
[9] H.-H. Knitter, Atomkernenergie **26**, 76 (1975), EXFOR-No. 20576
[10] J.P. Lestone et al., Nuclear Data Sheets **119**, 213 (2014)
[11] A. Chatillon et al., Physical Review C **89**, 014611 (2014)
[12] Th. Granier, Phys. Procedia **64**, 183 (2015)
[13] D. Neudecker et al., Nuclear Data Sheets, submitted (2016); Los Alamos National Laboratory Report LA-UR-15-20824 (2015)
[14] T.N. Taddeucci et al., Nuclear Data Sheets **123**, 135 (2015)
[15] T. Kawano et al., Journal of Nuclear Science and Technology **47**, 462 (2010)
[16] D. Neudecker et al., Nuclear Instruments and Methods in Physics Research A **791**, 80 (2015)
[17] D.G. Madland and J.R. Nix, Nuclear Science Engineering **81**, 213 (1982)
[18] J.P. Lestone et al., Nuclear Data Sheets **118**, 208 (2014)
[19] D.G. Madland, Nuclear Physics A **772**, 113 (2006)
[20] R.C. Haight et al., Nuclear Data Sheet **123** (2015)
[21] *International Handbook of Evaluated Criticality Safety Benchmark Experiments* (OECD Nuclear Energy Agency Report NEA/NSC/DOC(95)03, 2013)
[22] C. de Saint Jean et al., OECD Nuclear Energy Agency Report NEA/WPEC-34 (2015)
[23] M.E. Rising et al., ICNC 2015 proceeding, submitted (2015); Los Alamos National Laboratory Report LA-UR-15-24045 (2015)

EPJ Web of Conferences **111**, 06001 (2016)
DOI: 10.1051/epjconf/201611106001

Neutron Cross Section Processing Methods for Improved Integral Benchmarking of Unresolved Resonance Region Evaluations

Jonathan A. Walsh[1,2,a], Benoit Forget[1], Kord S. Smith[1], and Forrest B. Brown[2]

[1] *Department of Nuclear Science & Engineering*
Massachusetts Institute of Technology
77 Massachusetts Avenue, 24-107, Cambridge, MA, USA 02139

[2] *XCP-3, Monte Carlo Methods, Codes, & Applications*
Los Alamos National Laboratory
Los Alamos, NM, USA 87545

Abstract. In this work we describe the development and application of computational methods for processing neutron cross section data in the unresolved resonance region (URR). These methods are integrated with a continuous-energy Monte Carlo neutron transport code, thereby enabling their use in high-fidelity analyses. Enhanced understanding of the effects of URR evaluation representations on calculated results is then obtained through utilization of the methods in Monte Carlo integral benchmark simulations of fast spectrum critical assemblies. First, we present a so-called on-the-fly (OTF) method for calculating and Doppler broadening URR cross sections. This method proceeds directly from ENDF-6 average unresolved resonance parameters and, thus, eliminates any need for a probability table generation pre-processing step in which tables are constructed at several energies for all desired temperatures. Significant memory reduction may be realized with the OTF method relative to a probability table treatment if many temperatures are needed. Next, we examine the effects of using a multi-level resonance formalism for resonance reconstruction in the URR. A comparison of results obtained by using the same stochastically-generated realization of resonance parameters in both the single-level Breit-Wigner (SLBW) and multi-level Breit-Wigner (MLBW) formalisms allows for the quantification of level-level interference effects on integrated tallies such as k_{eff} and energy group reaction rates. Though, as is well-known, cross section values at any given incident energy may differ significantly between single-level and multi-level formulations, the observed effects on integral results are minimal in this investigation. Finally, we demonstrate the calculation of true expected values, and the statistical spread of those values, through independent Monte Carlo simulations, each using an independent realization of URR cross section structure throughout. It is observed that both probability table and OTF treatments reproduce the true expected values, calculated by averaging the results of many independent simulations, quite well. However, the spread of independent calculation results is shown to be relatively significant. The k_{eff} eigenvalues for fast spectrum systems can differ by more than 250 pcm from one simulation to the next.

[a] e-mail: walshjon@mit.edu

1 Introduction

At the intermediate energies between the well-defined resonance structure of neutron cross sections in the resolved resonance region and the broad, smoothed fluctuations in the fast energy region, resonances do exist, but are so closely-spaced in energy that their individual parameters (i.e. resonance energy and partial widths) cannot be determined. A nuclide's unresolved resonance region (URR) is composed of the energies characterized by this phenomenon [1].

Treatment of the URR is of particular importance to the modeling of fast spectrum reactors and critical assemblies which often have a significant neutron flux in the URR of many of their constituent nuclides. Because simply using averaged cross sections is tantamount to neglecting energy self-shielding effects and results in over-predicted reaction rates, URR resonance structure must be accounted for. Implemented in many nuclear data processing codes such as NJOY [2], CALENDF [3], and here, as part of this work, the probability table method [4] is the most widely-used approach for incorporating this structure. Using the average URR resonance parameters contained in an ENDF-6 format [5] evaluation file and the theoretical distributions of those parameters, the method proceeds by generating many realizations of resonance structure and then tabulating the probabilities of observing various cross section magnitudes. This tabulation is performed at several URR energies for each desired temperature.

This work examines alternate methods for processing URR neutron cross sections and their use in integral benchmark simulations. Specifically, in Sec. 2, we start by presenting applications of an on-the-fly (OTF) cross section calculation method which eliminates the need for the pre-generation of probability tables. Next, in Sec. 3, the effects of using a multi-level resonance formalism for representing URR cross sections are analyzed. Then, in Sec. 4, the use of single, independent URR realizations throughout their own independent Monte Carlo simulations is investigated to test the ability of the probability table method to produce true expected value tallies and the spread of independent tally realizations. Lastly, concluding remarks are offered in Sec. 5.

All of the presented methods are implemented in the continuous-energy neutron transport code OpenMC [6]. Except where noted, results are obtained from OpenMC simulations utilizing ENDF/B-VII.1 nuclear data evaluations [7].

2 On-the-Fly Cross Sections

One drawback of the probability table method is that data must be pre-generated at multiple energies for each desired temperature. In simulations of systems with many temperatures (e.g. fast reactor multiphysics analyses), this can result in a large memory requirement. Calculating URR cross sections OTF (i.e. within the Monte Carlo transport calculation) directly from temperature-independent average resonance parameters can reduce this burden significantly [8]. As an additional benefit, the need for performing the sensitivity studies that are typically required to ensure that probability tables are generated with a fine enough phase-space mesh is also eliminated because the OTF method is continuous in the energy, temperature, and cross section magnitude variables.

Here, we use the OTF method to illustrate the testing of pre-generated probability tables and also demonstrate the significance of varying the URR energy bounds in simulations of the Zero Energy Breeder Reactor Assembly, core 8H (ZEBRA-8H)[1,2]. Fig. 1a shows the flux computed with both the OTF method and the probability tables of the ENDF71x ACE-format processed nuclear data library [11]. Agreement between the two flux spectra is an indication that the probability tables are gen-

[1] MIX-MET-FAST-008 from the International Criticality Safety Benchmark Evaluation Project (ICSBEP) [9].

[2] ZEBRA-FUND-RESR-001 from the International Reactor Physics Experiment Evaluation Project (IRPhEP) [10].

erated with a phase-space mesh that is sufficiently fine to produce accurate tallies for the ECCO-33 energy group structure that is used in these simulations. Fig. 1b plots the k_∞ eigenvalues computed with varying URR upper energy bounds. The upward trend of k_∞ with increasing upper bound suggests that resonance structure may be important at energies above the current ^{238}U ENDF/B-VII.1 URR evaluation. As further indication of this, simulations utilizing Fröhner's complete evaluation [12], up to 300 keV, result in k_∞ eigenvalues that are 200 pcm closer to the benchmark value than those computed using the truncated ENDF/B-VII.1 evaluation.

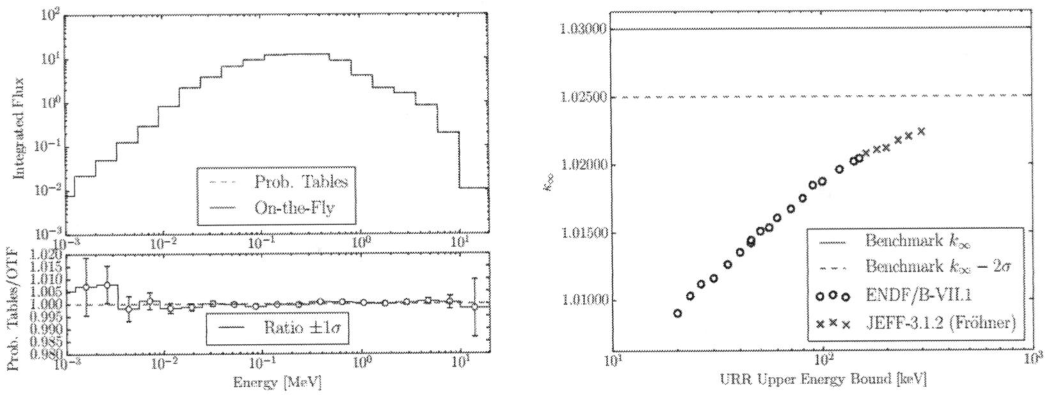

(a) Probability table mesh refinement studies

(b) Energy range sensitivity studies

Figure 1: URR evaluation benchmarking applications with ZEBRA-8H.

3 Multi-Level Resonance Representation

As is prescribed by the ENDF-6 format, the probability table method is typically implemented using the single-level Breit-Wigner (SLBW) resonance formalism to generate URR cross section realizations[3]. Here, we investigate the use of the multi-level Breit-Wigner (MLBW) resonance formalism for OTF URR cross section calculations[4] and compare the results to those obtained using the standard SLBW model in simulations of the Big Ten critical assembly[5]. Flux and ^{238}U scattering rate spectra for both formalisms are shown in Figs. 2a and 2b, respectively. Close agreement between the SLBW and MLBW spectra suggest that the use of a multi-level resonance formalism may not be necessary to reproduce many tallies of interest. This is further supported by the agreement between SLBW and MLBW k_{eff} values that is seen in Table 1.

Table 1: Big Ten k_{eff} eigenvalues.

Method	ENDF71x prob. tables	OpenMC prob. tables	OTF SLBW	OTF MLBW
$k_{\text{eff}}(1\sigma \times 10^5)$	1.00467(10)	1.00466(10)	1.00461(10)	1.00453(9)

[3]One exception is CALENDF which allows SLBW, multi-niveau Breit-Wigner, and Reich-Moore formalisms.

[4]In this preliminary study we investigate only the MLBW formalism as specified by the ENDF-6 format.

[5]IEU-MET-FAST-007, case 4 from the International Criticality Safety Benchmark Evaluation Project (ICSBEP) [9].

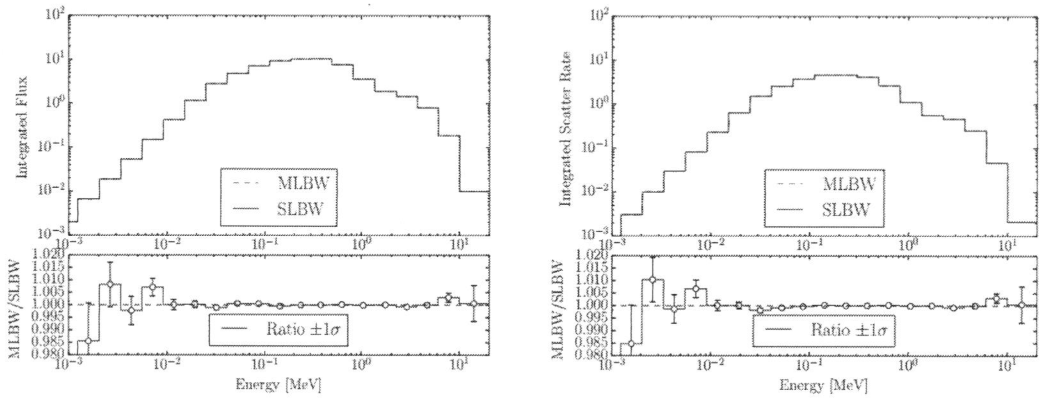

(a) Flux (b) ^{238}U scattering rate

Figure 2: SLBW and MLBW Big Ten tally spectra.

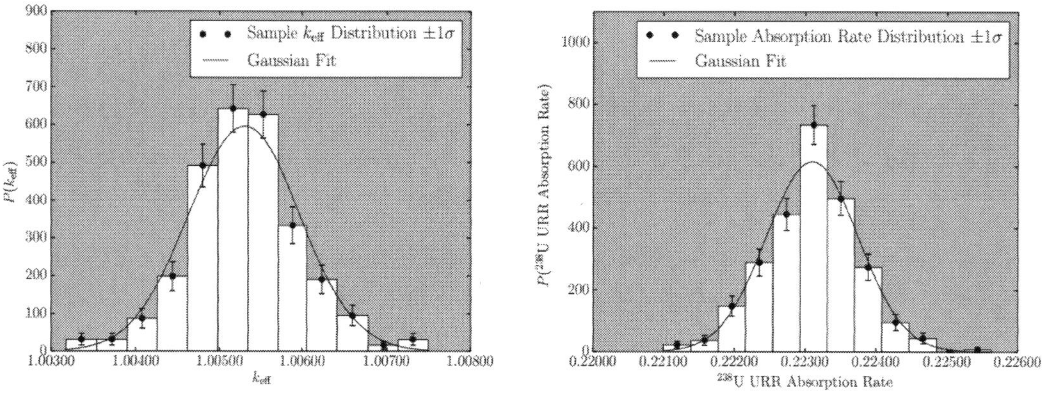

(a) k_{eff} eigenvalues (b) ^{238}U URR absorption rate

Figure 3: Big Ten tally distributions from independent realizations.

4 Independent Resonance Realizations

The typical probability table algorithm and the newly introduced OTF method both assume that neutrons, at a given energy, experience resonance structure that is independent of all structure previously encountered. The result is an artificial averaging over all possible resonance structures whereas, in nature, a single resonance structure exists. As a consequence, it is not guaranteed that simulations using probability tables or OTF calculations must reproduce true expected value tallies. However, by generating independent realizations of URR resonance structure and utilizing each of these realizations throughout its own independent Monte Carlo transport simulation, calculation of true expected value tallies is possible. Also, with each independent simulation now representing a viable realization of nature, it is possible to determine the spread of possible tally results. This spread represents the uncertainty induced by a lack of knowledge of precise URR resonance structure and cannot be ascertained using the probability table or OTF methods.

The distributions of k_{eff} and ^{238}U URR absorption rate from 350 simulations of the Big Ten critical assembly are displayed in Figs. 3a and 3b, respectively. The centers of these distributions, which are the true expected values, agree very well with the results produced using the probability table and OTF methods[6]. However, the spread of individual tally results, each corresponding to a valid realization of nature, is notable with the middle 95% of k_{eff} values spanning a range of 275 pcm.

5 Conclusions

This work demonstrates the use of various neutron cross section processing methods which allow for enhanced analyses of URR evaluations with integral benchmark simulations. An OTF cross section calculation approach is used to compute continuous phase-space results directly from resonance parameters. Further, it is used to demonstrate the variation of k_{eff} with increasing URR upper energy bound. This variation indicates a possible need for modeling resonance structure at energies above the upper bounds of the ^{238}U URR evaluations that are currently available. The ENDF-6 version of the MLBW resonance formalism is also used to calculate OTF URR cross sections. Differences in both k_{eff} and spectral tallies between simulations using the two formalisms are observed to be insignificant. Finally, independent simulations, each utilizing a single, independent URR realization, are used to compute true expected value tallies. The spread of tally realizations arising from unknown URR resonance structure is also demonstrated to be significant for a fast spectrum system with individual k_{eff} values showing differences greater than 250 pcm.

This work is funded by a U.S. Department of Energy (DOE) Nuclear Energy University Programs Graduate Fellowship, the Consortium for Advanced Simulation of Light Water Reactors (CASL), and the DOE National Nuclear Security Administration (NNSA) Advanced Simulation and Computing (ASC) program.

References

[1] A. Foderaro, *The Elements of Neutron Interaction Theory* (MIT Press, 1971)

[2] R. Macfarlane, D. Muir, Tech. Rep. LA-12740-M, Los Alamos National Laboratory (1994)

[3] J.C. Sublet, P. Ribon, M. Coste-Delclaux, Tech. Rep. CEA-R-6277, French Alternative Energies and Atomic Energy Commission (2011)

[4] L. Levitt, Nucl. Sci. Eng. **49**, 450 (1972)

[5] A. Trkov, M. Herman, D. Brown, Tech. Rep. BNL-90365-2009 Rev.2, Brookhaven National Laboratory (2012)

[6] P.K. Romano, B. Forget, Ann. Nucl. Energy **51**, 274 (2013)

[7] M. Chadwick, M. Herman, P. Obložinský, M. Dunn, Y. Danon, A. Kahler, D. Smith, B. Pritychenko, G. Arbanas, R. Arcilla et al., Nucl. Data Sheets **112**, 2887 (2011)

[8] J.A. Walsh, B. Forget, K.S. Smith, B.C. Kiedrowski, F.B. Brown, *Direct, On-the-fly Calculation of Unresolved Resonance Region Cross Sections in Monte Carlo Simulations*, in *Proc. M&C 2015,* Nashville (2015)

[9] NEA Nuclear Science Committee, Tech. Rep. NEA/NSC/DOC(95)03, OECD NEA (2013)

[10] NEA Nuclear Science Committee, Tech. Rep. NEA/NSC/DOC(2006)1, OECD NEA (2014)

[11] J.L. Conlin, D.K. Parsons, S.J. Gardiner, M. Gray, A. Kahler, M.B. Lee, M.C. White, Tech. Rep. LA-UR-13-20137, Los Alamos National Laboratory (2013)

[12] F.H. Fröhner, Nucl. Sci. Eng. **103**, 119 (1989)

[6] The difference between the OTF and true expected eigenvalues is 12 pcm with a 10 pcm 1σ.

EPJ Web of Conferences **111**, 06002 (2016)
DOI: 10.1051/epjconf/201611106002

Development of an automated platform for the verification, testing, processing and benchmarking of Evaluated Nuclear Data at the NEA Data Bank

Status of the NDEC system

F. Michel-Sendis[1,a], C.J. Díez[1], and O. Cabellos[1]

[1] *OECD Nuclear Energy Agency (NEA) - Data Bank, 92130, Issy-les-Moulineaux, France*

Abstract. Modern nuclear data Quality Assurance (QA) is, in practice, a multistage process that aims at establishing a thorough assessment of the validity of the physical information contained in an evaluated nuclear data file as compared to our best knowledge of available experimental data and theoretical models. It should systematically address the performance of the evaluated file against available pertinent integral experiments, with proper and prior verification that the information encoded in the evaluation is accurately processed and reconstructed for the application conditions. The aim of the NDEC (Nuclear Data Evaluation Cycle) platform currently being developed by the Data Bank is to provide a correct and automated handling of these diverse QA steps in order to facilitate the expert human assessment of evaluated nuclear data files, both by the evaluators and by the end users of nuclear data.

1 Introduction

Quality Assurance (QA) applied to nuclear data is designed to ensure that a comprehensive set of quality requirements are fulfilled for electronic evaluated nuclear data files. These requirements go well beyond the necessary verification of compliance with the ENDF-6 format for the electronic files. The verification that the information encoded in the evaluation is accurately processed and reconstructed for the application conditions (e.g. energy grouping and application temperatures) is necessary. Giving a systematic assessment of the performance of each file in pertinent (i.e. sensitive to the nuclide and reactions being tested) internationally evaluated integral benchmark experiments is necessary.

The Data Bank is currently developing an automated platform called NDEC (Nuclear Data Evaluation Cycle) that centralizes such QA processes in an user friendly approach. This paper aims to briefly present the current status of this development.

2 Development of the NDEC platform

In practice, evaluated nuclear data QA is a multistage process that implies having the appropriate computational framework and expertise to run different codes and corresponding input sequences,

[a] e-mail: franco.michel-sendis@oecd.org

with the capability to interpret the resulting output and handle the large numbers of secondary files generated per evaluated file tested. The problem of modern Evaluated Nuclear Data QA is an example where centralization and automation of sequenced processes become necessary in order to facilitate a thorough diagnose of the data. Such diagnoses aim at providing both evaluators and end-users of nuclear data with an in-depth understanding of the data's quality, domain of applicability and, importantly, its limitations. This information is essential when selecting candidate evaluations for a reference Nuclear Data Library.

2.1 Current Status

The Data Bank started in early 2015 to develop a platform for the systematization of a number of these QA processes, called NDEC (Nuclear Data Evaluation Cycle). Currently, NDEC is an operational prototype that automatically performs a number of steps following the submission of an ENDF-6 evaluated file. These steps will be described below. The longer aim of NDEC is to be available as an online NEA Data Bank service providing access to the output and diagnoses of these different QA steps. Currently the steps handled by NDEC are: evaluated format checking, processing, differential comparisons and integral benchmarking. The resulting output is organized in a user-friendly interface. The codes and tools presently used are briefly described below.

2.1.1 Evaluated format checking and processing of ENDF-6 files

ENDF-6 format compliance checks are performed with the utility codes CHECKR, FIZCON, PSYCHE, INTER. Their outputs are compiled. Processing of the submitted file is currently performed with NJOY-2012, PREPRO-2015 and AMPX-6.2. The use of the latter is being carried out by the Data Bank in collaboration with ORNL/SCALE development team under a beta-tester agreement, pending the official release of AMPX (see these proceedings on the processing of JEFF-3.2 with AMPX [1]).

As a general principle, NDEC provides access to all associated inputs used and the outputs obtained by constructing a modular XML organization accessible from a webpage, with modules that are visually color-coded according to the errors or warnings received (see Figure 1) in each sequence performed. A 'dictionary of errors' aims at directly providing a human readable description of the error found in order to facilitate the diagnostic.

2.1.2 Visualization of differential data

The user may visualize the comparison of the reconstructed cross-sections with the different processing codes, and includes the comparison to existing EXFOR data and major nuclear data libraries,for each reaction channel. For this NDEC integrates the functionalities of JANIS-4.0 [2] (see Figure 2).

2.1.3 Integral benchmarking

Assessing the integral performance of each evaluated nuclear data file in benchmark calculations is an essential step in the QA process, and an important criterion to consider when selecting candidate evaluations for a nuclear data library. Special care should be first given to the selection of benchmarks to run, which depend on the nuclide evaluation / reaction channel / energy domain and, in general, the application being tested for. In order to develop the relevant 'validation suites', the use of the different tools developed by the NEA for these purposes is foreseen. Currently in NDEC, this step is focusing exclusively in criticality benchmarking, with the aims to expand this scope in the future.

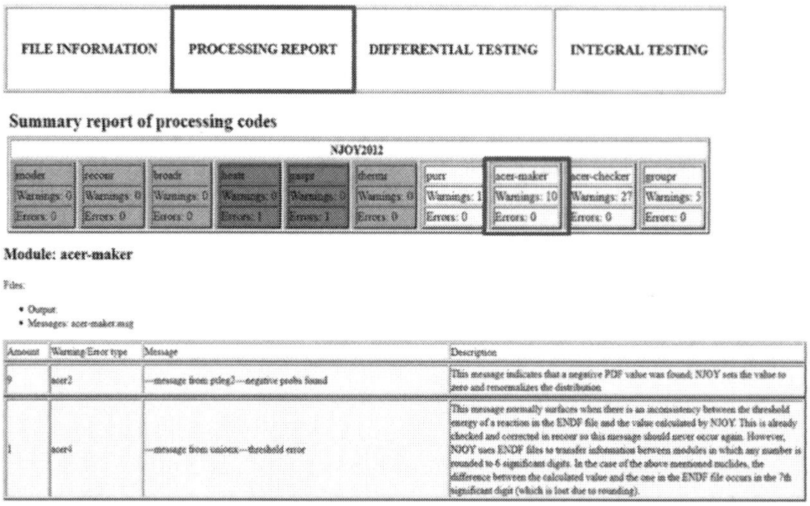

Figure 1. Screen-shot of the 'processing' panel of the NDEC prototype highlighting issues found during NJOY-2012 processing, modules are color-coded (*green*: no errors; *yellow*: only warnings; *red*: errors found) and clickable with respective detail of errors and explanations of errors given.

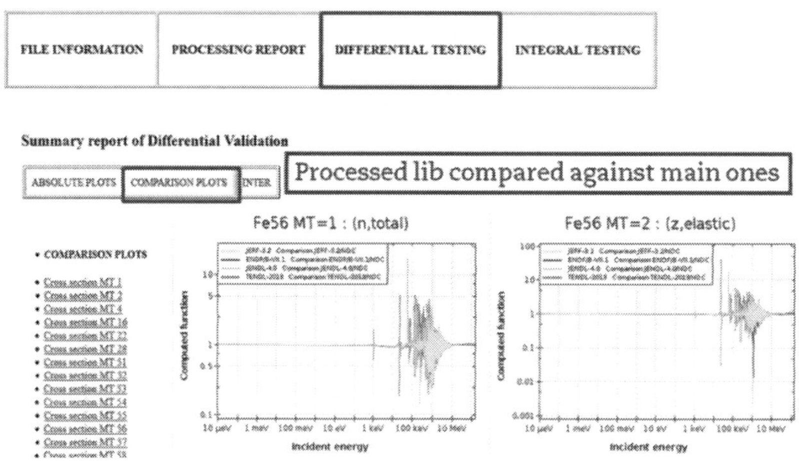

Figure 2. Screen-shot of the 'processing' panel of the NDEC prototype showing the visual comparison of ratio of processed cross sections with major nuclear data library releases (ENDF/B-VII.1; JENDL-4.0; TENDL-2013; JEFF-3.2).

A large computation and compilation effort of criticality sensitivity profiles for available ICSBEP[3] and IRPHE[4] benchmark experiments has been undertaken at the NEA in recent years, resulting in the creation of software searching tools that are able to interrogate the integral databases and select the pertinent benchmark cases (nuclide/reaction specific, above a given sensitivity threshold to k_{eff}); these are the DICE (Database for the International Criticality Safety Benchmark Evaluation Project) or the IDAT (International Reactor Physics Handbook Database and Analysis Tool) tools. A

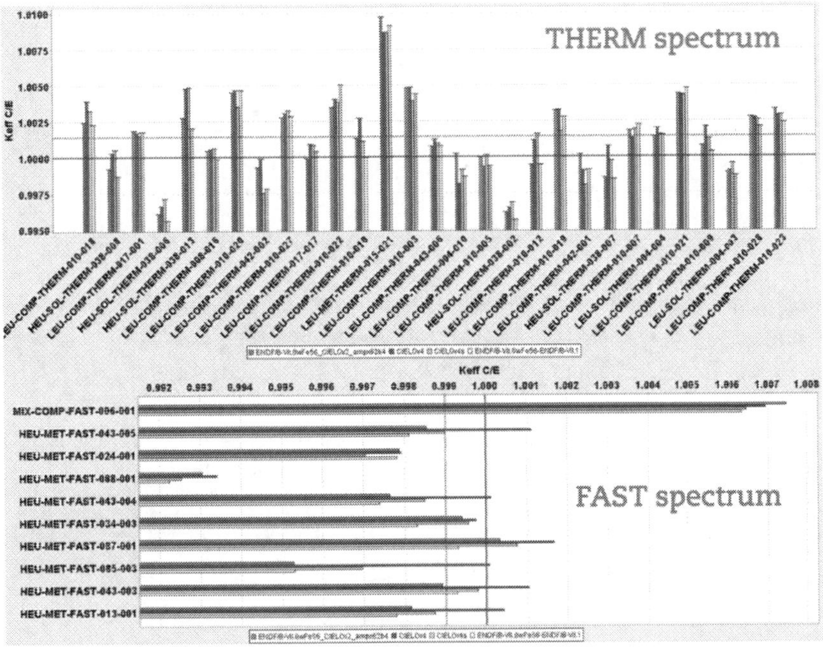

Figure 3. Screen-shot showing criticality benchmark results for 4 different Fe-56 evaluations being considered in the WPEC/CIELO initiative and tested with NDEC.

new tool called NDaST (Nuclear Data Sensitivity Tool)[5] has been recently developed[1] to make use of this sensitivity profile information and carry out sensitivity-based calculations; either performing broad perturbations to nuclear data as a function of nuclide, reaction and energy range or, through an interface JANIS, perform uncertainty propagation of nuclear covariance data to the given response parameter (e.g. keff). The calculated results of both methods allow the quantitative analysis of the impact upon a large number of selected benchmark cases and can be used to construct the pertinent benchmark validation suites per nuclide and application tested (see Figure 3).

2.2 Foreseen developments

The development of NDEC is still ongoing, and many developments could still be envisaged in the medium to longer term. Concerning integral validation, the enlarged scope beyond criticality benchmarking should be considered and could include the selection of relevant decay cases (decay heat, doses, reaction rates), the comparison to spectral average cross sections included in EXFOR, inclusion of transmission and shielding benchmarks from ICSBEP and SINBAD databases, or even performing simple computational benchmarks between evaluated files (e.g. critical mass, shielding transmission, particle emission, etc.) among others.

3 Conclusions

Centralization of processes and tools that allow for a thorough assessment of evaluated nuclear data and its performance in integral experiments is seen as a key necessity of modern nuclear data QA.

[1]NDaST is freely available online via the NEA webpage (http://www.oecd-nea.org/ndast/).

The development of the NDEC platform is at the center of the initiative undertaken by the NEA Data Bank to strengthen its role as a reference centre in verification, benchmarking and validation of nuclear data. These efforts aim at providing better nuclear data services to the Data Bank's member countries in general and to serve the Joint Evaluated Fission and Fusion (JEFF) Nuclear Data Library project in particular.

References

[1] C.J. Diez, F. Michel-Sendis, O. Cabellos, *On the processing of JEFF-3.2 neutron data library with AMPX 6.2 for its use with the SCALE tool suite*, these proceedings.

[2] N. Soppera, M. Bossant, E. Dupont, "JANIS 4: An Improved Version of the NEA Java-based Nuclear Data Information

[3] OECD NEA, *International Handbook of Evaluated Criticality Safety Benchmark Experiments*, NEA/NSC/DOC(95)03, OECD. And at www.oecd-nea.org/science/wpncs/icsbep/

[4] OECD NEA, *International Handbook of Evaluated Reactor Physics Benchmark Experiments*, NEA/NSC/DOC(2006)1, OECD. And at www.oecd-nea.org/science/wprs/irphe

[5] J. Dyrda, I. Hill, et al, *The New OECD-NEA Nuclear Data and Sensitivity Tool (NDaST)*, ANS Winter meeting proceedings, Washington DC, USA, Nov.8-12, 2015.

EPJ Web of Conferences **111**, 06003 (2016)
DOI: 10.1051/epjconf/201611106003

On the processing of JEFF-3.2 neutron data library with AMPX 6.2 for its use with the SCALE tool suite

C.J. Díez[1,a], F. Michel-Sendis[1], O. Cabellos[1], D. Wiarda[2], and M.E. Dunn[2]

[1] *OECD Nuclear Energy Agency (NEA) - Data Bank, 92130, Issy-les-Moulineaux, France*
[2] *Oak Ridge National Laboratory, Reactor and Nuclear Systems Division, 37831-6170, Oak Ridge, TN, USA*

Abstract. New processing capabilities are under development at the NEA Data Bank (DB) that aim to provide enlarged and enhanced nuclear data services to member countries in the framework of processing, verification and benchmarking of evaluated nuclear data. Within the context of the Joint Evaluated Fission and Fusion Nuclear Data Library Project (JEFF), we undertake to generate, with the latest version of AMPX processing code, a JEFF-3.2 incident-neutron nuclear data application library for the SCALE tool suite. In this paper we describe the requirements, in terms of nuclear data content, for new data to be considered and used in the SCALE tool suite. An overview of how to process and prepare JEFF-3.2 incident-neutron data with AMPX for its application in SCALE neutron transport codes is given. The resulting library is verified and tested at differential and integral level by comparing the performance of JEFF-3.2 in AMPX format with other processing and transport codes.

1 Introduction

The OECD Nuclear Energy Agency (NEA) Data Bank (DB) is host to the Joint Evaluated Fission and Fusion (JEFF) Nuclear Data Library project since its creation in 1981. JEFF official releases of nuclear data are distributed in ENDF-6 format [1]. Historically, the DB has also made available processed versions of neutron cross section libraries in ACE (A Compact ENDF) format for use with Continuous Energy (CE) Monte Carlo codes such as MCNP5 [2]. These processed libraries are available upon request from NEA or freely accessible from the web, e.g. for the latest release of JEFF: JEFF-3.2 library [3].

In order to expand JEFF's user-base to include other widely used neutron transport codes, the DB has undertaken to process neutron data of JEFF-3.2, into application libraries for the SCALE tool suite [4], for criticality and shielding calculations. This task is possible thanks to the AMPX processing code release since SCALE6.2β3.

In this paper, we describe how JEFF-3.2 neutron data files are processed with AMPX into CE libraries. Problems encountered during processing are reported and, whenever possible, their implemented solutions. Verification tests and numerical benchmarks are performed to check correct usage of the resulting processed data. Finally, the performance of these CE libraries is assessed with a selection of criticality benchmarks used in MCNP6 validation [5].

[a] e-mail: carlosjavier.diez@oecd.org

Article available at http://www.epj-conferences.org or http://dx.doi.org/10.1051/epjconf/201611106003

2 Processing evaluated cross section data

2.1 The JEFF-3.2 neutron library

The JEFF nuclear data library is a compilation of evaluated nuclear data files produced by a collaboration of Data Bank member countries which form the JEFF Project. The JEFF library consists in its diverse releases over the year of neutron and proton interaction data libraries, radioactive decay data libraries, fission yields, and thermal scattering law data libraries.

This work only addresses the treatment of JEFF-3.2 neutron data, which is the latest JEFF release of March 2014. The JEFF-3.2 neutron data library provides cross section data for 472 nuclides and thermal scattering libraries (TSL) for 9 different compounds in ENDF-6 formatted files.

2.2 Processing with AMPX

AMPX is a modular code system, developed by the Reactor and Nuclear Systems Division (RNSD) at Oak Ridge National Laboratory (ORNL), for processing ENDF-6 format evaluations into multi-group (MG) and CE cross sections using different formats that suit different particle transport codes.

In this paper, only how to generate CE cross section data is addressed. Such a task consists of generating the different parts of CE cross section data with the corresponding AMPX *module*:

- Point cross section data is the representation of cross section data as pairs of points (σ,E), cross section values at incident particle energy respectively, and linearly interpolating between two points for incident energies not represented. Contributions from the resonance region are added if present. The *Polident* module creates these point-wise data, reconstructing cross section values from resonance parameters, while *Broaden* performs Doppler-broadening. *Tgel* keeps the consistency between partial and total reactions.
- Unresolved resonance data (if present) represents the energy region for incident particles where resonances are so closely spaced in energy that it is either impossible or impractical to "resolve" each resonance. As a result, unresolved resonance parameters are averaged as resolved resonance parameters over energy. They can be treated in a statistical way with Monte Carlo techniques, using the so-called "probability tables". *Purm* generates such probability tables from the Unresolved resonance parameters at different temperatures.
- Scattering kinematics store differential cross section data: outgoing energy particles and angles as a function of incident particle energy. *Y12* is used to generate kinematics data for neutron scattering and gamma-production scattering, producing double-differential data, while *Jamaican* converts the double differential point-wise distributions into marginal probability distributions in angle and conditional probability distributions in exit energy. The same codes are used for processing thermal scattering libraries.

An additional module, *Platinum*, is required for merging all the parts into final CE libraries.

2.3 Processing JEFF-3.2 cross section data with AMPX

The JEFF-3.2 cross section data is processed with AMPX into CE libraries and TSL for their use in SCALE transport codes. Files for ^{49}V and ^{13}C are not considered as long as cross section files for natural isotopic compositions are given, ^{nat}V and ^{nat}C correspondingly. Regarding TSL, data for ^{1}H in CaH_2, Ca in CaH_2 and Mg in Mg metal are not processed as long as the presented bound states have not been considered nor included yet in SCALE.

Different issues arise during processing which should be reviewed, and then, communicated to evaluators if necessary:

- Negative cross section values found when reconstructing cross sections from resonance parameters (12 isotopes): ^{40}Ar ^{141}Ce, ^{54}Fe, ^{152,157}Gd, ^{22}Na, ^{128}Te, ^{46,49,50}Ti, ^{89}Y and ^{90}Zr. Any negative cross section value obtained during reconstruction is set to zero.
- Total resonance width given in the file differs from calculated (3 isotopes): ^{141}Ce, ^{22}Na and ^{90}Zr.
- The lower limit of the Unresolved Resonance Range (URR) does not include in its range some unresolved resonance parameters given (1 isotope): ^{239}U. Parameters are interpolated in order to have values within the boundaries of URR when processed with AMPX.
- Sampling negative cross sections when generating probability tables (1 isotope): ^{22}Na. It is not considered further in this work.

3 Verifying the processed CE libraries from JEFF-3.2

By verification, we mean that CE libraries generated can be used in the SCALE tool suite, and their performance in test cases are comparable with results from other codes/applications. Two different test cases are used for this purpose.

3.1 Infinite dilution calculations

They consist of criticality (k_{eff}) calculations of an infinite dilution of fissionable material (^{235}U as default) in a solution of the nuclide whose CE data is tested. Their results show that all CE libraries generated can be used except for two isotopes: ^{152m}Eu and ^{131m}Te. They have not yet been implemented in SCALE, in the modules where compositions and density mixtures are calculated. For the rest, the generated CE libraries successfully pass this test.

3.2 Transmission calculations

They consist in a mono-energetic and mono-directional neutron beam which impinges perpendicularly onto a slab of 25x25x5 cm of the nuclide to test, tallying the neutron spectrum behind the slab in a 2x2x2 cm cube situated 1.5 cm away from the slab (between the slab rear face and the cube's nearest face to the slab). The neutron energies used range from 14.0155 MeV to 0.1125 eV. Fission events are not considered in these calculations.

The total neutron flux observed in the volume tally is compared between MCNP5 [2], using cross section data processed with NJOY2012 [6], and SCALE. Due to issues arising during calculations, the isotopes (76 in total) in Table 1 were not assessed.

For the rest, differences larger than 10% are registered for 152 isotopes (see Table 1) when the statistical uncertainty associated with results is smaller than 10%. No discrepancies are found for incident neutrons at 0.1125 eV. Without probability tables, the number of isotopes with discrepancies decreases to 17 isotopes: 5 falling within statistical uncertainty, while the rest have different issues presented in Table 2.

Differences due to usage of probability tables come from the different handling of cross sections in URR during processing, where AMPX calculates values at the energy of reference and then interpolates on cross section to add additional points if needed, while NJOY calculates at the energy of reference and then interpolates on URR resonance parameters. The ENDF-6 manual says interpolation should be done on cross section. For isotopes whose cross sections are reconstructed from URR parameters, this could lead to differences between NJOY and AMPX. Also, MCNP and SCALE use probability tables in slightly different ways, especially while re-sampling cross sections in that range. In addition, differences can come from scattering data, when incident neutron energies are close to reaction thresholds which use Kalbach-Mann representation [1].

Table 1. Isotopes with differences between SCALE and MCNP larger than 10% in total neutron flux tallied in the cubic volume behind the slab for different mono-energetic neutron beams transmission calculations, not considering fission reactions. Below, those isotopes which could not be compared.

^{107}Ag	^{110m1}Ag	^{242m1}Am	^{36}Ar	^{38}Ar	^{197}Au	^{10}B	^{130}Ba	^{132}Ba	^{133}Ba	^{135}Ba
^{9}Be	^{247}Bk	^{249}Bk	^{106}Cd	^{108}Cd	^{110}Cd	^{111}Cd	^{112}Cd	^{114}Cd	^{136}Ce	^{138}Ce
^{139}Ce	^{143}Ce	^{144}Ce	^{249}Cf	^{250}Cf	^{251}Cf	^{252}Cf	^{242}Cm	^{246}Cm	^{247}Cm	^{248}Cm
^{250}Cm	^{133}Cs	^{135}Cs	^{137}Cs	^{156}Dy	^{158}Dy	^{160}Dy	^{161}Dy	^{162}Dy	^{163}Dy	^{164}Dy
^{253}Es	^{151}Eu	^{152}Eu	^{153}Eu	^{154}Eu	^{155}Eu	^{156}Eu	^{157}Eu	^{58}Fe	^{69}Ga	^{71}Ga
^{152}Gd	^{153}Gd	^{154}Gd	^{155}Gd	^{156}Gd	^{70}Ge	^{72}Ge	^{74}Ge	^{174}Hf	^{176}Hf	^{177}Hf
^{178}Hf	^{179}Hf	^{180}Hf	^{166m1}Ho	^{127}I	^{131}I	^{113}In	^{193}Ir	^{39}K	^{78}Kr	^{80}Kr
^{82}Kr	^{84}Kr	^{138}La	^{140}La	^{176}Lu	^{55}Mn	^{143}Nd	^{144}Nd	^{145}Nd	^{146}Nd	^{147}Nd
^{148}Nd	^{232}Pa	^{107}Pd	^{110}Pd	^{147}Pm	^{148}Pm	^{149}Pm	^{151}Pm	^{141}Pr	^{143}Pr	^{236}Pu
^{238}Pu	^{225}Ra	^{105}Rh	^{101}Ru	^{34}S	^{45}Sc	^{74}Se	^{76}Se	^{77}Se	^{78}Se	^{80}Se
^{82}Se	^{144}Sm	^{147}Sm	^{148}Sm	^{149}Sm	^{150}Sm	^{151}Sm	^{152}Sm	^{153}Sm	^{154}Sm	^{112}Sn
^{116}Sn	^{118}Sn	^{119}Sn	^{122}Sn	^{124}Sn	^{84}Sr	^{86}Sr	^{89}Sr	^{90}Sr	^{182}Ta	^{160}Tb
^{122}Te	^{124}Te	^{170}Tm	^{240}U	^{182}W	^{183}W	^{184}W	^{186}W	^{128}Xe	^{129}Xe	^{132}Xe
^{133}Xe	^{91}Y	^{64}Zn	^{66}Zn	^{68}Zn	^{70}Zn	^{90}Zr	^{92}Zr	^{94}Zr		

Failed to run in MCNP (x10):

^{31}P	^{62}Ni	^{76}Ge	^{109}Ag	^{115}In	^{134}Cs	^{162}Er	^{187}Re	^{233}Th	^{242}Am

Failed to run in SCALE (x66):

^{39}Ar	^{41}Ar	^{139}Ba	^{41}Ca	^{45}Ca	^{109}Cd	^{36}Cl	^{57}Co	^{60}Co	^{51}Cr	^{64}Cu
^{66}Cu	^{165}Dy	^{169}Er	^{171}Er	^{172}Er	^{148}Gd	^{161}Gd	^{181}Hf	^{126}I	^{128}I	^{192}Ir
^{177}Lu	^{27}Mg	^{53}Mn	^{54}Mn	^{20}Ne	^{21}Ne	^{22}Ne	^{63}Ni	^{18}O	^{184}Os	^{186}Os
^{205}Pb	^{209}Po	^{190}Pt	^{192}Pt	^{187}Os	^{188}Os	^{189}Os	^{190}Os	^{191}Os	^{192}Os	^{193}Os
^{194}Pt	^{195}Pt	^{196}Pt	^{198}Pt	^{88}Rb	^{188}Re	^{104}Rh	^{46}Sc	^{31}Si	^{32}Si	^{180m1}Ta
^{98}Tc	^{204}Tl	^{171}Tm	^{168}Yb	^{169}Yb	^{170}Yb	^{171}Yb	^{172}Yb	^{173}Yb	^{174}Yb	^{176}Yb

Table 2. Summary of sources for issues on isotopes in Table 1 when probabliy tables are not used.

Isotope	Energy	Region	Issue
^{107}Ag	1.10 keV	RRR → Fast	Different amount of interpolation points in AMPX and NJOY
^{197}Au	9.86 keV	URR	AMPX issue using scattering radius as function of energy (corrected)
^{9}Be	> 3.87 MeV	-	AMPX cannot handle (n,2n + excited residual) reactions (MT875+)
^{247}Bk	1.10 keV	RRR → Fast	Different amount of interpolation points in AMPX and NJOY
^{253}Es	90.1 eV	RRR	Discrepancies in RRR reconstruction, difference within statistical uncertainty
^{154}Gd	9.86 keV	URR	AMPX and NJOY generate different average cross section values in URR
^{147}Nd	1.10 keV	URR	AMPX and NJOY generate different average cross section values in URR
^{236}Pu	90.1 eV	URR	AMPX and NJOY generate different average cross section values in URR
^{225}Ra	90.1 eV	10 - 100 eV	Integral cross section value differs 0.6% and close to statistical uncertainty
^{45}Sc	104 keV	-	ENDF-6 evaluation provides two cross section values at 100 keV
^{150}Sm	2.94 MeV	Fast	Origin not found (probable issue with differential cross section)
^{90}Zr	104 keV	RRR	Different treatment of total width in RRR between AMPX and NJOY

4 Benchmarking the processed CE libraries from JEFF-3.2 for criticality

The processed libraries are benchmarked, comparing k_{eff} from SCALE6.2b3 and MCNP6.1 results [7]. The benchmarks to calculate are selected, due to input availability in both codes, from those in the criticality validation suite for MCNP6 [5], shown in Table 3 using the ICSBEP identifiers [8].

Results are presented in Table 3, marked in grey those cases where SCALE and MCNP k_{eff} values differ more than 60 pcm. The source of differences for the highlighted cases are discussed:

- HEU-COMP-INTER-003-006 (82 pcm) whose difference comes partially from ^{197}Au, due to differences in reconstruction of ^{197}Au cross sections in URR. However, not including ^{197}Au in calculations yields a difference between MCNP and SCALE of 154 pcm, so the equivalence of MCNP and SCALE models shall be checked.
- HEU-MET-FAST-026-021 (93 pcm) has differences between in MCNP and SCALE models, such as the amount of paraffin. Once equaled, the difference increases up to 343 pcm. Model representations in both cases shall be reviewed.
- U233-COMP-THERM-001-003 (204 pcm) has ^{90}Zr in its composition, for which large discrepancies have been observed in transmission tests. It has a large impact in k_{eff} as seen from sensitivity analysis, so when it is not included in calculations, differences are reduced to 23 pcm.
- U233-MET-FAST-005-002 (2210 pcm) is a ^{233}U sphere reflected by ^{9}Be which involves fast neutron spectrum. As seen in transmission cases, large differences will be encountered using ^{9}Be due to issues in handling MT875+ reaction channels with AMPX.

Therefore, despite HEU-COMP-INTER-003-006 and HEU-MET-FAST-026-021 cases where equivalence between MCNP and SCALE models shall be reviewed, a good agreement is achieved between MCNP and SCALE processing the JEFF-3.2 neutron cross section libraries with AMPX and NJOY.

Table 3. Results for k_{eff} in criticality benchmarks using SCALE6.2b3 and MCNP6.1, marked in grey are those with differences between MCNP and SCALE larger than 60 pcm.

	Experiments		MCNP6.1		SCALE6.2b3		Diff.
Identifier	k_{eff}	unc.	k_{eff}	unc.	k_{eff}	unc.	(pcm)
HEU-COMP-INTER-003-006	1.00000	0.00470	0.99793	0.00003	0.99875	0.000027	-82
HEU-MET-FAST-026-021	1.00000	0.00380	0.99885	0.00002	0.99792	0.000099	93
HEU-MET-FAST-028-001	1.00000	0.00300	1.00502	0.00002	1.00539	0.000035	-37
HEU-SOL-THERM-032-001	1.00150	0.00260	0.99903	0.00002	0.99908	0.000041	-5
IEU-MET-FAST-003-001 (Detailed)	1.00000	0.00170	1.00245	0.00002	1.00289	0.000099	-44
IEU-MET-FAST-004-001 (Detailed)	1.00000	0.00300	1.00705	0.00002	1.00748	0.000091	-43
LEU-COMP-THERM-008-002	1.00070	0.00120	1.00069	0.00005	1.00019	0.00005	50
LEU-SOL-THERM-002-002	1.00240	0.00370	0.99618	0.00002	0.99582	0.000099	36
LEU-SOL-THERM-007-004	0.99880	0.00110	0.99898	0.00002	0.99838	0.000043	60
MIX-COMP-THERM-002-004	1.00240	0.00240	1.00679	0.00002	1.00693	0.000097	-14
PU-COMP-INTER-001-001	1.00000	0.01100	1.00603	0.00002	1.00612	0.000073	-9
PU-MET-FAST-001-001 (Detailed)	0.99999	0.00129	1.00021	0.00002	1.00056	0.000029	-35
PU-MET-FAST-002-001	1.00000	0.00200	1.00258	0.00002	1.00296	0.000099	-38
PU-MET-FAST-003-003	1.00000	0.00300	0.99679	0.00002	0.99666	0.000063	13
PU-MET-FAST-006-001	1.00000	0.00300	1.00218	0.00002	1.00262	0.000099	-44
PU-MET-FAST-011-001	1.00000	0.00100	0.99786	0.00002	0.99795	0.00011	-9
U233-COMP-THERM-001-003	1.00000	0.00240	1.00288	0.00002	1.00493	0.0001	-204
U233-MET-FAST-001-001	1.00000	0.00100	0.99985	0.00002	1.00008	0.000099	-23
U233-MET-FAST-005-002	1.00000	0.00300	0.99515	0.00002	0.97363	0.000026	2210
U233-MET-FAST-006-001	1.00000	0.00140	1.0008	0.00002	1.00123	0.000049	-43

5 Conclusions

The JEFF-3.2 neutron cross section library has been processed with the AMPX version released in SCALE6.2b4. A continuous energy library for neutrons is generated for its use with SCALE transport calculations tools: KENO for criticality and MAVRIC-MONACO for shielding.

This work has identified different issues to address during the processing of JEFF-3.2 data with AMPX, which concern 17 isotopes (see Sec. 2.3). Those issues which are more relevant due to their impact in benchmarking are due to ^{22}Na and ^{90}Zr. Therefore, for these 17 isotopes, processing and evaluations shall be reviewed.

The verification of processed files is carried out by performing criticality calculations in infinite dilution cases and transmission tests. These tests show the need to implement the definition of some isotopes in SCALE: ^{152m}Eu, ^{180m}Ta and ^{131m}Te, as well as those marked in Table 1 for transmission tests. When comparing SCALE and MCNP results in transmissions tests, large differences were found for 150 isotopes (Table 1), due to a different treatment of the URR in processing and its usage in transport codes. The number of discrepant isotopes sharply decreases when probability tables are not used, where important differences are only found for 12 isotopes, with their likely issues being described in Table 2.

Finally, the generated CE library from JEFF-3.2 is used in a selection of benchmarks from the criticality validation suite for MCNP6. A comparison of SCALE and MCNP results is given. In general, a good agreement is obtained, within 60 pcm differences except for 4 cases. For 3 of these 4 cases, isotopes with large differences found in transmission tests are involved, such as ^{197}Au, ^{9}Be and ^{90}Zr. The differences found are however not completely explained, and would require the further review of both the MCNP and SCALE input models used.

Once the missing isotopes have been implemented in SCALE libraries, further study on transmission tests should be carried out with a wider selection of benchmarks in order to expand the application cases. Although, this latter work depends also in availability of inputs, for which checks on equivalence shall carried out.

This work has focused on processing the entire JEFF-3.2 neutron data library with the AMPX version released in SCALE6.2b4. Although most of the evaluations are processed successfully, the goal of releasing the thoroughly processed CE library from JEFF-3.2 library for SCALE would still need some necessary actions and decisions to be taken and which concern the issues mentioned. Nevertheless, this work has set the first milestones and established a successful collaboration for the future usage of JEFF libraries within SCALE.

References

[1] CSEWG, Tech. Rep. BNL-90365-2009 Rev.2, BNL (2012)

[2] X-5 Monte Carlo Team, Tech. Rep. LA-CP-03-0245, LANL (2003), Version 5

[3] OECD/NEA Data Bank, *JEFF-3.2 Evaluated Data Library - Neutron data* (2014), `http://www.oecd-nea.org/dbforms/data/eva/evatapes/jeff_32/`

[4] ORNL, *SCALE: A comprehensive modeling and simulation suite for nuclear safety analysis and design* (2011), ORNL/TM-2005/39, Version 6.1

[5] R.D. Mosteller, Tech. Rep. LA-UR-02-0878, LANL (2002)

[6] R.E. MacFarlane, D.W. Muir, R.M. Boicourt, A.C. Kahler, Tech. Rep. LA-UR-12-27079, LANL (2015)

[7] T. Goorley, M. James, T. Booth, F. Brown, J. Bull, L. Cox, J. Durkee, J. Elson, M. Fensin, R. Forster et al., Nuclear Technology **180**, 298 (2012)

[8] J.B.B. editor, Tech. Rep. NEA/NSC/DOC(95)03, OECD Nuclear Energy Agency (2010)

EPJ Web of Conferences **111**, 06004 (2016)
DOI: 10.1051/epjconf/201611106004

Use of integral experiments for the assessment of the ^{235}U capture cross section within the CIELO Project

Raphaelle Ichou[1,a], Nicolas Leclaire[1], Luiz Leal[1] and Evgeny Ivanov[1]

[1]*Institut de Radioprotection et de Sûreté Nucléaire (IRSN), 92262 Fontenay-aux-Roses Cedex, France*

Abstract. A new ^{235}U capture cross-section evaluation, evaluated by ORNL and the CEA Bruyères-le-Châtel (BRC) has been proposed within the CIELO project. IRSN, who participates in the CIELO project, contributes with data testing and has carried out benchmark calculations using few benchmarks, extracted from the ICSBEP database, for testing the new ^{235}U evaluation. The benchmarks have been selected by privileging the experiments showing small experimental uncertainties and a significant sensitivity to ^{235}U capture cross-section. The k_{eff} calculations were performed with both the MCNP 6 code and the 5.C.1 release of the MORET 5 code, using the ENDF/B-VII.1 library for all isotopes except ^{235}U, for which both the ENDF/B-VII.1 and the new ^{235}U evaluation was used. The benchmark selection allowed highlighting a significant effect on k_{eff} of the new ^{235}U capture cross-section. The results of this data testing, provided as input for the evaluators, are presented here.

1 Introduction

The *Collaborative International Evaluated Library Organization* Pilot Project, known as the CIELO project, is an initiative aimed at improving data evaluation by combining efforts from several institutions around the world. Over the last couple of years, evaluation work has been done to improve data evaluation for six isotopes, namely ^{1}H, ^{16}O, ^{56}Fe, $^{235\text{-}238}$U, and ^{239}Pu. The Working Party on International Nuclear Data Evaluation Co-operation (WPEC) subgroup 29 (SG 29) was established to investigate an issue with the ^{235}U capture cross-section in the energy range from 0.1 to 2.25 keV.1 The WPEC criticality calculation results indicated an overestimation of the ^{235}U capture cross-section of 10 % or more, starting around 100 eV. To understand and solve the problem, recommendation was made to perform new capture cross-section measurements followed by a resonance evaluation. Hence, time-of-flight capture cross-section measurements were done at the Rensselaer Polytechnic Institute (RPI) 2 and at the Los Alamos National Laboratory (LANL).3 Integral quantities, such as fission and capture resonance integral Westcott factors were also included in the evaluation. These new measurements were used together with the SAMMY computer code, developed by ORNL, to re-evaluate the ^{235}U resonance parameters in the energy range from thermal to 2.25 keV. The CEA/Bruyères-le-Châtel (BRC) was responsible to re-evaluate the ^{235}U resonance parameters in the fast energy region.

The Institut de Radioprotection et de Sûreté Nucléaire (IRSN) has carried out benchmark calculations, using few benchmarks extracted from the International Criticality Safety Benchmark Evaluation

[a] Corresponding author: raphaelle.ichou@irsn.fr

Article available at http://www.epj-conferences.org or http://dx.doi.org/10.1051/epjconf/201611106004

Project (ICSBEP) Handbook, 4 for testing the new ^{235}U capture cross-section evaluation over the whole energy range. The impact of the inclusion of the new evaluation was studied for selected critical benchmarks that give, when combining experiments, a sensitivity coverage to ^{235}U capture cross-section in each of the thermal, epithermal and fast energy range. The benchmark calculations, used to investigate the impact of the new ^{235}U evaluation, were performed using Monte-Carlo methods.

Table 1. Benchmarks integral sensitivities to the ^{235}U capture cross-section.

Series of experiments	Bench-mark k_{eff}	Benchmark k_{eff} uncertainty	k_{eff} sensitivity ^{235}U capt. < 0.625 eV	k_{eff} sensitivity 0.625 eV < ^{235}U capt. < 100 keV	k_{eff} sensitivity ^{235}U capt. > 100 keV	k_{eff} sensitivity ^{235}U capt. total
HEU-SOL-THERM-004-006	1.00000	0.00590	-10.76%	-6.03%	-0.04%	-16.82%
HEU-SOL-THERM-049-012	1.00120	0.00210	-8.72%	-8.67%	-0.14%	-17.53%
HEU-SOL-THERM-004-003	1.00000	0.00390	-7.14%	-13.91%	-0.16%	-21.21%
HEU-SOL-INTER-001-001	1.00000	0.00330	-5.33%	-15.73%	-0.34%	-21.40%
HEU-COMP-INTER-004	1.00000	0.00400	-0.25%	-16.01%	-0.31%	-16.56%
HEU-MET-INTER-007-031	0.99960	0.00220	-2.20%	-12.77%	-1.42%	-16.39%
HEU-MET-FAST-007-010	0.99810	0.00120	-0.77%	-9.64%	-2.07%	-12.48%
HEU-MET-FAST-042-002	0.99960	0.00090	-0.01%	-7.47%	-2.86%	-10.34%
HEU-MET-FAST-038-001	0.99990	0.00070	-0.01%	-6.31%	-3.00%	-9.31%
HEU-MET-FAST-070-003	1.00130	0.00130	-0.22%	-6.57%	-3.43%	-10.22%
HEU-MET-INTER-006-002	1.00010	0.00080	N.A	N.A	N.A	N.A
HEU-MET-INTER-001-001	0.99660	0.00260	N.A	N.A	N.A	N.A
IEU-MET-FAST-007-001	1.00460	0.00020	0.00%	-1.82%	-2.97%	-4.79%
HEU-MET-FAST-007-032	0.99410	0.00120	0.00%	-1.09%	-3.51%	-4.60%
HEU-MET-FAST-001-001	1.00000	0.00100	0.00%	-0.78%	-3.19%	-3.97%

2 Benchmark selection

2.1 Description of the method based on sensitivities

The DICE database, associated with the ICSBEP Handbook, was used to select the experiments sensitive to the ^{235}U capture. The available integral sensitivities, mainly calculated with the TSUNAMI-3D module of SCALE, were extracted from the database in each of the thermal (E < 0.625 eV), epithermal (0.625 eV < E < 100 keV), and fast (E < 100 keV) energy ranges, for both the ^{235}U capture cross-section and the ^{235}U fission cross-section, which is considered as the main background. The selection was based on several criteria: experiments showing too high uncertainty in the measured k_{eff} were discarded and, when possible, benchmarks showing the highest possible integral sensitivity to ^{235}U capture in each of the energy ranges and, at the same time, the lowest background cross-sections (^{235}U fission cross-section, H, O, .. etc.) sensitivity, have been selected. Table 1 shows the integral sensitivities to the ^{235}U capture cross-section in each energy range, as well as the total integral sensitivity to the ^{235}U capture, for the selected benchmarks. The DICE 2014 data also provide with the differential sensitivities of benchmarks to the different parameters as for instance nuclide, type of cross-section, etc. Figure 1 shows the sensitivity of k_{eff} to the ^{235}U capture and to the ^{235}U fission cross-section for the six most sensitive benchmarks (HST-004-006, HST-049-012, HCI-004, HMI-006-002, HMI-001 and HMF-007-032) among the selection.

2.2 Benchmark description

The experiments available in the ICSBEP Handbook are categorized by fissile media: Highly (HEU), Intermediate (IEU), or Low Enriched (LEU) Uranium; physical form of the fissile material (metal, solution and miscellaneous systems); and neutron energy range where the majority of the fissions occur (fast, intermediate, thermal spectra systems). The description of the selected benchmarks is presented in Table 2.

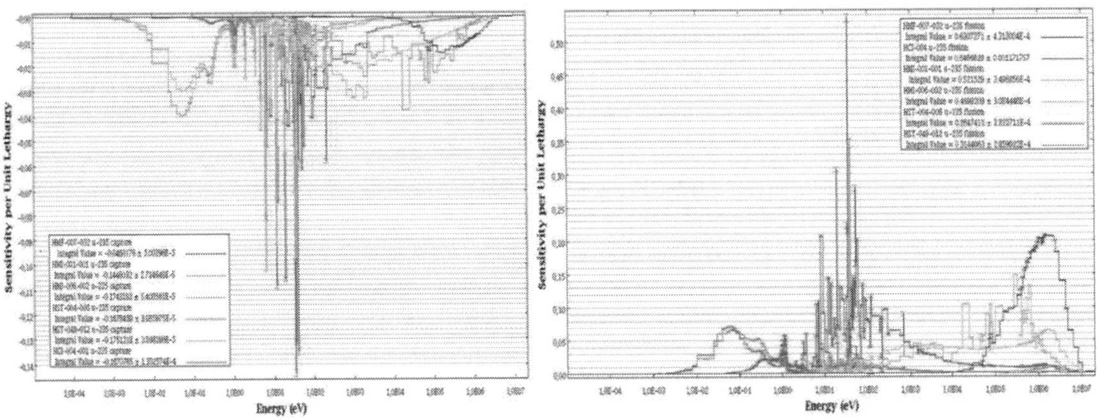

Figure 1. Sensitivity of k_{eff} to the ^{235}U capture (left) and to the ^{235}U fission (right) cross-section for the six most sensitive benchmarks, computed with the MORET 5 Monte Carlo code.

Table 2. Selected benchmarks description.

HEU-SOL-THERM-004-(003 and 006)	UO_2F_2 solution – U (93.6%^{235}U) – Spherical tank – Heavy water reflection
HEU-SOL-THERM-049-012	Uranyl Nitrate solution – U (85%^{235}U) – Spherical tank – Water reflection
HEU-SOL-INTER-001-001	UO_2F_2 solution – U (93.6%^{235}U) – Spherical tank – Heavy water reflection
HEU-COMP-INTER-004-001 (HECTOR)	Uranium/Boron/Graphite powder – U (92%^{235}U)
HEU-MET-INTER-007-031	Metallic Uranium plates (93.15%^{235}U) – Polyethylene moderated – No reflection
HEU-MET-FAST-007-(010 and 032)	Metallic Uranium plates (93.15%^{235}U) – Polyethylene (010), Teflon (032) moderated - No reflection
HEU-MET-FAST-042-002	Heterogeneous HEU cylinder –Vanadium diluted and Beryllium Oxide moderated
HEU-MET-FAST-038-001	Heterogeneous cylindrical assembly – (96%^{235}U), Beryllium and Beryllium Oxide disks – Depleted Uranium reflection
HEU-MET-FAST-070-003	Highly enriched Uranium, Tungsten, Aluminum Oxide plates – U (93%^{235}U) – Al_2O_3 reflected
HEU-MET-INTER-006-002 (ZEUS-002)	Cylindrical disks assemblies – metallic U (96%^{235}U) – Graphite moderated and Cu reflected
HEU-MET-INTER-001-001	Uranium/Iron cylinder – U (93%^{235}U) – Stainless steel reflected
IEU-MET-FAST-007-001 (BIGTEN)	Uranium (10%^{235}U) cylinders reflected by 15 cm depleted Uranium
HEU-MET-FAST-001-001 (GODIVA)	Metallic Uranium sphere Godiva – U (93.7%^{235}U) – No reflection

3 Results

The impact of the new preliminary ^{235}U resonance evaluation in benchmark calculations was investigated using the identified benchmarks listed in the ICSBEP compilation. The new set of ^{235}U resonance parameters was converted into the ENDF format and, for testing purpose, included in the ENDF/B-VII.1 ^{235}U library, where it replaced the existing ENDF/B-VII.1 resonance parameters. The new ^{235}U-v4 library, also labelled "*u235o4brc2*", was processed with the GAIA-NJOY 99 5 code to generate MCNP formatted cross-section. The k_{eff} calculations were performed, for the selected benchmarks, with both the MCNP 6 6 code and the 5.C.1 release of the MORET 5 7 code, using the ENDF/B-VII.1. In addition, k_{eff} calculations were also performed using ENDF/B-VII.1, together with the new ^{235}U evaluation. The MCNP and the MORET 5 k_{eff} results, obtained using the ENDF/B-VII.1 library, are shown in Figure 2, and compared to the benchmark k_{eff}. Except for a few cases, the calculation results are overall in good agreement with the experimental results, within the 3σ experimental uncertainty. The MCNP and MORET results are in good agreement, within 60 pcm (1 pcm = 10^{-5}), with a Monte-Carlo standard deviation of 10 pcm with the MORET code and varying from 5 to 25 pcm with the MCNP results.

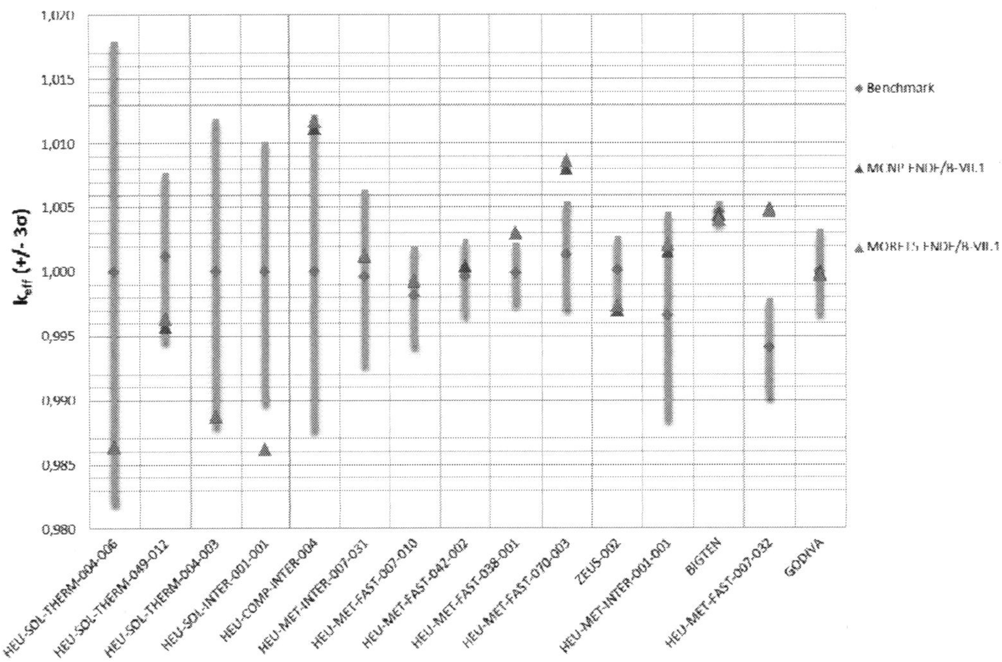

Figure 2. k_{eff} results obtained with MCNP 6 (purple) and MORET 5 (red), using ENDF/B-VII.1 library, together with the benchmark k_{eff} (experimental uncertainty in 3σ).

The C/E results, obtained with MORET 5, using the ENDF/B-VII.1 library, together with the new ^{235}U evaluation, are shown in Figure 3. The results of the new ^{235}U cross-section evaluation highlight a discrepancy compared to the previous evaluation results. It varies, as expected, as a function of the ratio in sensitivity ^{235}U capture/background reactions and according to the benchmarks characteristics (composition, energy spectrum, uranium enrichment, etc...). The impact of the new ^{235}U evaluation varies from about +370 pcm (HST-049-012) to about -1000 pcm (BIGTEN) for all the selected benchmarks. For the BIGTEN benchmark, which is an intermediate-enriched uranium benchmark with fast spectrum containing a large amount of ^{238}U, the new evaluation leads to a significant under-prediction of about 1000 pcm, whereas the previous one showed a good C/E. This phenomenon is due to compensations effects and the fact that ^{238}U and ^{235}U cross-sections are correlated, and highlights the fact that it would be necessary to re-evaluate ^{238}U together with ^{235}U. One can notice that, for the benchmarks sensitive to the intermediate energy range (HMF-038-001, HMF-070-003 and HMI-001), the k_{eff} tends to decrease compared to the old evaluation result and a trend in the C/E results can be observed. The impact of the new evaluation is smaller in the fast energy region (HMF-007-032 and GODIVA), except for BIGTEN, which involves intermediate-enriched uranium. Nevertheless, the benchmark selection allowed highlighting a significant effect on k_{eff} of the new ^{235}U capture cross-section, and the C/E results stay overall in good agreement, with, in some cases, a small improvement. It should be noted however that the results could be impacted by the nuclear data of other materials present in the ENDF/B-VII library. The C/E trend observed here may not be the same, if the ^{235}U evaluation is used with other cross section libraries for the other nuclides. Analysis work is ongoing to extract a tendency of the C/E results, which is important for the evaluation improvement.

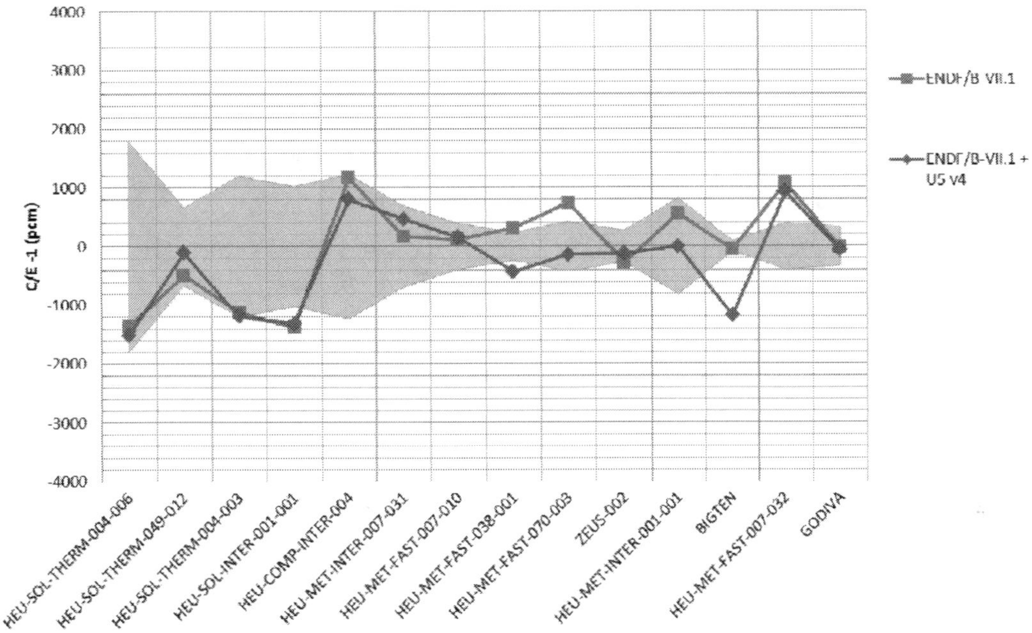

Figure 3. C/E results obtained with MORET 5 using ENDF/B-VII.1 library, for the ENDF/B-VII.1 ^{235}U (red) and the new ^{235}U-v4 (blue).

4 Conclusion

IRSN and ORNL have been working in collaboration with the CEA/BRC to produce a complete evaluation for ^{235}U. Under reserve of data testing and validation, the evaluation will be made available for possible inclusion in the Joint European Fusion and Fission (JEFF) cross-section library. IRSN, which is involved in the CIELO project in the data testing, especially for the new ^{235}U capture cross-section, has carried out key critical benchmarks selection, regarding the sensitivities to the ^{235}U capture cross-section, in each of the energy ranges, and performed MCNP and MORET calculations for the selected benchmarks, using the new evaluation. The benchmark selection allowed highlighting a significant effect on k_{eff} and C/E results with the new ^{235}U evaluation.

Acknowledgements

We would like to thank B. Morillon and P. Romain, from CEA/BRC, that provided the new ^{235}U evaluation in the fast energy region. Thanks to W. Haeck and I. Duhamel for the help and discussions.

References

1. NEA/NSC/WPEC/DOC(2011)433, "Final report of WPEC subgroup 29 on ^{235}U Capture Cross-section in the keV to MeV Energy Region", (2011).
2. Y. Danon, Private Communication, to be published, (2012).
3. M. Jandel et al., "New Precision Measurements of the ^{235}U (n,γ) Cross Section",PRL109, (2012).
4. International Handbook of Evaluated Criticality Safety Benchmark Experiments, NEA/NSC/DOC(95)03, OECD Nuclear Energy Agency, (2014).
5. W. Haeck, "GAIA User's Manual - Version 1.0.0", IRSN Report PSN-EXP/SNC/2015-165, Institut de Radioprotection et de Sûreté Nucléaire, France (2015).
6. MCNP6 User's Manual - Code Version 6.1.1beta, LA-CP-14-00745 (June 2014).
7. A. Jinaphanh, N. Leclaire, B. Cochet and B. Quaghebeur, "Validation of continuous energy sensitivity coefficients calculations in the MORET 5 code", Proc. ICNC 2015, Charlotte, (2015).

EPJ Web of Conferences **111**, 06005 (2016)
DOI: 10.1051/epjconf/201611106005

GALILÉE-1: a validation and processing system for ENDF-6 and GND evaluations

Mireille COSTE-DELCLAUX[1a], Cédric JOUANNE[1], Frédéric MOREAU[1] and Claude MOUNIER[1]

[1]*CEA Saclay, DEN, DM2S, SERMA, LLPR, F-91191 Gif-sur-Yvette, France*

Abstract: GALILÉE-1 is the new validation and processing system for evaluated data, developed at CEA. This system can handle evaluations stored either in the ENDF-6 format or in the new General Nuclear Data (GND) format. It consists of various components respectively dedicated to read/write the evaluated data whatever the format is, to diagnose inconsistencies in the evaluated data and to provide continuous-energy and multigroup data as well as probability tables for transport and depletion codes. All these components are written in C++ language and share the same objects. This paper describes the state of progress of the various parts of the system and gives some illustrations.

1 Introduction

GALILÉE-1 system is part of a CEA global development program dedicated to fine modelling of nuclear systems. It is carried out in synergy with the evaluation code CONRAD [1], and with the new generation of reactor physics codes, APOLLO3 [2], TRIPOLI-4 [3] and MENDEL [4]. It can handle evaluations stored either in the ENDF-6 or in the new General Nuclear Data (GND) format.

GALILÉE-1 system consists of:

- processing modules providing continuous-energy data (CE), multigroup data (MG) as well as probability tables (PT),
- interface modules creating application libraries (from these processed data) for deterministic or Monte Carlo transport codes, shielding codes or depletion codes,
- an automatic chain for creating these libraries.

GALILÉE-1 system originality lays in its two complementary running ways: an integrated one (Fig. 1) with a convivial and automatic chain for creating consistent and validated libraries for the various application codes and an open way for providing application codes with processing tools (Doppler on the fly for instance in a Monte Carlo transport code).

At the present, only the processing modules are under development. Cross-comparisons with other processing systems are systematically carried out at each step in order to fully master possible discrepancies.

[a] Corresponding author: mireille.coste-delclaux@cea.fr

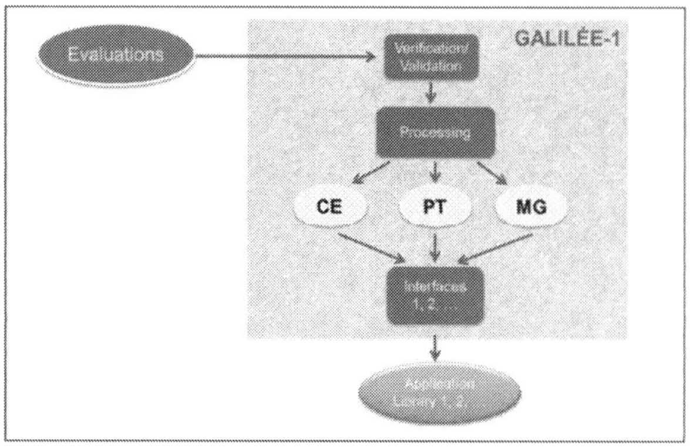

Figure 1. GALILÉE-1 Integrated running way.

2 GALILÉE-1 processing modules

Three main processing modules are under development in GALILÉE-1 system, a parser dedicated to read and write evaluated and processed data in ENDF-6 or GND format, a verification/validation tool described further and a processing code TREND (Treatment and Representation of Evaluated Nuclear Data). They are written in C++ language and share common objects grouped in GBASE component. GBASE objects are completely independent from the input data (evaluation) format and the output data (processed data) format.

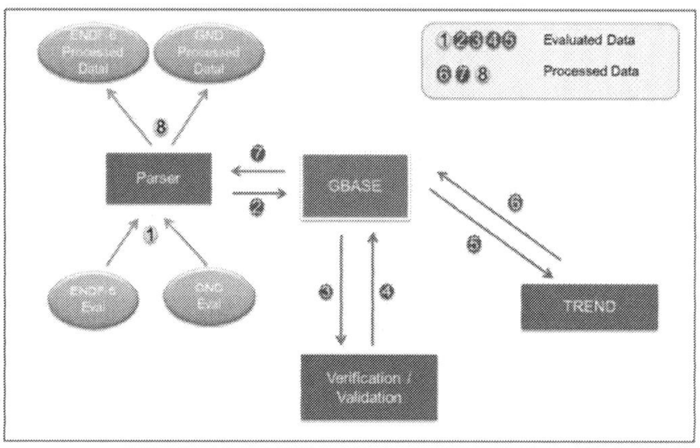

Figure 2. GALILÉE-1 Processing modules.

As shown in Figure 2, GBASE objects are initialized thanks to the Parser that reads the evaluation data. These objects are checked and eventually corrected by the Verification/Validation Tool and then processed data are created by TREND code. The objects storing processed data are also kept in GBASE and can be written on binary or character files by the Parser.

3 GBASE objects

The GBASE object hierarchy is very close to the GND object hierarchy. For each nucleus or element, we create a data base allowing us to store, in the same object, structure data and interaction data for a given projectile. GBASE structure data contain all the information needed to verify and eventually correct the evaluated data: masses, level scheme, spins, energy, half-life, decay modes, etc. GBASE interaction data contain the list of products that can be created by the interaction, all the information given by an evaluation (JEFF-3.2, ENDF/B-VII.1, JENDL4.0,…) but organized in such a way that processing is easier and processed data (CE data, PT, MG data, …).

Several GBASE structure data or several GBASE interaction data may exist in the same data base.

4 Parser

The Parser reads all the files of an evaluation in ENDF-6 format, and creates the GBASE objects corresponding to the resonance parameters, the reactions, and the kinematics of reaction products. Other GBASE objects are under construction. The reading of GND evaluations is under progress. Writing is not yet available.

5 Verification / Validation

One of the goals of GALILÉE-1 system is to test the consistency and the validity of nuclear data evaluations. We plan to perform a complete assessment of evaluated files before any treatment. Some checks can be performed by comparing data with the ones contained in structure databases, e.g. NUBASE or ENSDF. This is the case for:

- the masses of the nuclides, given in terms of neutron mass,
- the energies of the excited states reached in the inelastic scattering,
- gamma decay schemes of the excited states.

Some additional tests are designed to check the coherence between the data given in an evaluation:

- coherence between thresholds considered for various data of the same reaction,
- energy balance for reaction products,
- spin/parity of resonance parameters,
- normalization of distributions.

In this section, we present a list of test already considered in our code system allowing us to diagnose a list of incoherent values in an evaluation.

5.1 Target mass and Q-value reaction

The target mass is very important to determine the Q-value and the threshold of a reaction. In JEFF-3.2 library, the H2 evaluated mass, in neutron mass, is 1.995712 instead of 1.9968 (underestimation of 1 MeV). This discrepancy should impact radiative capture Q-value but it is not the case, the correct value being given in the evaluation. Other examples of important nucleus mass discrepancies can be found in JEFF-3.2 evaluations: for example, more than 4.5 MeV for Xe134 and 366 keV for Ti46.

In the case of transfer reactions, one can observe Q-value discrepancies higher than 1 MeV. Moreover, those values are not systematically given.

5.2 Resonance state spin (J value)

The spin of a resonance state, J, is calculated from the orbital angular momentum l, the neutron spin and the target spin I. For Bk247 which spin is 3/2, in agreement with Nubase, the J value given in JEFF-3.2 for *s* resonances ($l=0$), is *3/2*, out of the range of possible values, 1 or 2. This discrepancy has a significant impact on the calculated cross-section due to the spin statistical multiplicative factor.

5.3 Energy balance for emitted particles

Evaluations contain energetic distributions for emitted particles at various incident energies. From these distributions, we can reconstruct total and partial average energies and compare them to available energy for the reaction. Interpolation schemes for emitted energy distributions and for incident energies play an important role to determine energy release. By analyzing these distributions, we are able to highlight incoherent values.

5.4 Negative angular distribution

Angular distribution of emitted particles can be represented by Legendre polynomial expansion coefficients. This representation is particularly efficient in terms of compactness. However, the truncation at a given order can produce negative values for angular probabilities. Figure 3 deals with neutron elastic scattering on Au197 (for both JEFF-3.2 and ENDF/B-VII.1 libraries) at incident energy of 29 MeV. This is a difficult problem for Monte Carlo calculations. Indeed, the cumulative density function (CDF) is not continuously growing. In this case, for a sampling value, two or more scattering angles are possible.

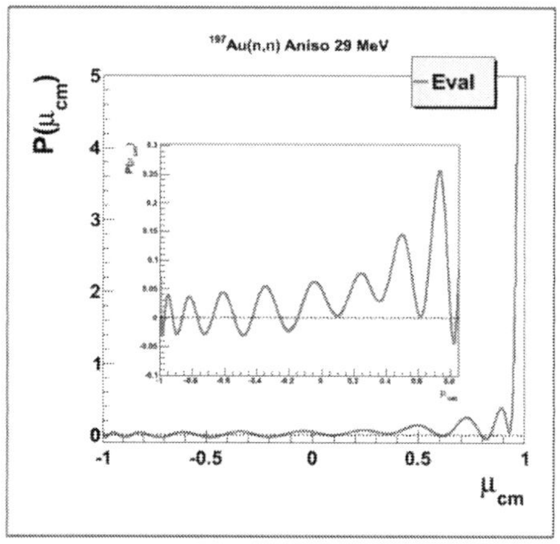

Figure 3. Angular distribution for ^{197}Au (n,n) at 29 MeV.

6 TREND

TREND code aims at replacing NJOY99 and CALENDF codes in CEA application library production. It consists of three main parts, TREND_CE corresponding to NJOY RECONR, BROADR, UNRESR, THERMR and HEATR modules, TREND_PT corresponding to CALENDF treatment and TREND_ MG corresponding to NJOY GROUPR module. At the present RECONR and TREND_PT functionalities are developed, BROADR /THERMR functionalities are under progress.

In order to validate TREND reconstruction at 0 Kelvin in the resolved resonance range, cross-comparisons with NJOY99 were carried out for all JEFF-3.2 nuclei (~ 470). In case of discrepancies, additional comparisons were made with NJOY2012 and PREPRO2015. About 25 nuclei show relative reconstruction discrepancies larger than 1.E-5, between NJOY99 and TREND, for MT1, MT2, MT102 and MT18 when it exists.

6.1 Discrepancies on MT2

For Fe54 and Os192, we observe relative discrepancies on MT2 (Table 1) not yet fully understood. For Fe54 these discrepancies appear between 3453.125 eV and 3497.071 eV and for Os192, between 2313.995eV and 2314.521eV.

Table 1. Maximum relative discrepancies on MT2 between NJOY99 and TREND.

	Maximum relative discrepancy	Energy (eV)
Fe54	1.26 E-3	3475.009
Os192	2.01 E-4	2314.337

6.2 Discrepancies on MT102

For C13, Kr86, O17, O18, Pb206, Pb208, Sr88 and Sr90, discrepancies occur on MT 102 reconstruction with, for all nuclei, the same "jump" between 0.01 eV and 0.1 eV (Fig.4). This problem disappears with NJOY2012 (except for O17 and O18) and with PREPRO15 for O18.

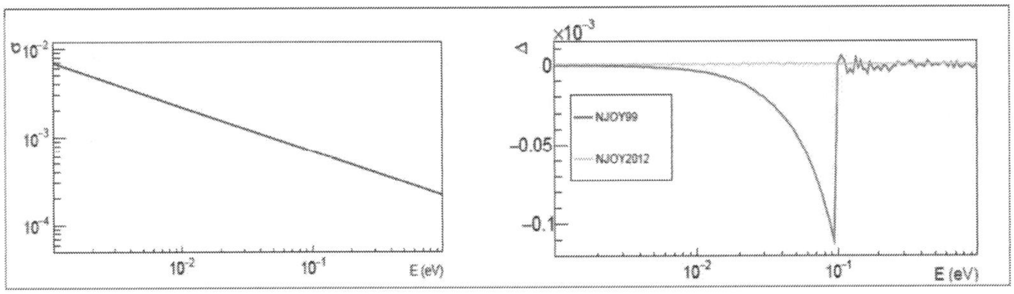

Figure 4. Relative discrepancies between NJOY and TREND on reconstructed MT102 file for C13.

6.3 Discrepancies on MT1

On other nuclei, discrepancies may appear on MT1, either when summing the background cross-sections or because of threshold problems. TREND uses, for the reconstruction, the partial cross-sections and not the redundant cross-sections. Moreover, as NJOY does, TREND modifies the threshold given in the evaluation when its value is not correct. But some differences remain.

As an example, for O16 in JEFF-3.2, the MT107 background file is not consistent with the sum of the MT800-MT803 files, above the threshold of the first excited state of C13 (MT801). That leads to discrepancies between NJOY 99 and TREND (Table 2). This problem is solved with NJOY2012.

Table 2. Maximum relative discrepancies on MT107 for O16.

	Maximum relative discrepancy	Energy (eV)
NJOY99/TREND	8.96 E-4	5910000.
NJOY2012/TREND	4.84 E-6	2380000.

7 Conclusion

GALILÉE-1 system allows R&D activity on new processing methods. It is also a powerful tool for nuclear data analysis. It will produce soon Monte Carlo transport libraries from validated evaluations, which is a first important step for our renewed library creation chain.

The authors acknowledge the Industrial Partners AREVA-NP and EDF for their partial financial support through SINET/DNT project.

References

1. P. Archier et al. "CONRAD evaluation code: Development status and Perspectives", *Nuclear Data Sheets*, **118**, 488-490, (2014)
2. H. Golfier et al., "APOLLO3: a common project of CEA, AREVA and EDF for the development of a new deterministic multi-purpose code for core physics analysis," *M&C 2009, Saratoga Springs*, New York, USA (2009)
3. E Brun et al., "Overview of TRIPOLI-4 version 7, Continuous-energy Monte Carlo Transport Code", *ICAPP 2011*, Nice, France (2011)
4. S. Lahaye et al., "First Verification and Validation Steps of MENDEL Release 1.0 Cycle Code System", *PHYSOR 2014*, Kyoto, Japan, (2014)

EPJ Web of Conferences **111**, 06006 (2016)
DOI: 10.1051/epjconf/201611106006

Need of a consistent and convenient nucleus identification in ENDF files for the automatic construction of the depletion chains

Pietro Mosca[1,a] and Claude Mounier[1,b]

[1] *CEA-Saclay, 91191 Gif-sur-Yvette, France*

Abstract. The automatic construction of evolution chains recently implemented in GALILEE system is based on the analysis of several ENDF files : the multigroup production cross sections present in the GENDF files processed by NJOY from the ENDF evaluation, the decay file and the fission product yields (FPY) file. In this context, this paper highlights the importance of the nucleus identification to properly interconnect the data mentioned above.

The first part of the paper describes the present status of the nucleus identification among the several ENDF files focusing, in particular, on the use of the excited state number and of the isomeric state number. The second part reviews the problems encountered during the automatic construction of the depletion chains using recent ENDF data. The processing of the JEFF-3.1.1, ENDF/B-VII.0 (decay and FPY) and the JEFF-3.2 (production cross section) points out problems about the compliance or not of the nucleus identifiers with the ENDF-6 format and sometimes the inconsistencies among the various ENDF files. In addition, the analysis of EAF-2003 and EAF-2010 shows some incoherence between the ZA product identifier and the reaction identifier MT for the reactions $(n, p\alpha)$ and $(n, 2np)$.

As a main result of this work, our suggestion is to change the ENDF format using systematically the isomeric state number to identify the nuclei. This proposal is already compliant to a huge amount ENDF data that are not in agreement with the present ENDF format. This choice is the most convenient because, ultimately, it allows one to give human readable names to the nuclei of the depletion chains.

1 Introduction

The automatic construction of evolution chains for depletion and transport codes like MENDEL[1] and APOLLO3® [2] is one of the last novelty in the nuclear data processing system GALILEE[3]. In this context, this paper highlights the importance of a correct nucleus identification to properly interconnect the depletion data (decay modes, production cross sections and fission product yields).

Normally, when producing depletion chains, the reactor physicists expect usual human readable names of nuclei based on the triplet (proton number Z, mass number A, isomeric state number). For instance, Am242M stands for the first isomeric state of Am242, its corresponding triplet is (95,242,1). The identifier (Z, A, isomeric state number) is in general preferred, because the isomeric states are

[a] e-mail: pietro.mosca@cea.fr
[b] e-mail: claude.mounier@cea.fr

the selection of the excited states of a nucleus with a significant half-life for depletion calculations and moreover the use of the excited state number is not convenient because roman alphabetical letters would be insufficient. Unfortunately, the various nuclear data in ENDF format[4], which are necessary to automatically construct an evolution chain, use different incomplete identifiers for products coming from the fission, from the nuclear reactions and from the decay. Depending on the kind of data, triplets such as (Z, A, excited state number) and (Z, A, isomeric state number) can be used. Moreover, even if the ENDF format specifies the excited state number for the reaction products, in some evaluations it can occur that the isomeric state number is provided in place.

The problems above mentioned are discussed in this paper according to the following structure. First, a description of the way to identify a nucleus in the depletion data processing is analyzed. Then some anomalies of nucleus identification while processing JEFF-3.1.1, JEFF-3.2, EAF-2003, EAF-2010 and ENDF/B-VII.0 evaluations are pointed out.

2 Nucleus Identification

In GALILEE system the production of the evolution chains is based on the analysis of:

- the decay data present in file (8,457) to recover the decay modes of each unstable nucleus (father) and the resulting stable or unstable nuclei (daughters),
- the independent fission product yields (FPY) that are contained in file (8,458), to extract the yields of the fission products (daughters) for each fissile nucleus (father),
- the production cross section of reaction products present in the GENDF files processed by NJOY [5] from the ENDF evaluations, which describe in files (8-9-10, any MT) the production reactions of a nucleus (father) and the corresponding products (daughters).

Table 1 summarizes the nucleus identifier for all this kind of data.

Table 1. Nucleus identifier used in ENDF evaluations.

Nucleus as ...	Target	Reaction product	Decay father	Decay Daughter	Fission product
Files	(1,451)	(8,MT) (9,MT) (10,MT) (6,MT)	(1,451)	(8,457)	(8,454)
proton number	Z	Z	Z	Z	Z
mass number	A	A	A	A	A
excited state number	LIS	LFS	LIS	-	FPS
isomeric state number	LISO	LIP only MF=6	LISO	RFS	-
excitation energy	ELIS	ELFS	ELIS	-	-

The identification of a nucleus-father can be carried out using the file (1,451) that is present in all these three types of data. This file characterizes a nucleus by: its proton number Z, its mass number A, its excited state number LIS, its isomeric state number LISO (LISO<=LIS) and its energy of excitation ELIS. The ELIS value can be hardly used as item of identification because the levels of nucleus excitation can be very close each other, but it is useful for checking purpose. The description of the nuclei-daughters is instead less rich.

The production cross section of a given product is identified by the Z, A and LIS parameters in ENDF format[4, p. 158]. However, according to the NJOY 2012 manual[5, p. 250] in the GENDF processing, the excited state number LFS is interpreted as the isomeric state number of the product.

In File 6 the isomeric state number LIP and the excitation energy ELIS are provided.

In decay files, the excited state of the decay products is defined by the isomeric state number RFS [4, p. 163].

Lastly, the fission products are identified by the excited state number FPS [4, p. 159].

In this context, the lack of homogeneity among all these files is potentially a source of errors when connecting these data together. If one disposes of an exhaustive description of each nucleus like in file 1, the identification can be carried out using the different keys (Z, A, excited state number) or (Z, A, isomeric state number). Nevertheless, a significant burden in the development would be avoided using a unique identifier everywhere like the triplet (Z, A, isomeric state number), which allows ultimately one to give useful names to nuclei.

Starting from these points, for the first version of GALILEE system, the JEFF-3.1.1 files that describe the reaction products have been modified substituting to the LIS the LISO parameter for some nuclei in order to allow the correct connection with the decay data.

For the following version of GALILEE system, some check tools have been developed in order to better track inconsistencies in nucleus identification.

A first control is now available between the Z and A parameters of the product and those obtained from the MT number of the reaction. If an inconsistency is tracked, the Z and A parameters deduced from the reaction are used in place of those of the product. Besides, a new control is now carried out among the daughters from cross section and FPY evaluations and those in decay. A warning is displayed on the console if a nucleus is not included in the decay data base and correspondingly the list of isomers of the decay data base is printed to help the identification of the product. This approach is based on the idea that the list of nuclei in decay data base is in principle the most complete available.

3 Decay data and Fission product yields evaluation

The coherence between decay and FPY evaluations is practically guaranteed by the fact that one needs the information contained in decay in order to calculate the evaluated cumulative fission product yields (MF=8,MT=459) from independent fission yields (8,454). While processing JEFF-3.1.1 and ENDF/B-VII.0 we observed that the fission products are identified using the triplet (Z,A,LISO) without compliance with the ENDF-6 format. Besides, in JEFF-3.1.1 all the decay products and all the fission products are included in the decay father list, while in the ENDF/B-VII.0 some radioactive products and more than one hundred fission products are not part of the decay father list.

4 Analysis of EAF-2003 and EAF-2010

In EAF-2003 evaluation all the reaction products derived from the file 8 are identified with the triplet (Z,A,LISO) instead of (Z,A,LIS) and the excitation energy of the isomeric state is set to zero. The same rule is adopted in EAF-2010.

Moreover, while processing the EAF-2003 evaluation, some incoherence between the ZA product identifier and the reaction identifier MT has been pointed out for the reactions $(n, p\alpha)$ and $(n, 2np)$. A systematic underestimation of the mass number of the products has been detected for these nuclei: Ar36, Ca40, Cl35, Co55, Co56, Co58, Er160, K39, K40, Mo93, Nb90, Nb92, Nb92M, Ni56, Ni57, Ni58, Ni59, Sc44M, Tb156M, Ti44, Ti45 and Y87M. Besides, we have found the absence from the decay data base of the following nuclei: Np246, Bi207M, U243, Sb131M, Ru109M and Am250.

The Np246, U243 and Am250 have been deleted because they are not included in the ENSDF [6] data base. The Bi207M, Sb131M and Ru109M have been substituted by the corresponding daughters Bi207, Sb131 and Ru109 resulting from γ-decay almost instantly (half-life of the order of the μs in accordance with the ENSDF data base) to account their contribution in the evolution chain.

5 Analysis of JEFF-3.2

The analysis of the JEFF-3.2 shows that more than one hundred reaction products are identified in compliance with the ENDF-6 norm using the LIS parameter and that the reaction product Os197 is not included in the decay data base.

To manage identification problem, we used the information present in decay (LIS, LISO and ELIS), in reaction product (LIS) and in the ENSDF database following this procedure. If two metastable nuclei one from decay and one from the files of production cross sections have the same triplet (Z, A, LIS) then the name of the nucleus in decay data base, which is defined using the LISO, is assigned. Otherwise, we identify the ELIS, which corresponds to the LIS of the reaction product, thanks to the ENSDF data base, and the isomer of the decay data base with the closest ELIS energy, names the nucleus (LIS = LISO+1 very often). In this way, most of the nucleus identification problems are removed. Moreover, to avoid any lack in the evolution chain, the Os197 has been substituted by its daughter Ir197, which is produced by β^- decay.

To illustrate the difficulty created mixing the nucleus identifiers, we consider the well-known case of the (n, γ) reaction of Am241. Two nuclei are produced by this reaction: the Am242 (LFS=2, LISO undefined) and the Am242 (LFS=0, LISO undefined). As target nuclei (MF=8), one can find the Am242 (LIS=3,LISO=1) and the Am242 (LIS=0,LISO=0) according to the file 451. In JEFF-3.1.1 decay data, these three father-nuclei Am242 (LIS=0,LISO=0), Am242 (LIS=2,LISO=1) and Am242 (LIS=141,LISO=2) are described. In this case, the correct decay modes can be attached to Am242 (LFS=2, LISO undefined) using the LFS identifier. No target can be found adopting the excited state number while one exists if the isomeric state number is used.

6 Conclusion

The importance of a homogeneous and direct identification of the nuclei among the different kind of data is crucial for the production of the evolution chains. Decay data should play a major role because it contains the most exhaustive nucleus description based on the quintuplet (Z,A,LIS,LISO,ELIS). For each target nucleus, all the reaction products and the fission products in ENDF evaluations should be included in the decay database as father.

To produce automatically depletion chain, the non compliant ENDF-6 evaluations of FPY and EAF that use (Z,A,LISO) combined with recent complete decay data are very practical. The nightmare begins when compliant and non compliant ENDF-6, which adopt different identification keys, are used together. The checking tool writen permits to overcome this situation but with a somewhat huge amount of human work.

Consequently, for the future, the isomeric state number for the nucleus identification of the product must be considered in order to simplify the post treatment analysis nowadays mandatory in the depletion data processing. The reaction products and fission products in the associated data file should be identified using the isomeric state number instead of the excited state number. In this way, any brief identification of a product, shown with a dash symbol in Table 2, will not cause any problem, provided that a complete decay data base are used in the processing. These changes are highlighted in red color in Table 2.

Table 2. Summary table of the proposal of identification of nucleus in the future release of ENDF evaluations.

Nucleus as ...	Target	Reaction product	Decay father	Decay product	Fission product
atomic number	Z	Z	Z	Z	Z
mass number	A	A	A	A	A
excited state number	LIS	-	LIS	-	-
isomeric state number	LISO	LFS or LIP (MF=6)	LISO	RFS	FPS
excitation energy	ELIS	ELFS	ELIS	-	-

With these modifications, no changes in NJOY are needed and the automatic chain construction can be carried out with high reliability. Moreover, the JEFF-3.1.1, ENDF/B-VII.0 files of fission product yields do not need to be changed as well as EAF-2003 and EAF-2010, because these evaluations are already compliant with our proposition. Ultimately, in recent evaluations, few reaction products files need to be modified using the same isomeric state number of the decay data.

References

[1] S. Lahaye et al., "First verification and validation steps of the MENDEL release 1.0 cycle code system", Proceedings of PHYSOR2014, Kyoto, Japan, September 28 – October 3 (2014).

[2] H. Golfier et al., "APOLLO3: a common project of CEA, AREVA and EDF for the development of a new deterministic multi-purpose code for core physics analysis", Proceedings of Mathematics and Computation Topical Meeting, Saratoga Springs, USA, May 3-7 (2009).

[3] M. Coste-Delclaux, "GALILEE: A nuclear data processing system for transport, depletion and shielding codes", Proceedings of PHYSOR 2008 Topical Meeting, Interlaken, Switzerland, September 14-19 (2008).

[4] Cross Section Working Group, "ENDF-6 formats manual, Data Formats and Procedures for the Evaluated Nuclear Data File", Technical Report BNL-90365-2009, Rev. 2, 2012. Edited by A. Trkov, M. Herman and D. A. Brown, October 24, 2012

[5] R. E. MacFarlane, D. W. Muir, R. M Boicourt and A. C. Kahler, "The NJOY Nuclear Data Processing System,Version 2012" , Technical Report LA-UR-12-27079 Rev, LANL (2012).

[6] "Evaluated Nuclear Structure Data File," http://www.nndc.bnl.gov/ensdf/ (2015).

EPJ Web of Conferences **111**, 07001 (2016)
DOI: 10.1051/epjconf/201611107001

Thermal neutron activation experiments on Ag, In, Cs, Eu, V, Mo, Zn, Sn and Zr in the MINERVE facility

Pierre Leconte[1,a], Benoit Geslot[2], Adrien Gruel[2], Alexandra Pépino[2]

[1]*CEA, DEN, DER/SPRC/LEPh Cadarache, F-13108 Saint Paul-Lez-Durance, France*
[2]*CEA, DEN, DER/SPEx/LPE Cadarache, F-13108 Saint Paul-Lez-Durance, France*

Abstract. The MAESTRO experimental program has been designed to improve nuclear data uncertainty on a large range of materials used for detection, absorption, moderation and structures in LWRs. It consists of pile-oscillation and neutron activation experiments, carried out in the MINERVE low power facility. For this program, the core configuration has been designed to be representative of HZP (Hot Zero Power) conditions of a typical PWR. Samples of high purity elements have been manufactured with severe technological constraints to reach a target accuracy of ±2% (1σ) on the measurement. This paper presents a preliminary analysis of activation experiments, based on TRIPOLI4 Monte-Carlo calculations and various nuclear data libraries.

1 Introduction

MAESTRO [1][2] is the follow-up in a series of experimental programs started more than thirty years ago in the MINERVE facility [3-4], which goal is to improve nuclear data uncertainties for Light Water Reactors (LWR) applications (GEN 2 and 3). Past programs mainly consist of pile-oscillation experiments on samples with small amount of separated isotopes, natural elements, fresh or irradiated fuels (UO_2, MOX).

Due to the lack of validation (or insufficient accuracy) for structural, moderating, detection and some absorbing materials, the MAESTRO experimental program was designed to improve the prediction of several neutronic parameters which are relevant for several LWR applications: core physics (reactivity of structural elements), instrumentation (EPR™ innovative instruments), reactivity control, criticality/safety (contributions of various elements in concrete), burn-up credit. Some of the experiments are also performed in support of the JHR irradiation reactor, currently being built in Cadarache. This reactor will use metallic fuel plates of U3Si2 cladded with aluminum, a beryllium reflector and hafnium control rods.

This paper describes the experiments and preliminary results on the neutron activation measurements that have been performed on the following natural materials: Ag, In, Cs, Eu, V, Mo, Zn, Sn and Zr.

[a] Corresponding author: pierre.leconte@cea.fr

Article available at http://www.epj-conferences.org or http://dx.doi.org/10.1051/epjconf/201611107001

2 Description of the experiment

2.1 MINERVE core configuration and related experimental techniques

MINERVE is a pool type reactor, located at CEA Cadarache. The core is submerged under 3 meters of light water at room temperature. The outside of the core is made of highly enriched UAl assemblies reflected by large graphite elements, being used as a driver zone. In a central square cavity of 70cm, various configurations can be loaded. For the MAESTRO program, a homogeneous core configuration consisting of ~800 UO_2 fuel pins (3% in ^{235}U) has been loaded. In the central position, the neutron spectrum is closed to the one of a UO_2-PWR in Hot Zero Power conditions (HZP).

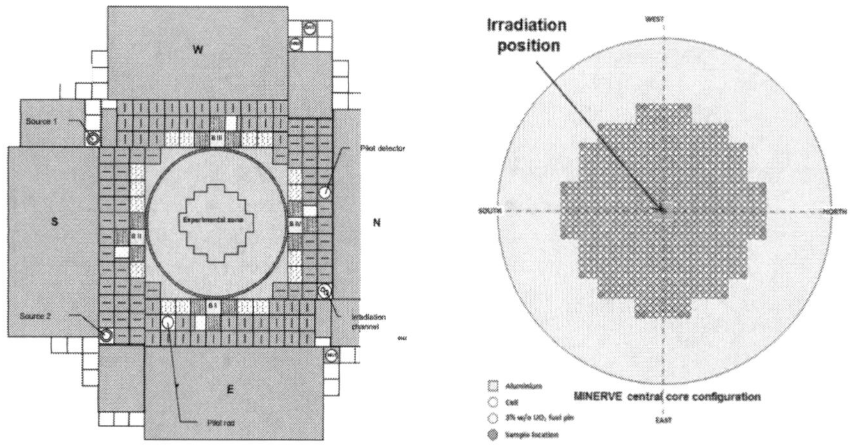

Figure 1. Overview of the MINERVE facility (left side) and central core configuration (right side).

Two kinds of experimental techniques have been used during the program. The first one is the pile-oscillation technique which consists in measuring the reactivity worth of a sample of pure material by successive translations between the center of the core and a position outside of the core. The technique is described in more details in [2] and [5]. The second one is the neutron activation techniques which consists in irradiating the sample in the center of the core and then to perform a γ-spectrometry analysis of the radionuclides which have been produced during the irradiation. This method was developed in previous programs to obtain complementary information on the capture rate of some absorbers and actinides, using the same sample in both pile-oscillation and activation techniques [6]. It is a relative method in which the reaction rate of a given isotope is normalized to the reaction rate of a reference isotope which is considered to be a neutron cross section standard.

2.2 Sample characteristics

All the samples used in MINERVE for the last fifty years consist of a mixture between a diluent material (DM) and a material of interest (MOI). For representativity considerations, UO_2 was often chosen as the DM. The MOI was mixed together with the DM, then crushed and sintered in pellets. Due to the possibility of material evaporation during the sintering process, additional pellets were prepared for destructive analysis by mass spectrometry. These analyses are quite expensive, time-consuming and can be complex for some insoluble elements.

For the MAESTRO program, the process was changed in order to avoid destructive analysis and to guarantee the material compositions just by accurate weighting and the provision of highly pure materials. Depending on physical and chemical characteristics of the samples, three different designs have been considered:

- Diluted solutions: this form uses a dilution of the MOI in 4-5% nitric acid. This form is well adapted for materials with high neutron cross section and for materials which are highly soluble. It is inserted in a double watertight cladding, one of welded Zircaloy-4 and one of aluminium alloy (Al-5754).
- Pure rods: this form uses cylinders of appropriated diameter for materials with low neutron cross section. The rod is inserted in a single watertight aluminium container.
- Powder mix: this form is a mixture between a DM (Al_2O_3 powder) and the MOI made of very fine powder (µm-sized powder). It is considered for materials which cannot be prepared from the two previous forms. The same cladding as for rod samples is used.

Table 1 summarises the characteristics of the samples used in the neutron activation experiments. Gold samples are used for calibration, as the capture of ^{197}Au is considered as a cross section standard.

Table 1. Characteristics of MAESTRO samples.

Sample name	Material	Physical form	MOI mass (g)	Diameter (mm)	Height (mm)	Reaction of interest
M-Ag	Ag	4% HNO_3 solution	1.4	8.4	82	$^{109}Ag(n,\gamma)^{110m}Ag$
M-In	In	4% HNO_3 solution	0.22	8.4	82	$^{113}In(n,\gamma)^{114m}In$ $^{115}In(n,\gamma)^{116m}In$
M-Cs	Cs	4% HNO_3 solution	0.75	8.4	82	$^{133}Cs(n,\gamma)^{134gs+m}Cs$ $^{133}Cs(n,\gamma)^{134m}Cs$
M-Eu	Eu	5% HNO_3 solution	0.045	8.4	96	$^{151}Eu(n,\gamma)^{152m}Eu$ $^{151}Eu(n,\gamma)^{152gs}Eu$ $^{153}Eu(n,\gamma)^{154}Eu$
M-V	V	Powder mix	7.2	10	100	$^{51}V(n,\gamma)^{52}V$
M-Mo	Mo	Pure rod	29	6	100	$^{98}Mo(n,\gamma)^{99}Mo$ $^{100}Mo(n,\gamma)^{101}Mo$
M-Zn	Zn	Pure rod	52	10	100	$^{64}Zn(n,\gamma)^{65}Zn$ $^{68}Zn(n,\gamma)^{69m}Zn$
M-Sn	Sn	Pure rod	173	10	300	$^{112}Sn(n,\gamma)^{113}Sn$ $^{117}Sn(n,\gamma)^{118m}Sn$ $^{122}Sn(n,\gamma)^{123m}Sn$
M-Zr	Zy4 alloy	Pure rod	152	10	300	$^{94}Zr(n,\gamma)^{95}Zr$ $^{96}Zr(n,\gamma)^{97}Zr$
M-Au-1 M-Au-2 M-Au-3	Au	Pure rod	1.5 3.9 6.0	1 1.6 2	10	$^{197}Au(n,\gamma)^{198}Au$
M-AlAu	Al-0.1%Au alloy	Alloy rod	0.006	1	10	$^{197}Au(n,\gamma)^{198}Au$

2.3 Neutron activation experiments

The measurement procedure follows two steps. Firstly, the sample is introduced in a watertight stainless steel tube between aluminium spacers in order to have the sample at the radial and axial central position of the core. The reactor is set to an average power of 80W during 30 minutes to 6 hours, depending on the target activity to reach. Secondly, after waiting for a few minutes to a few hours (depending on the sample dose rate after the irradiation), the sample is placed in front of the HPGe detector to start a γ-spectrometry analysis (Fig. 2). This process is repeated at least twice for each sample in order to check the measurement reproducibility. For long-lived nuclides, a background subtraction is required to account for previous irradiations in the total activity.

In order to extrapolate the measured count rate to the saturated activity, the γ-counting rate is modified using several factors:
- the radioactive decay during the cooling and the measurement phases,
- the detector efficiency,
- the γ-emission probability of the γ-ray of interest.

A correction factor, called "efficiency transfer correction", is also required to account for two effects:
- A spatial effect which is related to the volumic and heterogeneous distribution of the γ-source within the sample, while the detector efficiency was calibrated with a point source.
- An attenuation effect which accounts for the self-absorption within the sample, especially for the dense materials like gold.

This factor is evaluated by MCNP5 calculations (Fig. 2). More details can be found in [2] and [6].

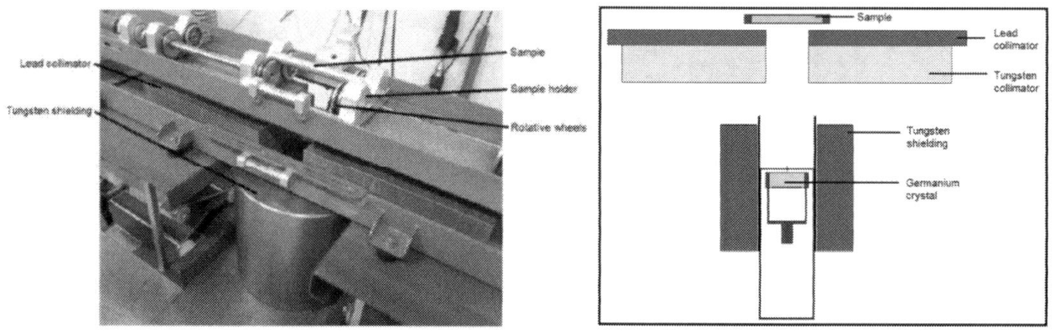

Figure 2. γ-spectrometry device (left side) and MCNP model for the efficiency transfer correction (right side).

The experimental results are presented as ratios of effective macroscopic capture rate relative to the gold macroscopic capture rate:

$$R_i = \frac{<N_i \sigma_{c,i}\, \phi>}{<N_{Au} \sigma_{c,Au}\, \phi>} \quad (1)$$

Four different gold samples were used and combined together. The combination of pure gold rods and aluminum-gold alloy wires is useful to validate the efficiency transfer correction between one with negligible (M-AlAu) and one with large (M-Au-3) self-absorption effect.

3 Interpretation of the neutron activation experiments

3.1 Calculation code and model

Neutron activation experiments were analysed by Monte-Carlo calculations based on TRIPOLI4 [7]. A detailed 3D model (Fig. 3) was settled including the graphite reflector, the driver zone (with the fuel assemblies explicitly described) and the experimental zone with the sample located in the core centre.

1 to 10 billion histories were simulated for each sample, leading to a convergence uncertainty of less than 0.5% (1σ). Three different neutron cross section libraries were considered for the calculation: JEFF-3.1.1 [8], JEFF-3.2 and ENDF/B-VII.0 [9]. As some reactions are partial captures, i.e. capture to a metastable stable, it was required to correct the computed total capture rate by the isomeric branching ratio. The value was taken from the EAF-2010 library [10] at thermal energy (25 meV).

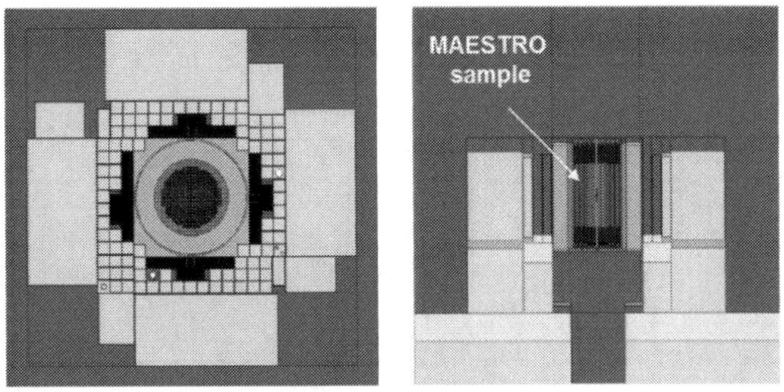

Figure 3. TRIPOLI4 3D detailed model of the MINERVE core.

3.2 Calculation vs Experiment

Table 2 presents the relative differences between calculated (C) and measured (E) capture rate ratios. The total uncertainty is a combination of the Monte-Carlo convergence (<0.5%), counting statistics (0.8 to 3%), technological uncertainty related to dimensions and compositions of the materials (<0.5%) and isotopic abundance in the natural element (0-3%).

Table 2. Calculation over measurement comparisons of macroscopic capture cross section ratios.

Type/Interest of the material	**Reaction**	**Total uncertainty (1σ)**	**TRIPOLI4.9 [C/E-1]**		
			JEFF-3.1.1	**JEFF-3.2**	**ENDF/B-VII.0**
Control rod (AgInCd) and absorbing FP)	$^{109}Ag(n,\gamma)^{110m}Ag$	1.1%	0.0%	0.6%	-0.5%
Absorber (AgInCd)	$^{113}In(n,\gamma)^{114m}In$	2.0%	-12.6%	-12.1%	-11.4%
	$^{115}In(n,\gamma)^{116m}In$	1.6%	-3.3%	-2.7%	-3.7%
Absorbing Fission Product	$^{133}Cs(n,\gamma)^{134gs+m}Cs$	0.8%	-2.3%	-1.7%	-3.1%
	$^{133}Cs(n,\gamma)^{134m}Cs$	3.5%	-0.1%	0.5%	-0.9%
Absorber (possible burnable poison, Burn-up Credit)	$^{151}Eu(n,\gamma)^{152m1}Eu$	1.3%	-12.5%	-12.0%	-9.1%
	$^{151}Eu(n,\gamma)^{152gs}Eu$	2.1%	-12.1%	-11.6%	-8.7%
	$^{153}Eu(n,\gamma)^{154}Eu$	1.3%	-8.5%	-8.0%	-9.5%
Instrumentation	$^{51}V(n,\gamma)^{52}V$	1.3%	-1.4%	-0.8%	-2.2%
Structure (clads, grids)	$^{98}Mo(n,\gamma)^{99}Mo$	2.0%	-1.2%	-0.3%	-1.9%
	$^{100}Mo(n,\gamma)^{101}Mo$	3.5%	-0.5%	-3.2%	-1.3%
	$^{64}Zn(n,\gamma)^{65}Zn$	2.1%		2.3%	
	$^{68}Zn(n,\gamma)^{69m}Zn$	3.6%		7.0%	
Structure (Zy clads)	$^{112}Sn(n,\gamma)^{113}Sn$	2.0%	27.7%	28.8%	22.6%
	$^{117}Sn(n,\gamma)^{118m}Sn$	1.8%	-82.5%	-81.9%	-82.5%
	$^{122}Sn(n,\gamma)^{123m}Sn$	1.8%	18.7%	-8.3%	-11.7%
	$^{94}Zr(n,\gamma)^{95}Zr$	2.0%	17.3%	6.9%	17.0%
	$^{96}Zr(n,\gamma)^{97}Zr$	3.8%	-4.9%	-4.3%	-5.1%

Here are the conclusions that can be derived from this table:

- *Silver*: the C/E agreement is within 1σ uncertainty for the three nuclear data libraries. This trend is consistent with previously published results obtained from the BUC program in MINERVE, but with a significantly reduced uncertainty (from 4% to 1%) [11].

- *Indium*: the C/E agreement is within 2σ uncertainty for ^{115}In capture for the three libraries. All of them are also consistent with a 11-13% underestimation of ^{113}In capture. Additional results would be required to separate the contribution of the isomeric ratio to the capture cross section.

- *Cesium*: the C/E agreement is higher than 2σ uncertainty for ^{133}Cs total capture for the three libraries, with a systematic underestimation of 2-3%. The partial capture to ^{133m}Cs is predicted within 1σ, so it can be concluded than the isomeric ratio from EAF-2010 is correctly evaluated within 3.5%.

- *Europium*: the C/E agreement of the two ^{151}Eu partial capture is consistent for the three libraries, indicating an underestimation of 12% for JEFF-3.1.1 and JEFF-3.2 and 9% with ENDF/B-VII.0. Just like for caesium, one can conclude that the isomeric ratio to the first metastable state is correctly evaluated in EAF-2010. Moreover, it seems that a revision of the capture cross sections would be required to solve the underestimation of this capture rate ratio. Additional information will be provided by the analysis of the pile-oscillation experiments which will give information on the total capture rate of ^{nat}Eu (mostly depend on the one of ^{151}Eu). For ^{153}Eu, there is also an 8-9% capture cross section underestimation in the three libraries. Previous pile-oscillation experiments [10] and irradiation experiments [12] led to the same conclusions.

- *Vanadium*: the C/E agreement for ^{51}V capture is within 1σ for the JEFF libraries and within 2σ for the ENDF/B-VII.0. This trend confirms preliminary results previously published from the MAESTRO program, based on a pure rod sample [1].

- *Molybdenum*: the C/E agreement is within 2σ for both ^{98}Mo and ^{100}Mo capture cross section and for the three nuclear data libraries. These results complete the feedback obtained in the past on the ^{95}Mo and on the overall ^{nat}Mo captures [11].

- *Zinc*: no isotopic evaluations are available in JEFF-3.1.1 and ENDF/B-VII.0 so only the JEFF-3.2 can be computed. The C/E agreement is within 2σ for both ^{64}Zn and ^{68}Zn capture cross section.

- *Tin*: the C/E agreement for ^{112}Sn and ^{117}Sn capture is significantly outside of the 2σ uncertainty range. Further investigations are required to understand such large differences, reaching up to 80% for ^{117}Sn. For ^{122}Sn capture, the strong overestimation of 19% observed in JEFF-3.1.1 was reduced to -9% in JEFF-3.2 but it is still outside of the 2σ uncertainty, with the same trend for ENDF/B-VII.0. Additional work would be required to reach the 2σ uncertainty range.

- *Zirconium*: the C/E agreement for ^{96}Zr capture is within 2σ uncertainty for the three nuclear data libraries. For ^{94}Zr capture, a strong overestimation of 17% is observed with both JEFF-3.1.1 and ENDF/B-VII.0 but it is significantly reduced to 7% in JEFF-3.2. Nevertheless, additional work would be required to reach the 2σ uncertainty range.

4 Conclusions

The MAESTRO experiment benefits from the feedbacks of previous programs in the MINERVE reactor. A careful design of the samples with severe constraints on the material mass, purity and homogeneity, just like some improvements on the irradiation conditions, allows reaching a target accuracy of less than 2% on the experimental results (previously 3-4%).

Preliminary results on neutron activation experiments performed on natural elements were obtained with TRIPOLI4 Monte-Carlo calculations. They confirm JEFF-3.1.1 evaluations of ^{109}Ag, ^{115}In, ^{51}V, ^{98,100}Mo, ^{64,68}Zn and ^{96}Zr capture cross sections within 2σ uncertainty, while some other indicate that a possible re-evaluation should be required for ^{113}In, ^{133}Cs, ^{151,153}Eu, ^{112,117,122}Sn and ^{94}Zr. Moreover, it was possible to validate the isomeric ratio to the metastable state in the capture of ^{133}Cs and ^{151}Eu, thanks to the measurement of both ground and metastable states. All these results will have to be confirmed by the analysis of the pile-oscillation measurements that were performed on the same sample, which is currently ongoing.

Acknowledgments

Authors would like to acknowledge their industrial partners AREVA and EDF for their financial support and also the operators and experimentalists of the MINERVE facility.

References

1. P. Leconte et al., MAESTRO: an ambitious experimental programme for the improvement of nuclear data of structural, detection, moderating and absorbing materials - First results for ^{nat}V, ^{55}Mn, ^{59}Co and ^{103}Rh, Proceedings of the ANIMMA2013 Conference, paper #1376, May 2013, Marseille, France.
2. A. Gruel et al., Integral (n,γ) Cross-sections Measurement of Standard Light Water Reactors Materials, Proceedings of the IGORR2014 conference, Bariloche, Argentina, November 2014.
3. S. Cathalau et al., More than forty years of neutronics experiments in critical facilities of Cadarache center: from EOLE and MINERVE to the future ZEPHYR reactor, Proceedings of the IGORR2014 conference, Bariloche, Argentina, Nov. 2014.
4. G. Bignan et al., Reactor Physics Experiments on Zero Power Reactors, *Handbook of Nuclear Engineering,* Springer Science, pp 2053-2184, 2010.
5. A. Santamarina et al., Reactivity Worth Measurement of Major Fission Products in MINERVE LWR Lattice Experiment, *Nuclear Science and Engineering,* Vol. 178, 4, December 2014, 562-581.
6. P. Leconte, J.P. Hudelot, M. Antony, Integral Capture rate Measurements of Neutron Absorbers and Minor Actinides by Gamma-Ray Spectrometry on Specific Fuel Samples, Proceedings of the IYNC 2006 Conference, Stockholm, Sweden, June 2006.
7. E. Brun et al., TRIPOLI-4®, CEA, EDF and AREVA reference Monte Carlo code, Joint International Conference on Supercomputing in Nuclear Applications and Monte Carlo 2013, SNA + MC 2013, Volume 82, p 151–160, August 2015.
8. A. Santamarina et al., The JEFF-3.1.1 Nuclear Data Library, JEFF Report 22, OECD, NEA No. 6807, 2009.
9. M.B. Chadwick et al., ENDF/B-VII.0: Next Generation Evaluated Nuclear Data Library for Nuclear Science and Technology, *Nuclear Data Sheets*, Vol. 107, 12, p. 2931-3060, December 2006.
10. J.C. Sublet et al., The European Activation File: EAF2010 neutron-induced cross section library, EASY Documentation Series, CCFE-R (10) 05, 2010.
11. A. Gruel et al., Interpretation of Fission Product Oscillations in the MINERVE Reactor, from Thermal to Epithermal Spectra, *Nuclear Science and Engineering*, 169, 229–244, 2011.
12. D. Bernard et al., Validation of JEFF-3.1.1 Thermal and Epithermal Neutron-Induced Capture Cross Sections Through MELUSINE Experiment Analysis, *Nuclear Science and Engineering*, 179, 302–312, 2015.

EPJ Web of Conferences **111**, 07002 (2016)
DOI: 10.1051/epjconf/201611107002

Fission Rate Ratios of FCA-IX Assemblies as Integral Experiment for Assessment of TRU's Fission Cross Sections

Masahiro Fukushima[1,a], Kazufumi Tsujimoto[1], and Shigeaki Okajima[1]

[1]*Japan Atomic Energy Agency (JAEA), 2-4 Shirakata, Tokai-mura, Naka-gun, Ibaraki, Japan, 319-1195*

Abstract. At the fast critical assembly (FCA) of JAEA, central fission rate ratios for TRU such as ^{237}Np, ^{238}Pu, ^{239}Pu, ^{242}Pu, ^{241}Am, ^{243}Am, and ^{244}Cm were measured in the seven uranium-fueled assemblies (FCA-IX assemblies) with systematically changed neutron spectra. The FCA-IX assemblies were constructed with simplicity both in geometry and composition. By virtue of these FCA-IX assemblies where the simple combinations of uranium fuel and diluent (graphite and stainless steel) in their core regions were systematically varied, the neutron spectra of them cover from the intermediate to fast one. Taking their advantages, benchmark models with respect to the central fission rate ratios had been recently developed for the evaluation of the TRU's fission cross sections. As an application of these benchmark models, the Japanese Evaluated Nuclear Data Library JENDL-4.0 was utilized by a Monte Carlo calculation code. Several results show large discrepancies between the calculation and experimental values. The benchmark models would be well suited for the evaluation and modification of the nuclear data for the TRU's fission cross sections.

1 Introduction

In studies on the evaluation and modification of nuclear data, integral and differential experiments have complementarity relations. The measured data in integral experiments apparently lack information of neutron energy since they are integrated out by weighting of neutron spectra. However, the set of integral data measured in a variety of neutron spectra has a potential to be useful for the evaluation of cross sections since the sensitivities of the measured data show energy dependence.

Seven uranium-fueled assemblies (FCA-IX assemblies) with systematically changed neutron spectra were constructed at the FCA in the 1980s, and the central fission rate ratios (CFRRs) were measured for seven TRU nuclides containing minor actinides, ^{237}Np, ^{238}Pu, ^{239}Pu, ^{242}Pu, ^{241}Am, ^{243}Am, and ^{244}Cm [1-2]. Each assembly is identified by a number (1–7) after the "IX," e.g., IX-1 for the first one. Each was a coupled system comprising a central core region and a surrounding blanket region with depleted uranium metal blocks. They were simple, both in geometry and composition. Each core region was composed of a simple combination of high enriched uranium (HEU) and graphite in the assemblies IX-1 to IX-3 and that of HEU and stainless steel in the assemblies IX-4 to IX-6. Their neutron spectra cover from the intermediate to fast ones by varying the combinations of fuel and loaded diluent. In addition, the assembly IX-7 was constructed as a low enriched uranium (LEU)

[a] Corresponding author: fukushima.masahiro@jaea.go.jp

assembly without diluent. The three-dimensional benchmark models of the FCA-IX assemblies were recently developed with respect to the criticalities [3] and the CFRRs [4]. The purpose of this paper is to introduce the analysis results of CFRRs of the FCA-IX assemblies with using the new benchmark models, from a viewpoint of integral experiment for the nuclear data evaluation. In Section 2, we briefly review the benchmark models with respect to the CFRRs. In Section 3, we discuss the Monte Carlo calculation results by JENDL-4.0 [5]. Finally, the conclusion is shown in Section 4.

2 Brief review of CFRRs of FCA-IX assemblies

2.1 Experiments

The fission rates for seven nuclides, ^{237}Np (F37), ^{238}Pu (F48), ^{239}Pu (F49), ^{242}Pu (F42), ^{241}Am (F51), ^{243}Am (F53), and ^{244}Cm (F64), were measured at the core center of each FCA-IX assembly [1-2]. Seven fission chambers of parallel-plate type [6] were used. **Figures 1** and **2** illustrate the structure of the fission chamber and the loading of fission chamber into the central core drawer, respectively. The deposit mass was determined within ±1.5% error (±3.0% for ^{244}Cm) by α-spectrometry using silicon surface barrier detector mounted in a vacuum chamber. The uncertainties in the CFRRs measurements are due to those in the α-assay of the deposit mass, the statistical counting errors, errors introduced by the impurity isotope corrections. As a typical example, **Table 1** shows uncertainties of F37, F49, and F64 in the assembly IX-1. The uncertainties in the α-assay mainly contribute to the total ones.

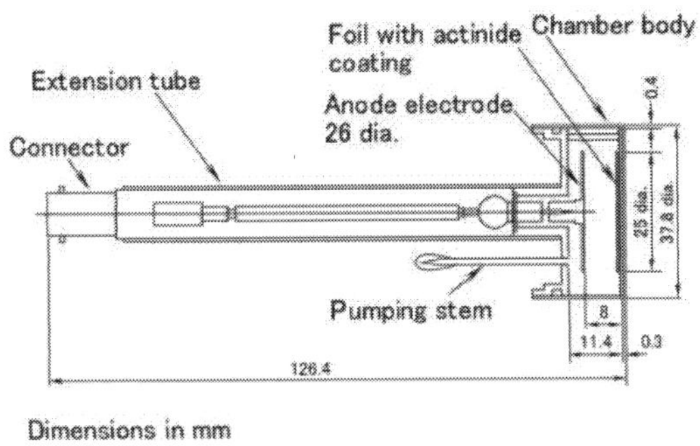

Figure 1. Structure of fission chamber.

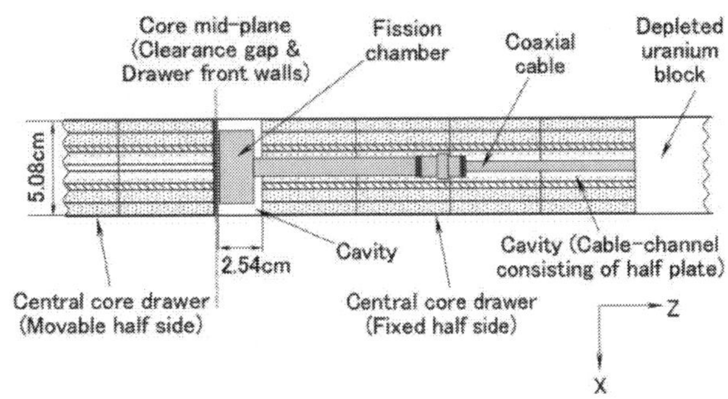

Figure 2. Loading of fission chamber.

Table 1. Components of uncertainties of typical fission rates in the assembly IX-1.

	F37	F49	F64
α-assay of the deposit mass	1.5%	1.5%	3.0%
Statistical counting error	0.4%	0.4%	0.8%
Impurity isotope correction	—	0.3%	1.0%
Total uncertainty	1.6%	1.6%	3.3%

2.2 Benchmark models and Calculations

Even though the FCA-IX assemblies are simple in geometry and composition, we should be taken account for the geometric conditions during the measurement in analysis of the CFRRs. To develop the benchmark models as simple as possible, we took care not to increase the total uncertainty far beyond the original experimental one. In reference [4], the new three-dimensional benchmark models for the CFRR analysis were developed by simplifying geometric details while keeping the plate-wise heterogeneity due to the plate arrangement in the core drawers. The CFRRs calculated in this new benchmark models, C_{bench}, were compared with the values (E_{bench}) corresponding to the measured ones (E_{exp}) with taking account for the following two simplification effects: (a) the first one is relating to geometric modelling of the CFRR measurement conditions, i.e. the fission chamber and the cavities occupied in the central core drawer and so on, and (b) the second one is relating to that of other complex conditions specific to the FCA facility as mentioned in the reference [3]. Each simplification effect was estimated by the Monte-Carlo code with continuous neutron energy, MVP [7], as described below.

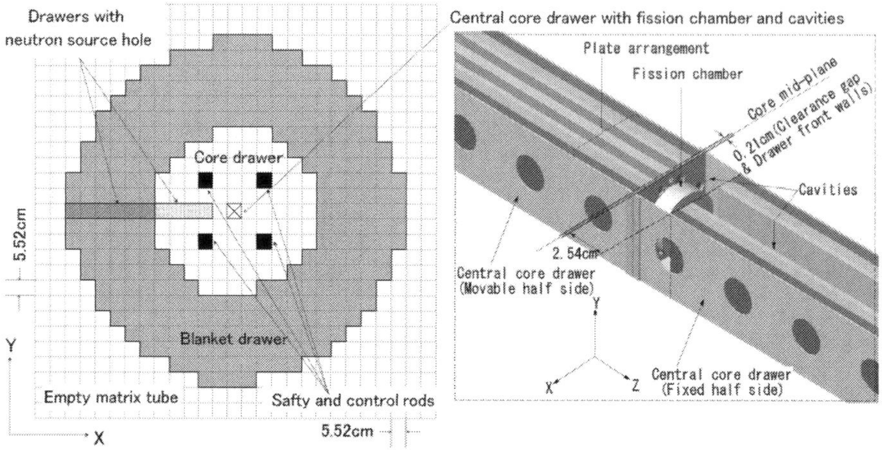

Figure 3. Configurations of as-built model for the assembly IX-1.

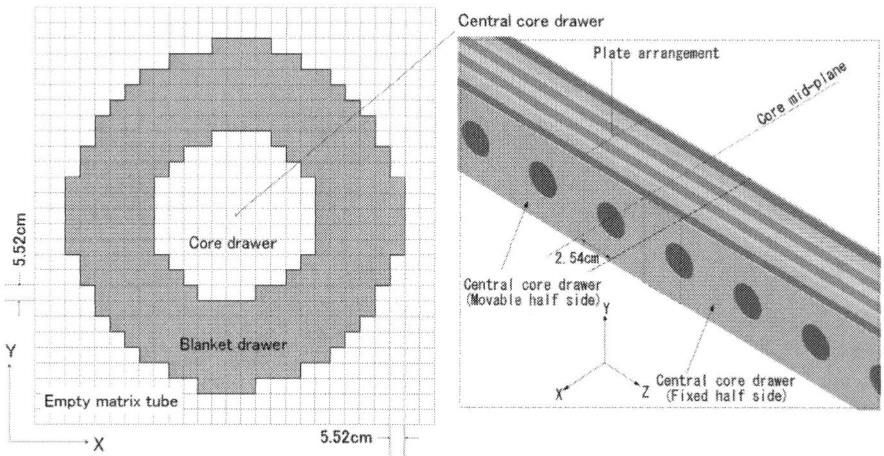

Figure 4. Configurations of benchmark model for the assembly IX-1.

(1) Step 1
As the as-built calculations, the experimental configuration with the fission chamber and the cavities and so on was modelled as detailed as possible by MVP. **Figure 3** illustrates the XY-configuration and the bird's eye view of the central core drawer of the as-built model for the assembly IX-1. To calculate the CFRR, the cylindrical space inside the fission chamber was adopted as the tally. The calculated CFRR with this model is denoted as $C_{as\text{-}built}$.
(2) Step 2
As for the CFRR measurement conditions, the central drawer with the fission chamber and the cavities was simplified to be same as the core drawers in the core region. To calculate the CFRR, the amounts of fissionable nuclide corresponding to the fission chamber were doped into the matrix tube and drawer at the core center ranging from the core mid-plane to 2.54 cm along the Z-direction. This model could be used in deterministic calculations. The calculated CFRR in this model is denotes as $C_{simple(a)}$.
(3) Step 3
The experimental configuration includes other complex conditions such as the clearance gap and the drawer front walls at the core mid-plane, the structure of the safety control rods, and the hole for

neutron source insertion as also shown in **Figure 3**. As the second simplification, these geometries were simplified while keeping the plate-wise heterogeneity due to the plate arrangement in the core drawers. **Figure 4** illustrates the XY-configuration and the bird's eye view of the central core drawer of the benchmark model for the assembly IX-1. The calculated CFRR in this simple model is denoted as $C_{simple(b)}$.

These simplification effects were indicated as the ratios, $f_a=C_{simple(a)}/C_{as\text{-}built}$ and $f_b=C_{simple(b)}/C_{simple(a)}$. For each ratio calculation by MVP, the history number was 200 million. The benchmark CFRR value (E_{bench}) was finally obtained by multiplying these effects, f_a and f_b, to the measured one (E_{exp}). As a typical example of the results, **Table 2** summarizes the measured CFRRs (E_{exp}), the simplification effects, and the benchmark CFRRs (E_{bench}) in the assembly IX-1. The results for the other assemblies IX-2 to IX-7 are given in reference [4]. The simplification effects are within a few percent in total and the increase of uncertainty of the CFRR is negligible small for all the assemblies.

The further information related to the homogenization and its effect were provided in reference [4] together with the homogenous atomic number densities and the three-dimensional configuration of each FCA-IX assembly.

Table 2. Measured CFRRs, simplification effects, and benchmark CFRRs of the assembly IX-1.

Type	Measured CFRR (E_{exp})	Simplification effects		Benchmark CFRR (E_{bench})
		$f_a= C_{simple(a)}/C_{as\text{-}built}$ *1	$f_b= C_{simple(b)}/C_{simple(a)}$ *2	
F37/F49	0.2048 ±2.2%	0.9842±0.0046	1.0031±0.0032	0.2022 ±2.3% *3
F48/F49	0.5856 ±2.4%	0.9948±0.0050	1.0008±0.0034	0.5830 ±2.5%
F42/F49	0.1584±2.2%	0.9834±0.0046	1.0031±0.0032	0.1562 ±2.3%
F51/F49	0.1961±2.2%	0.9831±0.0047	1.0037±0.0033	0.1935 ±2.3%
F53/F49	0.1509±2.2%	0.9822±0.0047	1.0038±0.0033	0.1488 ±2.3%
F64/F49	0.2614±3.6%	0.9866±0.0045	1.0022±0.0031	0.2585 ±3.7%

*1 Simplification effect relating to modelling of the CFRR measurement conditions. *2 Simplification effect relating to modelling of the other complex geometries. *3 Estimated by error propagation of the measurement uncertainties and the Monte Carlo statistical errors in $C_{as\text{-}built}$ and $C_{simple(b)}$.

3 Discussions of benchmark results

The benchmark calculations by JENDL-4.0 were carried out and the results are shown in **Table 3**. The calculations for F37/F49 agreed with the experimental values (E_{bench}) for all the assemblies. On the other hand, the calculations for F64/F49 obviously overestimate by 11%-18% and the C_{bench}/E_{bench} values slightly increase with neutron spectrum hardening from the assemblies IX-1 to IX-4. For the other CFRRs, the results are summarized below:

- For F48/F49, the calculations underestimate about by 5% for all the assemblies,
- For F42/F49, the calculations overestimate about by 8% for all the assemblies,
- For F51/F49 and F53/F49, the calculations underestimate about by 4% and 6%, respectively.

Table 3. Benchmark calculation results (C_{bench}/E_{bench}) with MVP and JENDL-4.0 [4].

Type	IX-1	IX-2	IX-3	IX-4	IX-5	IX-6	IX-7
F37/F49	0.994±0.023*	0.996±0.027	0.991±0.025	1.004±0.025	0.994±0.025	0.992±0.020	0.999±0.025
F48/F49	0.953±0.024	Not measured	0.961±0.026	0.956±0.026	0.955±0.026	0.954±0.026	0.955±0.026
F42/F49	1.077±0.025	1.071±0.029	1.081±0.027	1.086±0.022	1.090±0.022	1.081±0.022	1.090±0.027
F51/F49	0.968±0.022	0.947±0.027	0.963±0.024	0.961±0.024	0.957±0.024	0.962±0.024	0.958±0.024
F53/F49	0.939±0.022	0.933±0.025	0.945±0.024	0.934±0.023	0.938±0.023	0.943±0.024	0.948±0.024
F64/F49	1.112±0.041	1.135±0.044	1.152±0.043	1.165±0.043	1.168±0.043	1.175±0.046	1.159±0.044

* Sum of squares of measurement uncertainties and MVP statistical error

As a typical case in the results, we discuss the large overestimations of F64/F49 as following. The previous work [2] shows that the C/E values of F64/F49 by JENDL-3.3 range from 1.01 to 1.05. From these results, the calculation values of F64/F49 become about 10% larger by changing from JENDL-3.3 to JENDL-4.0. **Figure 5** shows the sensitivity coefficient of F64/F49 to ^{244}Cm fission cross section. The sensitivity coefficient is mainly distributed above the fission threshold energies around 1 MeV for all the assemblies. **Figure 6** compares the ^{244}Cm fission cross sections between JENDL-4.0 and JENDL-3.3. The ^{244}Cm fission cross section of JENDL-4.0 around 1 MeV is about 10% larger than that of JENDL-3.3, which results in the large overestimations of F64/F49 by JENDL-4.0.

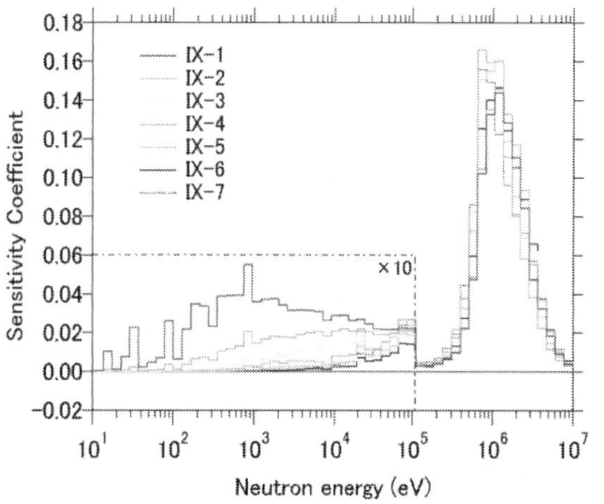

Figure 5. Sensitivity of F64/F49 to ^{244}Cm-fission.

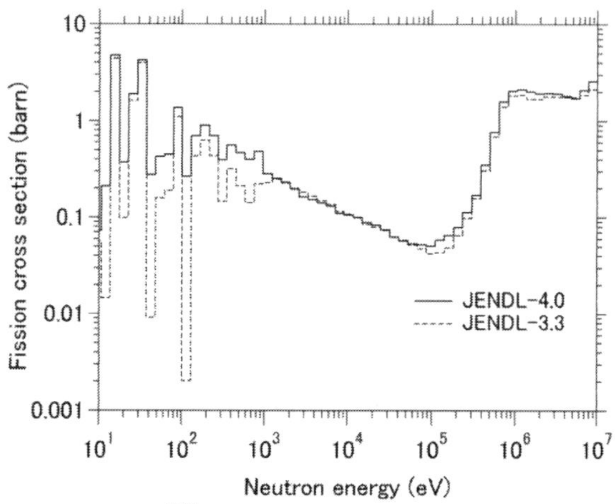

Figure 6. ^{244}Cm-fission cross sections.

4 Conclusion

In this paper, we reviewed the three-dimensional benchmark models for the CFRRs of the FCA-IX assemblies with systematically changed neutron spectra and discussed benchmark results by JENDL-4.0. The new benchmark models were developed without increasing the total uncertainty far beyond that of the original experiment. The benchmark results for F37/F49 show that the calculations by JENDL-4.0 agreed with the experimental values for all the assemblies. For F64/F49, the large overestimations by JENDL-4.0 are found for all the assemblies, which is caused by the large fission cross section of ^{244}Cm above the fission threshold energies around 1 MeV. For the other CFRRs, F48/F49, F42/F49, F51/F49, and F53/F49, the results by JENDL-4.0 show discrepancies between the calculation and experimental values. The benchmark models would be well suited for the evaluation and modification of the nuclear data for these fission cross sections owing to their advantages of simplicity and systematically changed neutron spectra.

Acknowledgements

We would like to thank Mr. T. Mukaiyama for valuable information about the CFRR measurements of the FCA-IX assemblies.

References

1. T. Mukaiyama, M. Obu, M. Nakano, et al., *Proc. of Int. Conf. on Nuclear Data for Basic and Applied Science*, Santa Fe, USA, 483 (1985)
2. S. Okajima, M. Fukushima, T. Mukaiyama, *Proc. of International Conference on the Physics of Reactors (PHYSOR2008)*, Interlaken, Switzerland (2008)
3. M. Fukushima, Y. Kitamura, T. Kugo, et al., J. Nucl. Sci. Technol., Published online, (2015)
4. M. Fukushima, A. Oizumi, H. Iwamoto, et al., JAEA-Data/Code 2014-030 (2015) [in Japanese]
5. K. Shibata, O. Iwamoto, T. Nakagawa, et al., J. Nucl. Sci. Technol., **48**, 1 (2011)
6. M. Obu, JAERI-M 9757 (1981) [in Japanese]
7. Y. Nagaya, K. Okumura, T. Mori, et al., JAERI 1348 (2005)

EPJ Web of Conferences **111**, 07003 (2016)
DOI: 10.1051/epjconf/201611107003

Feasibility study of the AOSTA experimental campaign

M. Carta[1,a], P. Blaise[2], C. Bethaz[3], F. Boccia[4], V. Fabrizio[1], B. Geslot[2], A. Grossi[1], and A. Gruel[2]

[1]*ENEA, C.R. Casaccia, Via Anguillarese 301, S. M. Galeria, Rome, Italy*
[2]*CEA Cadarache Centre, St Paul Lez Durance. France*
[3]*Politecnic of Turin, Corso Duca degli Abruzzi 24, Turin, Italy*
[4]*University of Rome'La Sapienza', Via Eudossiana 18, Rome, Italy*

Abstract. The reduction of the nuclear waste is one of the most important nuclear issues. The high radiotoxicity of the spent fuel is due to plutonium and some minor actinides (MAs) such as neptunium, americium and curium, above all. One way to reduce their hazard is to destroy by fission MAs in appropriate nuclear reactors. To allow the MAs destruction an important effort have been done on the nuclear data due to the poor knowledge in this field. In the framework of one of the NEA Expert Group on Integral Experiments for Minor Actinide Management an analysis of the feasibility of MAs irradiation campaign in the TAPIRO fast research reactor is carried out. This paper provides preliminary results obtained by calculations modelling the irradiation, in different TAPIRO irradiation channels, of some CEA samples coming from the French experimental campaign OSMOSE, loaded with different contents of MAs, in order to access, through particular peak spectrometry, to their capture cross section. On the basis of neutron transport calculation results, obtained by both deterministic and Monte Carlo methods, an estimate of the irradiated samples counting levels from the AOSTA (Activation of OSMOSE Samples in TAPIRO) experimental campaign is provided.

1 Introduction

Among the foundations of a sustainable use of nuclear fission energy certainly there are a reliable fuel cycle and a safe manage of radioactive waste. The spent fuel discharged from nuclear power plants constitutes the main contribution to nuclear waste. Most of the hazard from the spent fuel stems from only a few chemical elements such as plutonium and some Minor Actinides (MAs) such as neptunium, americium and curium, plus some long-lived fission products such as iodine and technetium. In this frame, one of the key aspects is the appropriate management of MAs. A technological route to reduce the risks associated with MAs is to transmute them in nuclear systems. However, due to a lack of experimental data, which has meant an absence of precision in MAs nuclear data, it remains difficult to

[a] Corresponding author: mario.carta@enea.it

establish a detailed design of transmutation systems with reliable accuracy and the capacity to precisely predict the composition of spent fuel. Several NEA and IAEA Working Groups addressed these issues, in particular recommending integral measurements, complementary to parallel efforts for differential measurements, for the following nuclides of MAs from viewpoints of design of transmutation systems and of fuel cycles: ^{237}Np, ^{241}Am, ^{242m}Am, ^{243}Am, ^{242}Cm, ^{243}Cm, ^{244}Cm and ^{245}Cm. In the frame of the NEA Expert Group on Integral Experiments for Minor Actinide Management [1] a joint collaboration between ENEA (Italian National Agency for New Technologies, Energy and Sustainable Economic Development) and CEA (French Alternative Energies and Atomic Energy Commission) was established with the aim to study the feasibility of a MAs irradiation campaign in the TAPIRO fast neutron source research reactor located at the ENEA Casaccia center near Rome [2]. This paper provides preliminary results obtained by calculations modelling the irradiation, in different TAPIRO irradiation channels, of some CEA samples coming from the French experimental campaign OSMOSE [3] (Oscillation in MINERVE of isotope in 'Eupraxic' Spectra), loaded with different contents of MAs. On the basis of the neutron transport calculations results, obtained by both the deterministic ERANOS [4] and the Monte Carlo Serpent [5] codes, an estimate of the irradiated samples counting levels from the AOSTA (Activation of Osmose Samples in TApiro) experimental campaign, taking into account both geometry and efficiency of the counting system, is provided.

2 The TAPIRO fast neutron source research reactor

TAPIRO (**TA**ratura **PI**la **R**apida a potenza zer**O** - Fast Pile Calibration at 0 Power) is a fast neutrons source research reactor located at C.R ENEA-CASACCIA (Italy). The project, entirely developed by ENEA's staff, is based on the general concept of AFSR (Argonne Fast Source Reactor - Idaho Falls). It was built to support an experimental program on fast reactors and it is in operation since 1971.TAPIRO is currently used, in addition to education and training, for experimental programs in support of different research fields like nuclear data, nuclear fusion, aerospace industry.

It has a maximum power of 5 kW with a neutron flux around $4 \cdot 10^{12}$ n/(cm^2·s) in the centre of the core. The core is cylindrical with a diameter of about 12 cm and a similar height. It is made by metallic uranium (98.5 % uranium and 1.5 % molybdenum) with an enrichment of 93.5 % in ^{235}U. It consists of 2 parts: the upper part is fixed while the lower one is movable. The core is surrounded by a double layer of a copper reflector and by an external borate concrete biological shield. The core is cooled by helium. The reactor is equipped with 2 shim rods, 2 safety rods and a regulating rod. These rods are made of the same material of the reflector, i.e. copper, and the reactor is controlled increasing or reducing the neutron leakage. The system has different experimental channels with various diameters. A horizontal section of the reactor is shown in Fig. 1.

3 Energy averaged capture cross sections evaluation

Neutron transport calculations have been performed by two different methodologies, deterministic by the ERANOS code, and stochastic by the Monte Carlo Serpent code. In Fig. 2 the calculation models are shown.

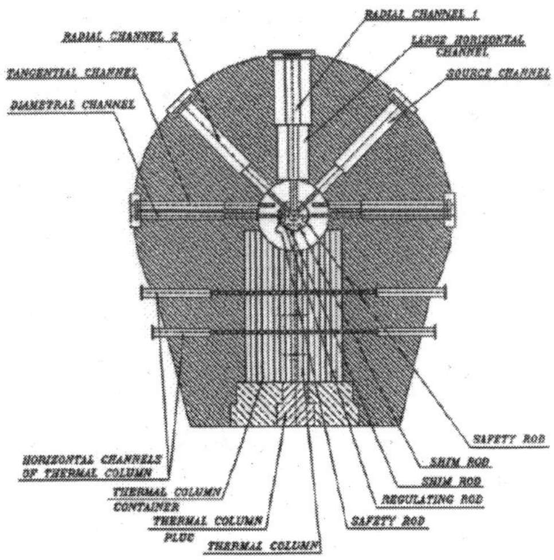

Figure 1. TAPIRO horizontal section.

Neutron fluxes, energy spectra and reaction rates have been evaluated for several minor actinides in correspondence of different radial positions in the diametral experimental channel (cf. Fig. 1).

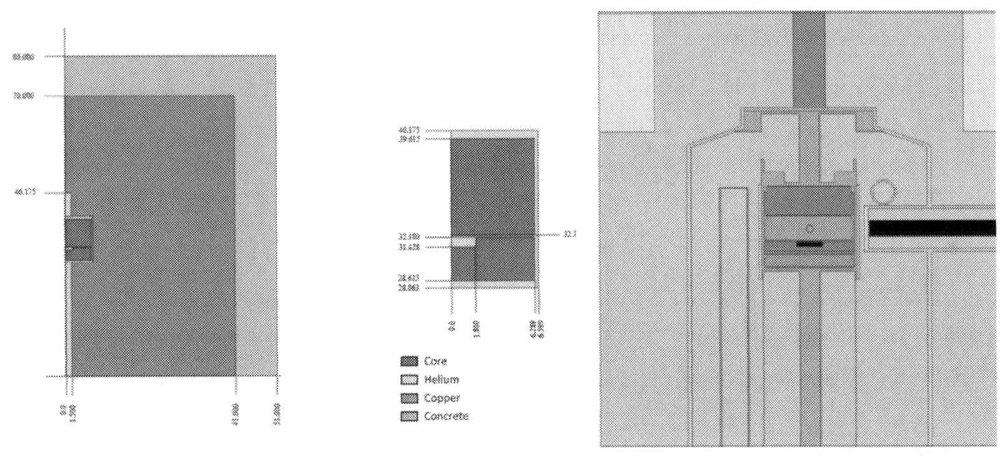

(a) - RZ ERANOS model (b) - Serpent model

Figure 2. ERANOS and Serpent TAPIRO calculation models.

Subsequently position dependent average microscopic capture cross sections have been calculated to predict the impact on their values of the spectral variations across the system. Average capture microscopic cross sections $\bar{\sigma}_c(\mathbf{r})$ are defined as:

$$\bar{\sigma}_c(\mathbf{r}) = \frac{\int \sigma_c(\mathbf{r},E)\varphi(\mathbf{r},E)dE}{\int \varphi(\mathbf{r},E)dE} \tag{1}$$

with φ neutron flux. As an example, the behavior across the system of the average capture microscopic cross sections for the minor actinide ^{241}Am, calculated with both ERANOS and Serpent codes, is shown in Fig. 3.

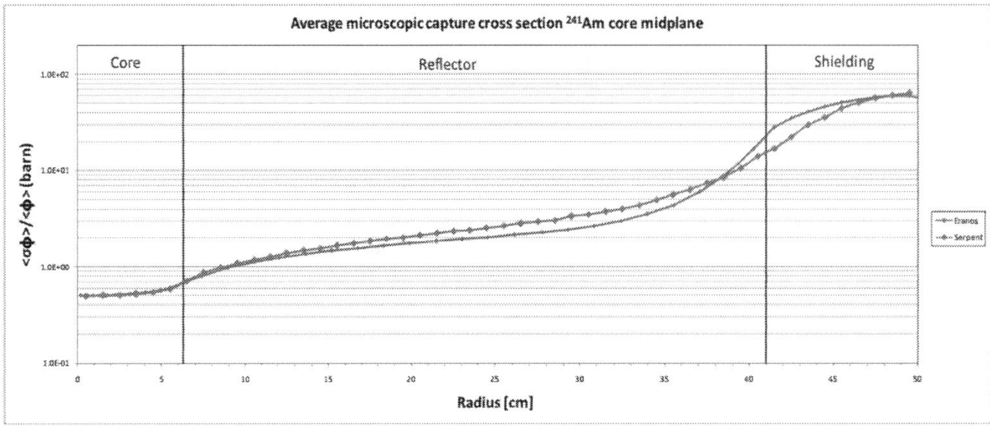

Figure 3. ^{241}Am microscopic capture cross section.

A satisfactory agreement can be observed, also for the other samples, among ERANOS and Serpent codes results.

4 Evaluation of the counting rates after irradiation of the OSMOSE samples

The selected OSMOSE samples contain ^{237}Np, ^{242}Pu, ^{241}Am, ^{243}Am in a double Zircaloy sheath (Fig. 4). The matrix for all the samples is composed by natural uranium.

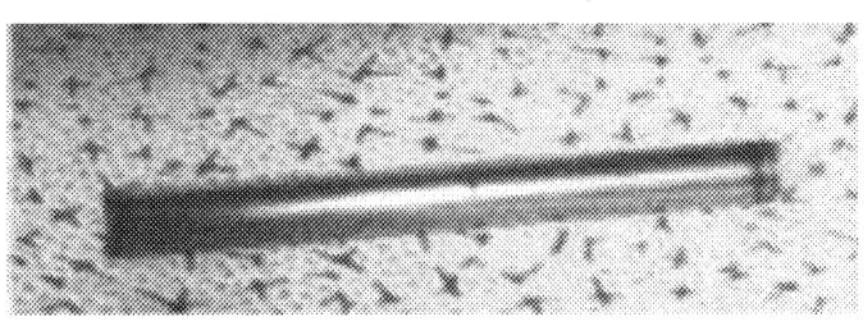

Internal sheath [mm]	9.56
External sheath [mm]	10.6
Length [mm]	103.5

Figure 4. OSMOSE samples.

It was assumed to irradiate these samples following a weekly scheme characterized by 5 hours of irradiation and 19 hours of cooling repeated for 4 times and then 5 hours of irradiation and 2 hours of cooling. Activity values have been evaluated after these last 2 hours of cooling (Fig. 5). Three sample positions have been considered along the diametral channel: r=12.07 cm (near the core), 24.58 cm (about reflector center) and 45.4 cm (entrance of the thermal column), see Fig. 2 (a). In correspondence of such positions have been calculated by the FISPACT code [6] the activity values, for each OSMOSE sample, associated to the irradiation scheme shown in Fig. 5. The obtained results, by using the ERANOS neutron flux results for this preliminary analysis, are shown in Table 1.

The counting rate [counts/s] for each detector have been evaluated by the relationship:

$$C = A \cdot I_\gamma \cdot \varepsilon$$

with A activity level [Bq], I_γ intensity of the γ or X emission [%], ε geometric efficiency of the detector [%]. The geometric efficiency ε depends by the scintillator type used for detection and by the overall experimental geometrical arrangement. In our case it was considered as scintillator n-type coaxial HPGe detector made of high purity germanium, showing high precision and efficiency for both γ and X rays in the energy range 3 keV÷10

MeV. The geometric efficiency has been evaluated by Monte Carlo MCNP code [7] modeling the arrangement of the counting system (Fig. 6).

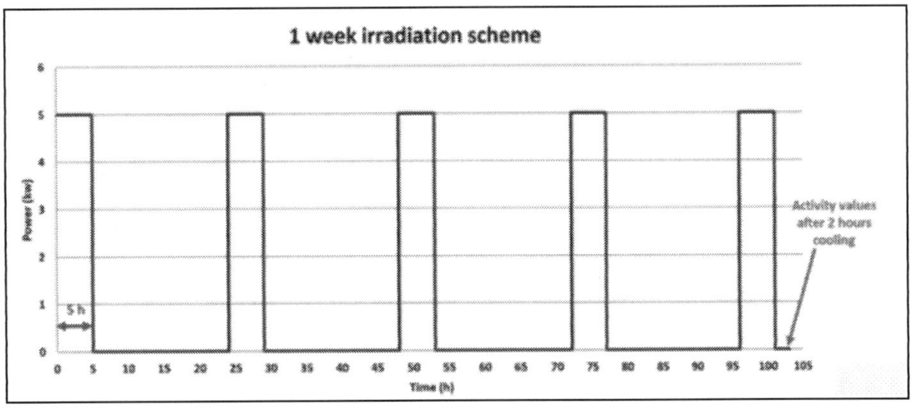

Figure 5. Irradiation scheme.

Table 1. Energy average cross sections and activity levels for each sample at different position in TAPIRO.

	Position	r = 12.07 cm	r = 24 .58 cm	r = 45.5 cm
OSMOSE Samples	ϕ (n·cm^{-2}·s^{-1})	6.94E+11	1.74E+11	8.79E+09
Np237	$\sigma_{c,Np237}$ (barn)	1.04	1.73	19.57
	A (Bq)	2.04E+08	8.49E+07	4.85E+07
Pu242	$\sigma_{c,Pu242}$ (barn)	0.34	0.62	22.48
	A (Bq)	1.15E+08	5.30E+07	9.68E+07
Am241	$\sigma_{c,Am241}$ (barn)	1.32	2.06	32.61
	A (Bq)	1.15E+08	4.50E+07	3.61E+07
Am243	$\sigma_{c,Am243}$ (barn)	1.13	1.83	33.40
	A (Bq)	2.10E+07	8.10E+06	7.14E+06

Table 2 summarizes the results obtained. In particular, in Table 1 are shown, for each sample, the γ or X rays characteristics (energy and intensity), the geometric efficiency and the counting level obtained by the relationship (1). The counting rate levels shown in Tab. 2 seem to predict the feasibility of the AOSTA experimental campaign although, as evident, a more detailed analysis is needed in the next future to confirm these promising preliminary results.

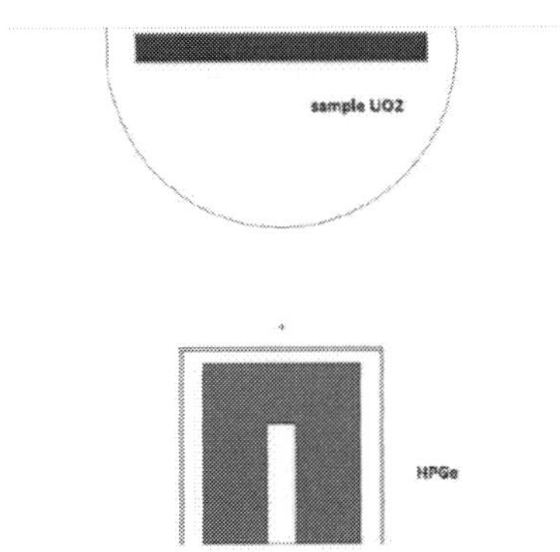

Figure 6. MCNP modeling of the counting system.

Table 2. Counting rates for the selected OSMOSE samples in TAPIRO (ERANOS neutron fluxes).

	Position	r = 12.07 cm	r = 24 .58 cm	r = 45.5 cm
OSMOSE Samples	ϕ (n·cm^{-2}·s^{-1})	6.94E+11	1.74E+11	8.79E+09
Np237	Np238 E γ (keV)	984.45	984.45	984.45
	γ Intensity (%)	25.19	25.19	25.19
	ε Detection (%)	0.186	0.186	0.186
	C (cps)	95487	39779	22738
Pu242	Pu243 E γ (keV)	84	84	84
	γ Intensity (%)	23.10	23.10	23.10
	ε Detection (%)	0.021	0.021	0.021
	C (cps)	5559	2572	4698
Am241	Am 242 E $X_{K\alpha 1}$ (keV)	103.374	103.374	103.374
	$X_{K\alpha 1}$ Intensity (%)	5.70	5.70	5.70
	ε Detection (%)	0.107	0.107	0.107
	C (cps)	7014	2745	2204
Am243	Am244 E γ (keV)	743.971	743.971	743.971
	γ Intensity (%)	66.00	66.00	66.00
	ε Detection (%)	0.213	0.213	0.213
	C (cps)	29466	11391	10032

Conclusions

In the frame of the NEA Expert Group on Integral Experiments for Minor Actinide Management, a joint collaboration between ENEA and CEA was established with the aim to study the feasibility of a MAs irradiation campaign, named AOSTA (Activation of Osmose Samples in TApiro), in the TAPIRO fast neutron source research reactor located at the ENEA Casaccia center near Rome. Preliminary neutronic analyses of the AOSTA experimental campaign have been performed by both deterministic and stochastic methods, and a good agreement has been observed among the results provided by the two calculation approaches.The counting rate levels obtained by these simulations seem to predict the feasibility of the AOSTA experimental campaign although, as evident, a more detailed analysis is needed in the next future to confirm these promising preliminary results.

References

1. https://www.oecd-nea.org/science/ma/.
2. http://www.enea.it/en/research-development/documents/nuclear-fission/tapiro-eng-pdf.
3. JP. Hudelot at al., "OSMOSE: An Experimental Program for the Qualification of Integral Cross Sections of Actinides", Proc. PHYSOR-2004, Chicago, Illinois, April 25-29, 2004.
4. G. Rimpault et al., 'The ERANOS code and data system for fast reactor neutronic analyses', Proc. PHYSOR 2002, Seoul, Korea, October 7-10, 2002.
5. http://montecarlo.vtt.fi.
6. R. A. Forrest, "FISPACT-2007: User manual", UKAEA FUS 534, EURATOM/UKAEA Fusion Association, Culham Science Centre, March 2007.
7. X-5 Monte Carlo Team "MCNP: A General Monte Carlo Transport Code" LA-CP-03-245, April 24, 2003 (Revised 10/3/05).

EPJ Web of Conferences **111**, 08001 (2016)
DOI: 10.1051/epjconf/201611108001

Accurate measurements of fission-fragment yields in ^{234,235,236,238}U(γ,f) with the SOFIA set-up

A. Chatillon[1,a], J. Taïeb[1], J.-F. Martin[1], E. Pellereau[1], G. Boutoux[1], T. Gorbinet[1], L. Grente[1], G. Bélier[1], B. Laurent[1], H. Alvarez-Pol[2], Y. Ayyad[2], J. Benlliure[2], M. Caamaño[2], L. Audouin[3], E. Casarejos[4], D. Cortina-Gil[2], F. Farget[5], B. Fernández-Domínguez[2], A. Heinz[6], B. Jurado[7], A. Kelić-Heil[8], N. Kurz[8], S. Lindberg[6], B. Löher[8], C. Nociforo[8], C. Paradela[2], S. Pietri[8], D. Ramos[2], J.-L. Rodriguez-Sanchez[2], C. Rodrìguez-Tajes[5], D. Rossi[8], K.-H. Schmidt[8], H. Simon[8], L. Tassan-Got[3], H. Törnqvist[9], J. Vargas[2], B. Voss[8], H. Weick[8], and Y. Yan[3]

[1] *CEA,DAM,DIF, F-91297 Arpajon, France*
[2] *Dpt. of Particle Physics, University of Santiago de Compostela, E-15782 Santiago de Compostela, Spain*
[3] *Institut de Physique Nucléaire d'Orsay, Université Paris-Sud, CNRS, IN2P3, 91406 Orsay, France*
[4] *University of Vigo, E-36310 Vigo, Spain*
[5] *CNRS, GANIL, Bd. H. Becquerel, 14076 Caen, France*
[6] *Chalmers University of Technology, 41296 Gothenburg, Sweden*
[7] *CNRS, CENBG, F-33175 Gradignan, France*
[8] *GSI, Planckstr. 1, 64291 Darmstadt, Germany*
[9] *Institut für Kernphysik, Technische Universität Darmstadt, 64289 Darmstadt, Germany*

Abstract. SOFIA (Studies On Fission with Aladin) is a new experimental set-up dedicated to accurate measurement of fission-fragments isotopic yields. It is located at GSI, the only place to use inverse kinematics at relativistic energies in order to study the (γ,f) electromagnetic-induced fission. The SOFIA set-up is a large-acceptance magnetic spectrometer, which allows to fully identify both fission fragments in coincidence on the whole fission-fragment range. This paper will report on fission yields obtained in ^{234,235,236,238}U(γ,f) reactions.

1 Introduction

Data on fission-fragment yields are still up to now incomplete and often inaccurate despite decades of advanced investigations. Even for the most studied reaction, i.e., thermal-neutron-induced fission of ^{235}U, uncertainties attached to isotopic fission yields are mainly above 30 % [1]. This lack of high-resolution data constitutes an obstacle to the development of predictive and reliable models, and yet, accurate yields data of isotopic fission fragments are crucial for nuclear-reactor applications, to simulate the nuclear fuel burn-up associated with the accumulation of fission products in the core, the neutron flux, or, the decay heat after a core shutdown.

[a] e-mail: audrey.chatillon@cea.fr

1.1 Experimental limitations due to direct kinematics

Experimental constraints in direct kinematics, where a neutron, light charged particle or γ beam impinges on an actinide target, prevent from measuring unambiguously the mass and the nuclear charge of the fission fragments especially for the heavy group. From radiochemical methods, high-resolution data of fission-product isotopic yields, after β-decay, can be obtained but only for few nuclei [2]. From physical methods, ionic charge states and masses can be respectively deduced from the energy-loss and total energy measurements. The nuclear-charge identification is then limited to the light fragments [3–5] due to the strong fluctuation of ionic charge states preventing a clear assignment of the atomic number in the heavy group. Moreover, fission fragments having a low recoil energy, their energy straggling in dead layers limits the resolution to $\sigma \sim 1.5$ MeV [6]. Therefore the mass resolution is around 4 mass unit (FWHM). With the advent of recoil spectrometers, such as Lohengrin [7] or Hiawatha [8], high-resolution data on isobaric yields are finally reached with an uncertainty below 5%. But still, isotopic identification is limited to the light group, for example Refs. [9–11], except during experiments where β-delayed γ-rays are measured [12]. This latter method allows to measure only some isotopic yields in the heavy group, and results on isotopic yields remain partial.

1.2 An alternative approach: Surrogate reactions using inverse kinematics

A new generation of experiments based on inverse kinematics coupled with a spectrometer has been developed. Actinides are accelerated, and the compound nuclei are produced using surrogate reactions. Thanks to the kinematical boost, fission fragments are emitted at forward angles with a higher recoil energy, and their elemental distribution is measured with an improved resolution. Using this technique, isotopic yields are reachable even for the heavy group.

At GANIL, inverse kinematics is used to produce fissioning nuclei by transfer or fusion reactions of ^{238}U beam at 6 A.MeV on a ^{12}C target [13]. In this experiment, the VAMOS spectrometer [14] is used to isotopically identify one fission fragment. It is coupled to a ΔE-E telescope to select the reaction channel, i.e. the fissioning nucleus (from U to Cf), and to deduce the excitation energy. Mass distribution is obtained with a resolution below 0.8 mass unit (FWHM) and elemental distribution with a resolution around $\Delta Z/Z$=1.5% (FWHM) [15]. For the first time, isotopic yields depending on excitation energy were obtained over the whole fission-fragments range. Nevertheless, the kinematical boost is still too weak for the heavy group, and uncertainties on isotopic yields for the most produced fission fragments of this group are around 20% (statistical) and 10% (systematic).

Accurate measurement of nuclear charge over a large dynamics can only be obtained using inverse kinematics at relativistic energy. At such high energy, ions are fully stripped, thus the ionic charge obtained by energy-loss (ΔE), gives a direct measurement of the nuclear charge (Z). The main difficulty lies in the measurement of the mass number since it requires a large-scale detection system to combine the energy-loss measurement of the heavy ion with its time-of-flight (ToF) and its tracking through a magnetic field ($B\rho$). With these three observables, the mass A of the ion can be deduced using the so-called ΔE-$B\rho$-ToF method, based on the following equation: $A/Z \propto B\rho/\beta\gamma$.

This method was already applied in the 90's at GSI in pionneering experiments. In a first type of experiment, one fragment only, produced after fragmentation or fission of primary beam, was identified in mass and charge [16] thanks to the fragment separator (FRS) [17], but with a limited transmission due to the $\pm$15 mrad angular acceptance. In a second type of experiments, the FRS was used to select and identify, on an event-by-event basis, secondary radioactive actinide beams, and their fission was induced at its final focal plane by Coulomb excitation. Both fission-fragment charges were measured in coincidence with a good resolution over the whole fragment range [18, 19], but the masses were missing.

The SOFIA experiment comes within the scope of this last experiment. The breakthrough compared to previous experiments, is that the ΔE-$B\rho$-ToF method was applied at two levels: first for the identification of the secondary beam and then, for each fission fragment in coincidence. For this purpose, the experiment can only take place at the GSI facility, the only heavy-ion accelerator to provide and unambigously identify secondary actinide beams at relativistic energy up to neptunium isotopes. Moreover, the spectrometer was conceived to fit with the already existing ALADIN (A Large Acceptance DIpole magNet) magnet, located at the cave C, mandatory to get the masses of the fission fragments.
We will report on results obtained on $^{234,235,238}U(\gamma,f)$ from the first experiment in August 2012, and on $^{236}U(\gamma,f)$ from extra-shifts performed with an upgraded set-up in October 2014.

2 Experimental set-up

2.1 Radioactive secondary beam

All secondary actinide beams are produced by fragmentation of ^{238}U primary beam at 1 A.GeV in a Be target coupled with a Nb stripper. Secondary beams are selected by the FRS operated as a momentum-loss achromat spectrometer [20]. The first two dipoles make a selection in $B\rho$ and the last two dipoles, in energy loss in the intermediate focal plane degrador. Figure 1 (left) shows the 2014 set-up. Event by event, actinides are identified thanks to a Triple-MUSIC (MUltiple Sample Ionization Chamber) to get not only the ΔE measurement from the energy loss collected on the anode plane, but also the horizontal angle from the electron drift times. The time of flight is measured by two plastic scintillators, one located at S2 and the other one at cave C. The horizontal measurements are provided at S2 by the scintillator, and at cave C by a MWPC [21]. Figure 1 (right) shows the identification plot for the ^{236}U setting. It illustrates that the actinides are unambigously identified and can be selected with a graphical cut.

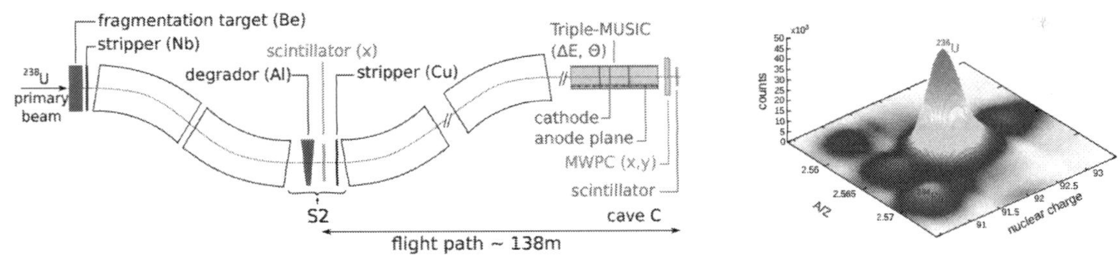

Figure 1. Left: schematic view of the FRS and the setup built in October 2014 to apply the ΔE-Bρ-ToF method for the secondary-beam identification. Right: Secondary-beam identification for the ^{236}U FRS setting.

2.2 Electromagnetic versus nuclear fission

Depending on the nuclear charge of the target and on the impact parameter, different reaction channels are opened: Nuclear reactions (for small impact parameters) and Coulomb excitations (for high-Z targets and large impact parameters). Coulomb excitation populates the giant dipole resonance (GDR) with an excitation energy around 12 MeV. The excited compound nucleus may decay via fission. As an example, electromagnetic-induced fission of ^{236}U is a surrogate reaction of 6 MeV neutron induced fission of ^{235}U. To maximize this low-energy fission induced by Coulomb excitation, high-Z targets are used. Two uranium targets and one lead target are mounted in an active target as cathode whose principle is detailed in [19], whereas the anodes are made of aluminium foils. In 2014, one more thick aluminium target dedicated to nuclear-fission measurement was mounted into the chamber. The nuclear-induced fission contribution, calculated from fission in anodes and alumium target, is subtracted at the last stage of the analysis. The number of electromagnetic-induced fissions for the different uranium isotopes is reported in Table 1.

Table 1. Statistics for the different electromagnetic-induced fission reactions.

reaction	statistics	year	reaction	statistics	year
^{234}U(γ,f)	1.9×10^5	2012	^{236}U(γ,f)	2×10^6	2014
^{235}U(γ,f)	1.1×10^6	2012	^{238}U(γ,f)	1×10^6	2012

2.3 Fission Fragments

Figure 2 shows on the left side a scheme of the dedicated set-up developed for the ALADIN magnet, and on the right side the results obtained for the nuclear charge (up) and mass (down) distributions. Fission fragments go through a Twin-MUSIC made of two identical MUSICs with a common vertical cathode. Each part has a segmented anode plane in order to provide for each fission fragment its nuclear charge from the energy-loss signals and its horizontal angle from the electron drift times. To complete the tracking, two MWPC detectors [21] located up- and downstream ALADIN give the (x,y) coordinates. Finally, the time of flight of each fragment is measured between the plastic scintillator located prior to the active target, and the time-of-flight wall [22]. The elemental distribution was measured with a resolution (FWHM) of 0.4 charge unit in 2012 and 0.35 charge unit in 2014. The mass distribution was measured in 2012 with a resolution (FWHM) of 0.6 mass unit in the light fragments group and 0.8 mass unit in the heavy fragments group. Analysis is still in progress for the data taken in 2014.

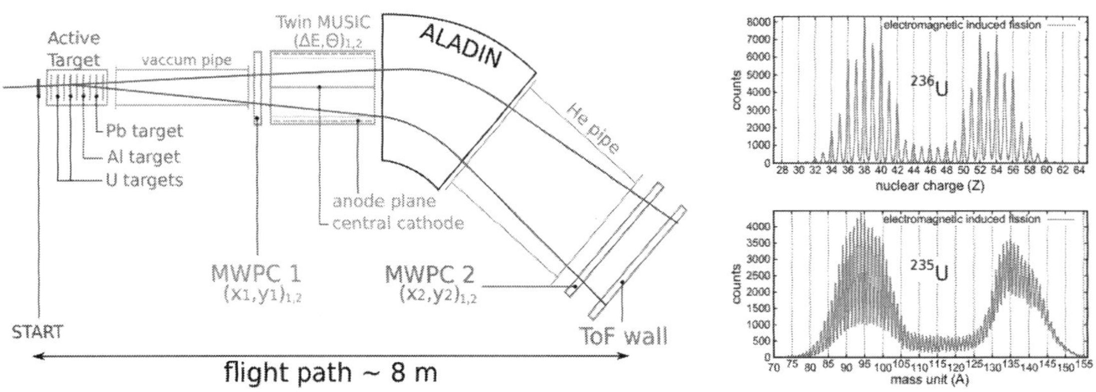

Figure 2. Left: schematic drawing of the setup built in October 2014 to apply the ΔE-Bρ-ToF method for each fission fragment. Right up: elemental distribution obtained in 2014 for ^{236}U(γ,f) reaction. Right down: mass distribution obtained in 2012 for ^{235}U(γ,f) reaction [23].

3 Elemental, isobaric and isotopic fission yields

Yields are calculated from the different distributions. Figure 3 shows the comparison of the experimental results for the elemental yields (left) and mass yields (right) obtained for the ^{234,235,236,238}U(γ,f) reactions. Error bars are reported in the figure. Except for the ^{234}U case, where the statistics is a factor 10 less than for the ^{235,236,238}U cases, uncertainties on Y(Z) and Y(A) are better than 2 % (σ) even for the symmetric valley which is less populated. Mass yields are a good probe of the influence of the nascent heavy fission-fragments shell structure in the scission process. Whereas the heavy fragment seems to be stabilized around A =138-140, the average mass of the light fragments increases with the mass of the fissionning nuclei. More precisely, the elemental yields show that the standard 1 (ST1) and standard 2 (ST2) asymmetric fission modes [24], have different weights, depending on the mass of the fissioning nucleus. The ST1 mode, characterized by a heavy fragment strongly influenced by the spherical shell gap Z=50, is predominent for ^{238}U. Then, for lighter compound nuclei, the ST2 mode, characterized by a heavy fragment strongly influenced by shell effects leading to an enhancement of Z=54, is more and more populated.

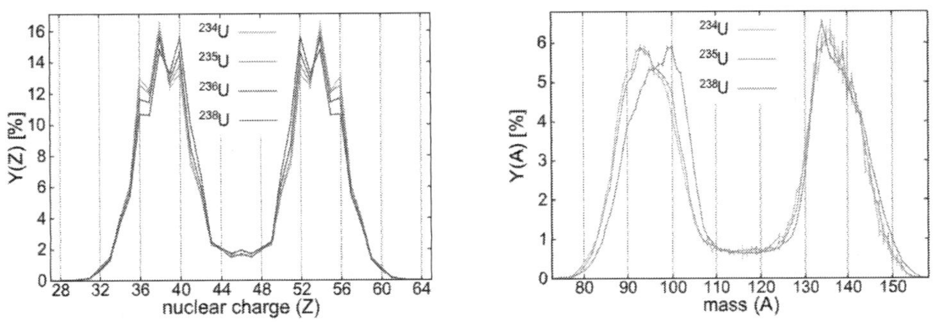

Figure 3. Elemental yields (left) and isobaric yields (right) measured for ^{234,235}U(γ,f) [23], ^{236}U(γ,f) and ^{238}U(γ,f) [25].

Figure 4 presents the comparison between the experimental elemental yields for ^{236}U(γ,f) and the JEFF-3.1.1 and BVII-1 evaluated elemental yields for ^{235}U(n,f) for 3 different neutron-beam energies.

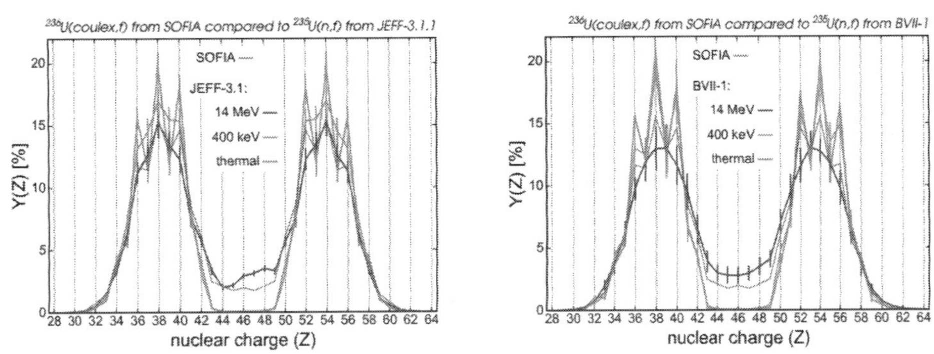

Figure 4. Comparison of elemental yields between the experimental data for ^{236}U(γ,f), and the evaluated data from JEFF-3.1.1 (left) and BVII-1 (right) for ^{235}U(n,f) reactions for 3 different neutron-beam energies.

As previously underlined, the excitation energy is not measured but is expected to be around 12 MeV in average. With increasing excitation energy, the symmetric valley is more populated, and the even-odd staggering is decreasing. The comparison of the SOFIA experimental data with evaluated data shows that the peak-over-valley ratio is consistent with an excitation energy slightly above 12 MeV, and that the JEFF-3.1.1 evaluated file is underestimating the even-odd staggering for the 400 keV incident neutrons.

Finally, isotopic fission yields obtained for $^{235}U(\gamma,f)$ [26] are presented in Fig. 5. The whole fission-fragments dynamics are covered, from Z=32 to Z=60. Error bars are reported in the figure. For the heavy fission-fragment group and at symmetry, uncertainties are below 7% for the most probable isotopes. At strong asymmetry, always for heavy fission fragments ($Z \geq 57$) uncertainties are below 20%. For the light fission-fragments group, uncertainties are mainly below 2.5%.

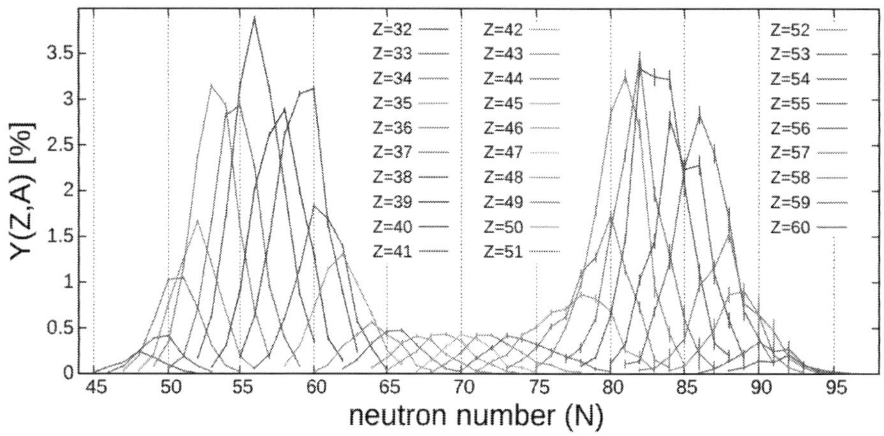

Figure 5. Isotopic yields measured for $^{235}U(\gamma,f)$ [26].

4 Summary and outlooks

Nuclear charge of heavy ions over a large dynamics, up to transuranium nuclei, can only be measured with a good accuracy using inverse kinematics at relativistic energy. This method first used in the 90's at the GSI facility to study fission yields, was recently applied again with a novel experimental set-up, the SOFIA spectrometer. For the first time, the isotopic identification was obtained for both fission fragments in coincidence and for a large range of fissioning nuclei. Fission was induced using Coulomb excitation, with a mean excitation energy around 12 MeV.

For high-statistics data, elemental yields and isobaric yields can be obtained with an uncertainty better than 2%, even for the heavy fission-fragment group, which is the most difficult but most challenging measurement. Even if the excitation energy is not measured, the comparison of the experimental elemental yields with the evaluated files shows that the JEFF-3.1.1 library underestimates the even-odd staggering for the 400 keV neutron-induced fission. Concerning the isotopic fission yields, unprecedented resolution has been reached. For the most produced fission fragments, in the light group, uncertainties are below 2.5%. It is getting a bit worse in the heavy group and at symmetry, but still, it mostly remains below 7%.

Besides the results presented in this proceeding, since the fissioning nuclei and the fission fragments are unambigously isotopically identified event by event, the mean neutron multiplicity can be obtained. This can be correlated with the total kinetic energy, which is another observable of the SOFIA experiment. Such results can be found in [26] and will be developed in a further publication [23].

The SOFIA experiment is part of the R^3B (Reactions with Relativistic Radioactive Beams) [27] research program at the FAIR (Facility for Antiproton and Ion Research) facility. Coupling the SOFIA set-up with standard R^3B detection systems, new experimental observables will be reachable (see Ref. [26] for more details). Some of these future data will be of high interest for the nuclear-reactor applications, as the neutron multiplicity per fragment and the evolution of the yields with the excitation energy. Finally, if a primary beam of ^{242}Pu could be produced, accurate fission yields for ^{238,239}Np, ^{239,240,241,242}Pu and ^{241,242}Am fissioning nuclei could be obtained.

References

[1] https://www-nds.iaea.org/exfor/endf.htm
[2] A.C Wahl *et al.*, *Proc. 2nd IAEA Symp. on physics and chemistry of fission*, 813 (1969).
[3] G. Siegert *et al.*, Phys. Lett. B **53**, 45 (1974).
[4] H.G. Clerc *et al.*, Nucl. Instrum. Methods **124**, 607 (1975).
[5] U. Quade *et al.*, Nucl. Instrum. Methods **164**, 435 (1979).
[6] S.I. Mulgin *et al.*, Nucl. Phys. A **824**, 1 (2009).
[7] E. Moll *et al.*, Nucl. Instrum. Methods **123**, 615 (1975).
[8] G. Diiorio and B.W. Wehring, Nucl. Instrum. Methods **147**, 487 (1977).
[9] W. Lang *et al.*, Nucl. Phys. A **345**, 34 (1980).
[10] U. Quade *et al.*, Nucl. Phys. A **487**, 1 (1988).
[11] D. Rochman *et al.*, Nucl. Phys. A **710**, 3 (2002).
[12] A. Bail *et al.*, Phys. Rev. C **84**, 034605 (2011).
[13] C. Rodríguez-Tajes *et. al*, Phys. Rev. C **89**, 024614 (2014).
[14] H. Savajols *et al.*, Nucl. Instrum. Methods B **204**, 146 (2003).
[15] M. Caamaño, O. Delaune *et al.*, Phys. Rev. C **88**, 024605 (2013).
[16] P. Armbruster *et al.*, Z Phys. A **355**, 191 (1996).
[17] H. Geissel *et al.*, Nucl. Instrum. Methods B **70**, 286 (1992).
[18] K.-H. Schmidt *et al.*, Phys. Lett. B **325**, 313 (1994).
[19] K.-H. Schmidt *et al.*, Nucl. Phys. A **665**, 221 (2000).
[20] K.-H. Schmidt *et al.*, Nucl. Instrum. Methods A **260**, 287 (1987).
[21] ALICE collaboration, J. Phys. G: Nucl. Part. Phys. **30**, 1517 (2004).
[22] A. Ebran *et al.*, Nucl. Instrum. Methods A **728**, 40 (2013).
[23] J.-F. Martin *et al.*, in preparation
[24] U. Brosa, Phys. Rev. C **38**, 1944 (1988).
[25] E. Pellereau *et al.*, in preparation
[26] J.-F. Martin *et al.*, *Proc. Perspectives on Nuclear Data in the next Decade Workshop* to be published in Eur. Phys. Jour. Web of Conferences
[27] https://www-alt.gsi.de/forschung/kp/kp2/collaborations/R3B/R3B-TP-Dec05.pdf

EPJ Web of Conferences **111**, 08002 (2016)
DOI: 10.1051/epjconf/201611108002

^{233}U mass yield measurements around and within the symmetry region with the ILL Lohengrin spectrometer

A. Chebboubi[1], G. Kessedjian[1], C. Sage[1,a], D. Bernard[2], A. Blanc[3], H. Faust[3], U. Köster[3], O. Litaize[2], P. Mutti[3], and O. Serot[2]

[1] *LPSC, Université Grenoble-Alpes, CNRS/IN2P3, F-38026 Grenoble Cedex, France*
[2] *CEA, DEN, DER, SPRC, Cadarache, Physics Studies Laboratory, F-13108 Saint-Paul-lès-Durance, France*
[3] *Institut Laue-Langevin, F-38042 Grenoble Cedex 9, France*

Abstract. The study of fission yields has a major impact on the characterization and understanding of the fission process and is mandatory for reactor applications. The LPSC in collaboration with ILL and CEA has developed a measurement program on fission fragment distributions at the Lohengrin spectrometer of the ILL, with a special focus on the masses constituting the heavy peak. We will present in this paper our measurement of the very low fission yields in the symmetry mass region and the heavy mass wing of the distribution for ^{233}U thermal neutron induced fission. The difficulty due to the strong contamination by other masses with much higher yields will be addressed in the form of a new analysis method featuring the required contaminant correction. The apparition of structures in the kinetic energy distributions and possible interpretations will be discussed, such as a possible evidence of fission modes.

1 Introduction

The accurate knowledge of the fission data in the actinide region is important for studies of innovative nuclear reactor concepts. Today, in the framework of nuclear data evaluation, fission models are necessary to inscrease the consistency and the precision of the libraries. For instance, post neutron fission yields are actually needed in the current and innovative fuel cycles for the calculation of the inventory and the radiotoxicity of the spent fuel, and the estimation of the residual power after shutdown.

Moreover, fission yield measurements supply experimental data to put constraints on fission models and improve their predictive power. In this context, since 2007, various experiments from our collaboration have been performed to investigate fission yields at the Lohengrin spectrometer of the ILL with a special focus on the heavy mass and symmetry region, where can be found an inconsistency between models or evaluations and the scarce experimental data.

Recent measurements from [1, 2] of ^{233}U fission yields in the heavy region with the Lohengrin spectrometer showed an important discrepancy near the symmetry region when compared to evaluations. This is shown in Fig. 1 where these measurements are displayed with the JEFF-3.1.1 evaluations [3]. Firstly, our work presented here constitutes a deep investigation around and within the symmetry region to explain this disagreement and get accurate fission yields, which is a key element

[a] e-mail: sage@lpsc.in2p3.fr

to ensure the self-normalisation of our data [1, 2]. A second important goal of our work here is to take this opportunity to study the structures in the fission fragment kinetic energy distributions and use symmetric fission as a laboratory to test fission models.

2 Experimental setup: the Lohengrin spectrometer

The Lohengrin mass spectrometer [4] is a nuclear physics instrument from the ILL research reactor facility which allows to study fragment distributions from thermal neutron induced fission with a very good resolution. A fissile actinide target is placed close to the reactor core, in a thermal neutron flux reaching 5×10^{14} neutron.cm^{-2}.s^{-1}.

Fission fragments emerge from the target with an ionic charge distributed around an average ionic charge state of about 21. Those fragments that are emitted along the beam tube axis undergo a horizontal deflection in a magnetic field, directly followed by a vertical deflection in an electric field. These combined fields separate ions according to their A/q and E/q ratios, with A, q and E the mass, ionic charge state and kinetic energy of the ions respectively.

At the spectrometer exit, different detection systems can be installed, such as an ionisation chamber for mass yield measurements, or Ge clovers that are used with an additional magnet whose aim is to focus the ion beam. A schematic view of the spectrometer is shown in Fig 1.

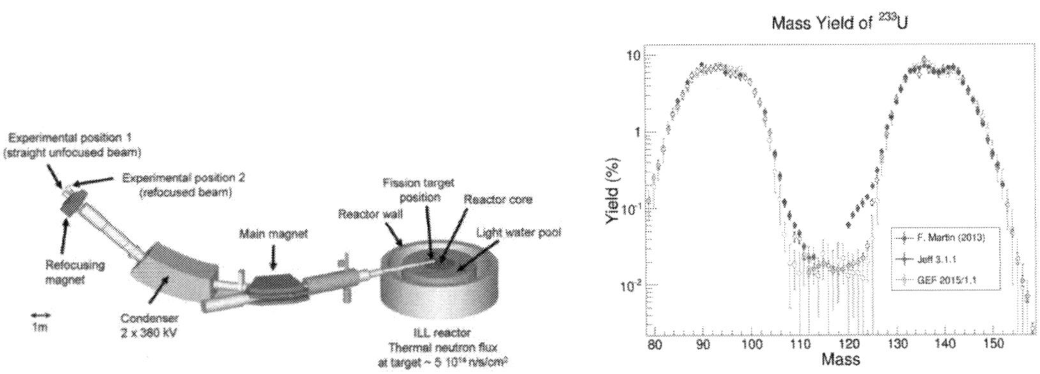

Figure 1. Schematic overview of the Lohengrin spectrometer at the ILL (left). ^{233}U(n_{th}, f) fission yields from F. Martin *et al.*[2] (right).

3 Yield measurements around and within the symmetry region

3.1 Absolute mass yields in the heavy region

High yields in the heavy region are obtained after an integration over the kinetic energy and the ionic charge distributions of the count rates measured with the ionisation chamber. A new measurement method and consequent analysis path have been developed and are detailed in Ref [1, 2]. Among the special features of this method are the self-normalisation of our data and the calculation of the experimental covariance matrices. Provided that all the heavy mass rates are measured, it is possible to self-normalise the data by defining to 100 % the sum of the whole heavy peak yields. As a consequence, these new measurements are independent from another experiment or assessment and may be compared directly with the existing data and evaluations. The results for $^{233}U(n_{th}, f)$ are shown in

Fig. 1 and present a clear difference of the yields on the way down to the symmetry region. The goal of this work is the complete analysis of the ionic charge and the kinetic energy distributions which present structures that had not been seen in past works [5].

3.2 Contamination highlighting

In the descent to the symmetry region ($A < 130$), a second component appears at low kinetic energy. We previously observed this component to be sensitive to the pressure of the spectrometer. The evolution of this effect with pressure is shown in Fig. 2. This phenomenon is explained as a consequence of ionic charge exchange between the fission fragment and the residual gas of the Lohengrin vacuum. If we consider that the main magnet selected the fragments with mass A_0 and A_1 with the condition $\frac{A_0 v_0}{q_0} = \frac{A_1 v_1}{q_1}$ followed by a charge change for the second fragment $q_1 \rightarrow q_1^{'}$, then the condenser will select the kinetic energies following the equation :

$$\frac{E_0}{q_0} = \frac{E_1}{q_1^{'}} \Leftrightarrow \frac{A_0 v_0^2}{q_0} = \frac{A_1 v_1^2}{q_1^{'}} \tag{1}$$

Since this contaminant will appear at the same energy in the ionisation chamber, we get the following condition on the charge : $E_1 = E_0 \Leftrightarrow q_1^{'} = q_0$. Consequently the contaminant mass follows the relation $A_1 = \left(\frac{v_0}{v_1}\right)^2 A_0$. If we compare to the initial condition from the magnet we get the final relation for the contaminant :

$$A_1 = \left(\frac{q_1}{q_0}\right)^2 A_0 \tag{2}$$

This hypothesis was investigated with the building of an indicator based on equation 2. Fig. 2 shows the indicator results for mass 124. The masses with discrepant count rates out of the distribution shaped by the other ones are typically the masses with the highest contamination indicator values. This indicator was validated by gamma spectrometry, where the presence of the expected contaminant was proven.

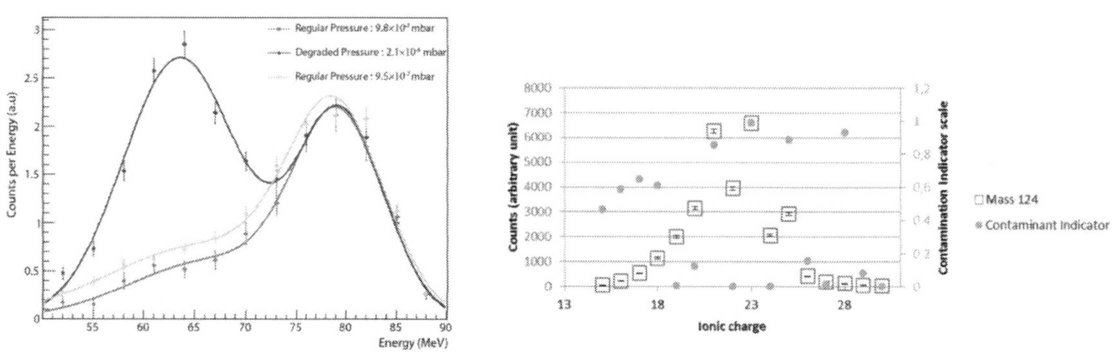

Figure 2. Influence of the condenser pressure on the kinetic energy distribution (left). Contamination indicator for the mass 124 as a function of ionic charge (right).

3.3 Analysis method for the contamination correction

The contaminant subtraction was achieved on the kinetic energy distributions $P(E_k)$ following equation 3.

$$P(E_k)^{sym} = P(E_k)^{meas} - \sum_{cont} P_{qq'} \times P_{ind} \times \frac{Y(A_{cont})}{Y(A_{sym})} \times P(E_k)^{cont} \quad (3)$$

where $P_{qq'}$ represents the measured charge changing probabilities using γ spectroscopy. These probabilities have the same order of magnitude for several nuclei. In the following, it will be supposed identical for all isotopes : $P_{q\to q+1} \sim 0.02$, $P_{q\to q+2} \sim 0.0013$ and $P_{q\to q-1} \sim 0.02$.

P_{ind} is linked to the presence of the contaminant in the ionisation chamber and is based on the indicator derived from equation 2. The correction factor is weighted by the yield ratio of the contaminant mass (with a higher yield) over the measured symmetry mass, multiplied by the kinetic energy distribution of the contaminant mass $P(E_k)^{cont}$.

Figure 3 shows the results of such a correction for the kinetic energy distributions of masses 116 and 118. We can notice that the presence of two components in the distribution remains after the correction, and this is the case for all masses around and within the symmetry region. Moreover, even if we consider $P_{qq'}$ as free parameters, it is not possible to fully suppress the low component.

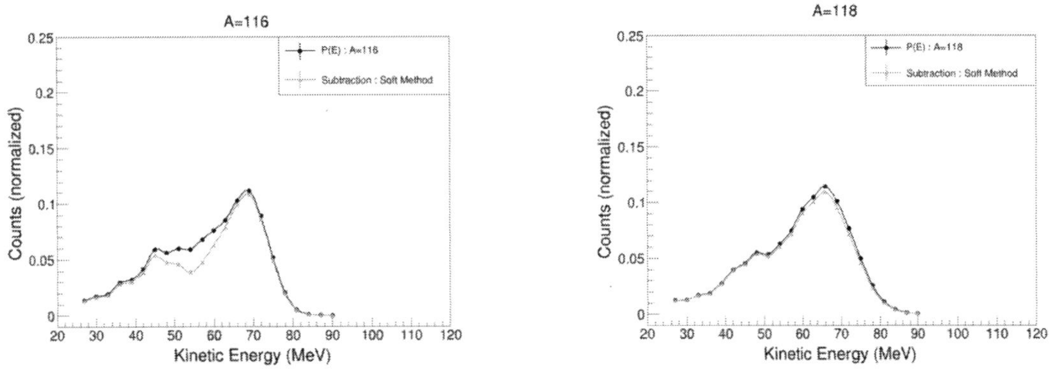

Figure 3. Kinetic energy distributions before (black) and after (blue) contaminant correction for masses 116 (left) and 118 (right).

4 Preliminary results and discussion

Figure 4 shows the preliminary results for our symmetry yield measurements. The yields are shown in the case of no correction (black dots), corrected (red dots), but also in the hypothesis where only one of the two components in the kinetic energy distribution is considered to estimate the yield (blue and green dots). The disagreement of our corrected values with the JEFF-3.1.1 evaluation is still present. The mean kinetic energy for the two components and for the whole kinetic energy distribution is also shown on the right figure, after correction from the energy loss in the target using a set of reference data (orange dots) with a very thin target of 30 μg.cm^{-2} without any cover.

An important result here is the apparition of two modes after effective decontamination. We remind that the modality of the fission process has already been observed for the Fermium region [6] and in the case of fast neutron induced fission [7]. In the ponctual approximation of the Coulomb repulsion, the total kinetic energy of fragments is related to the scission configuration by the following equation :

$$TKE = E_L + E_H \cong E_L\left(1 + \frac{m_L}{m_H}\right) = \frac{Z_L Z_H e^2}{4\pi\epsilon_0 d} \tag{4}$$

with m_i and Z_i the mass and nuclear charge of each fragment and d the distance between the two centers of charge of the fragments at the scission point. The results of this calculation, considering additional hypothesis such as the Unchanged Charge Density (UCD), are shown in Tab. 1 and compared to estimations from the Brosa approach [8] and from microscopic model calculations [9]. Our values for the low and high kinetic energy components are rather consistent with the theoretical predictions from the two models considered here. These results tend to prove the modality of the $^{233}U(n_{th}, f)$ fission process.

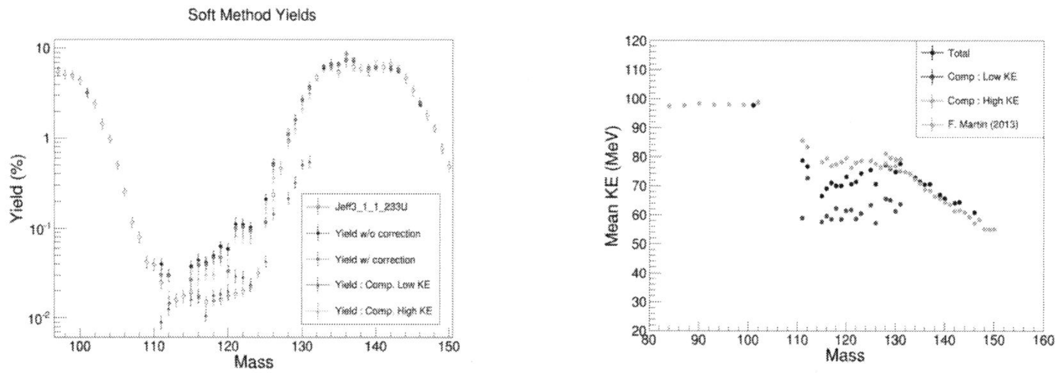

Figure 4. Mass yields per component in kinetic energy distributions compared to the Jeff-3.1.1 library (left). Mean kinetic energy of the extracted modes after energy loss corrections in the target (right).

Table 1. Comparison of scission configurations d, between our work and different fission models. Estimation of neutron emission $\overline{\nu}$ and comparison with dedicated measurement (Nishio [10]).

Component	This work	Brosa	Microscopic	Configuration	This work	Nishio
Low KE	$d \cong 24\ fm$	$d \cong 21\ fm$	$d \cong 20\ fm$	Symmetric	$\overline{\nu} \cong 7.0$	$\overline{\nu} \cong 8.0$
High KE	$d \cong 19\ fm$	$d \cong 18\ fm$	$d \cong 16\ fm$	Asymmetric	$\overline{\nu} \cong 3.6$	$\overline{\nu} \cong 4.0$

5 Conclusion and perspectives

This work presents an investigation of the $^{233}U(n_{th}, f)$ yields in the symmetry region, with a focus on the measurement of the fragment kinetic energy distributions. After a proven contamination presence and its necessary correction, two components in the distributions are observed for all of the masses around and within the symmetry but not int the high yield region.

A further study relative to possible biases in the correction process and an estimation of the uncertainties are still needed to conclude on the direct evidence of the modality of thermal neutron induced fission in the region of Uraniums. The development of a Time Of Flight setup is under development to get a triple coincidence between a double Frisch grid ionisation chamber and the velocity of the incoming fragments.

References

[1] F. Martin *et al.*, *Proc. 2nd ANIMMA Conference*, June 2011, Ghent
[2] F. Martin *et al.*, *Nucl. Data Sheets* **119**, 328-330 (2014)
[3] The JEFF-3.1.1 nuclear data library, OECED, NEA, JEFF Report 21, ISBN 92-64-02314-3 (2006)
[4] P. Armbruster *et al.*, *Nucl. Instrum. Methods* **139** 213 (1976)
[5] R. Brissot *et al.*, *Physics and Chemistry of Fission 1979*, IAEA, 99-110 (1980)
[6] E.K. Hulet *et al.*, *Phys. Rev. C* **40**, 770 (1989)
[7] E. Pfeiffer *et al.*, *Z. Phys.* **240**, 403 (1970)
[8] U. Brosa *et al.*, *Phys. Reports* **197**, 167 (1990)
[9] H. Goutte *et al.*, *Phys. Rev. C* **71**, 024316 (2005)
[10] K. Nishio *et al.*, J. *Nucl. Sc. Tech.* **35**, 631 (1998)

EPJ Web of Conferences **111**, 08003 (2016)
DOI: 10.1051/epjconf/201611108003

Isomeric ratio measurements with the ILL LOHENGRIN spectrometer

A. Chebboubi[1], G. Kessedjian[1,a], O. Litaize[2], O. Serot[2], H. Faust[3], D. Bernard[2], A. Blanc[3], U. Köster[3], O. Méplan[1], P. Mutti[3], and C. Sage[1]

[1]*LPSC, Université Grenoble-Alpes, CNRS/IN2P3, F-38026 Grenoble Cedex, France*
[2]*CEA, DEN, DER, SPRC, Cadarache, Physics Studies Laboratory, F-13108 Saint-Paul-lès-Durance, France*
[3]*Institut Laue-Langevin, F-38042 Grenoble Cedex 9, France*

Abstract. The modelling of γ heating and neutron damage inside a nuclear reactor is essential to design the next generation of nuclear reactors. The determination of the fission fragment momentum is a key element to perform accurate calculations of the γ heating. One way to assess this information is to look at the isomeric ratio of different nuclei. According to the lifetime of the isomeric state, different experimental techniques were developed at the LOHENGRIN spectrometer. A focus on the measurement of isomeric ratios of ^{136}I in neutron induced fission of ^{241}Pu is presented. A discussion with the current assumptions used in the evaluation process for isomeric ratio is also shown.

1 Introduction

Seven decades ago, nuclear fission was discovered [1, 2]. Even though the understanding of fission has improved over all these decades, some aspects such as the fission fragment angular momentum were barely studied. Today, this quantity is major for the determination of the prompt γ spectra which are essential in the calculation of γ heating and damage of nuclear reactor components [3]. In order to design the next generation of nuclear reactors with a higher level of confidence, a high accuracy of the prompt γ spectra is required. One probe to study fission fragment angular momentum is to look at isomeric ratio (IR) [4–6]. In this work, a new experimental method is presented to measure IRs at the LOHENGRIN spectrometer [7]. The validation of the assumptions used in the modelling of fission fragment angular momentum in the evaluation process is also discussed.

2 Experimental setup and analysis path

The measurement of IRs were done at the LOHENGRIN mass separator of the Institut Laue-Langevin (ILL). The different parts of the LOHENGRIN spectrometer are shown on Fig. 1 (left). The fissile isotope target is placed near the core of the reactor under a thermal neutron flux of $\approx 5\times10^{14}$ n.cm^{-2}.s^{-1}. The combination of a magnetic and electric field allows a separation of the produced fission fragments according to their ratios mass A over ionic charge q and kinetic energy E_k over ionic charge. Finally a refocusing magnet [8] permits to increase the particle density at the focal position 2.

[a]e-mail: kessedjian@lpsc.in2p3.fr

According to the lifetime of the isomeric state, different experimental methods were developed. Ref. [9] presents the case of μs isomers whereas in this article a focus on min isomers is shown.

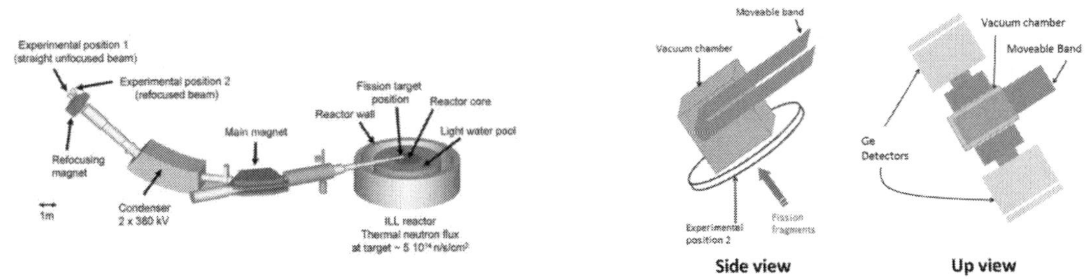

Figure 1. Scheme of the LOHENGRIN spectrometer (left). Detection system placed at the focal position 2 (right).

2.1 Beamcut method

The incoming fission fragments are implemented into a tape surrounded by two clover detectors consisting of four high purity germanium crystals, placed at the focal position 2 of the spectrometer. The band can be moved after each measurement in order to reduce the ambient background coming from the impure selection of the LOHENGRIN spectrometer. Figure 1 (right) shows the scheme of the detection system.

However, for certain nuclei, the detected γ-ray cascade is the same for the ground and the isomeric states. To remove the degeneration, for a given setting of the mass spectrometer, two measurements are performed. When the beam is on, the γ-rays are fed by both states, whereas when the beam is off, because of the difference in the lifetime between states, the γ-rays are fed mostly by one state. Figure 2 summarizes the beamcut method and its principle.

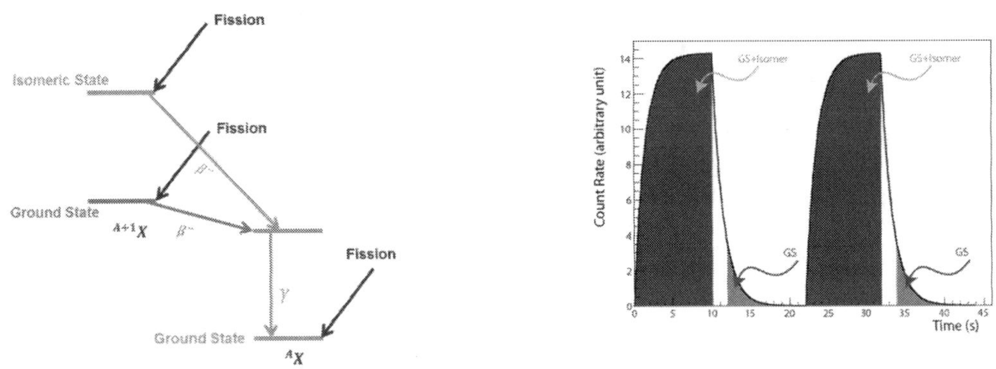

Figure 2. Principle of the beamcut method. When a γ-ray is fed by different states, the combination of two measurements permits to determine the feeding from both states. The blue area is associated to the beam on and the red area corresponds to the beam off.

To test this new procedure, the IR of ^{136}I in neutron induced fission of ^{241}Pu was measured. Indeed, some γ-rays of the isomeric state are contaminated by the isomeric state of ^{136}Xe. A double analysis, one using independent γ-rays and the other using the common γ-rays through the beamcut method was done. Table 1 shows the different γ-rays used in both methods.

Table 1. List of γ-rays used in both methods.

Isotope	Independent γ-rays (keV)	Common γ-rays (keV)
^{136m}I	369.8 / 750	197.3 / 381.4 / 1313
^{136}I	1321.08 / 2289.6 / 2414.6 / 2634.2	1313
^{136m}Xe	∅	197.3 / 381.4 / 1313

In this work, the isomeric ratio is defined as following:

$$IR = \frac{\eta(^{m}X)}{\eta(^{m}X) + \eta(^{gs}X)} \quad (1)$$

with $\eta(X)$ the fission rate of the isomeric state (m) or the ground state (gs) respectively. The determination of the IRs from the γ-ray count rate involved several corrections. The current analysis path was presented in Ref. [9]. IRs were measured as a function of the fission fragment ionic charge because the final objective is to determine the isomeric yield. Figure 3 shows the result of both analysis and the correlation matrix calculated for the mean IR. Both methods are in good agreement, which permits to validate the beamcut technique.

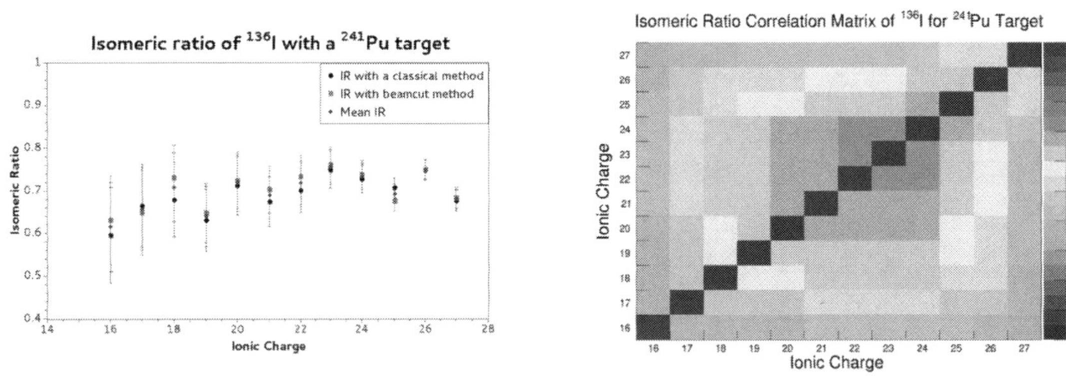

Figure 3. Validation of the beamcut method to determine the IR as a function of the ionic charge (left). Experimental correlation matrix which shows the weight of the systematic uncertainty over the total uncertainty.

3 Results and discussion

Currently, the isomeric ratio determination in the evaluation process is achieved by using the approach developed by Madland and England [10]. Their assumptions are the following : the isomeric ratio depends only on the spin of the isomeric and ground states. Also, all fission fragments have the same spin distribution characterized by a spin-cutoff $J_{rms} = 7.5\ \hbar$. Figure 4 presents the comparison of our result, corresponding to the averaged IR over ionic charge, with previous experiments performed with different fissioning systems, the JEFF evaluation based on the England-Madland hypothesis and a calculation from the code GEF [11].

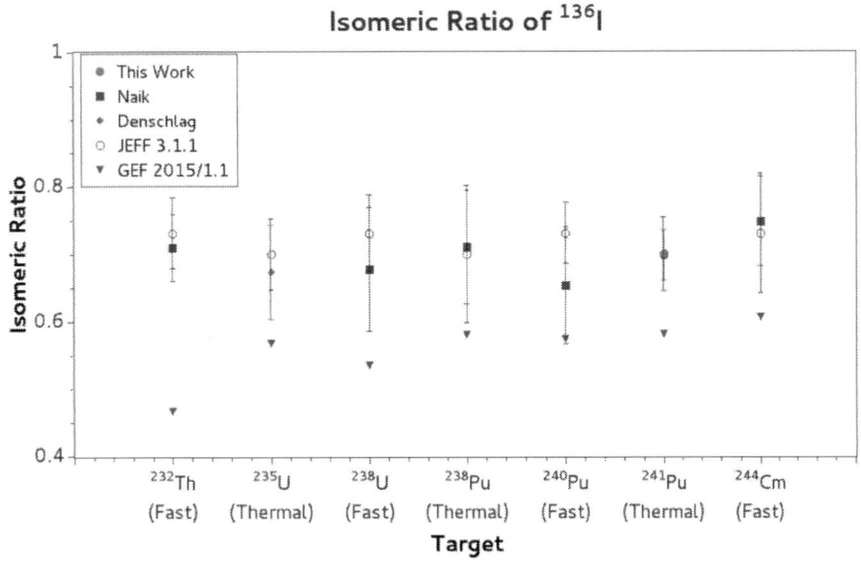

Figure 4. Comparison of the isomeric ratio of ^{136}I with other experimental measurements, the JEFF-3.1.1 evaluation and a GEF calculation. The IR measured doesn't seem to be sensitive to the fissioning system. Also, the Madland-England modelling involved in the evaluation process is in good agreement with experimental values.

The first observation is that the Madland-England modelling seems to be in agreement with all experimental measurements. The second observation is that the isomeric ratio of ^{136}I doesn't seem to be sensitive to the fissioning system, which is in disagreement with the prediction of the GEF calculation.

To go further, an investigation of the isomeric ratio of ^{132}Sn and the extraction of its angular momentum distribution was performed. To determine the spin-cutoff of this nucleus, we used a γ de-excitation code : FIFRELIN [12, 13]. For each state (E^*, J^π) an IR was calculated. More details about the analysis are presented in Ref [9]. Then an average according to the most commonly used spin distribution was performed :

$$IR_{calc}(E^*) = \sum_J \sum_\pi P(\pi)P(J)IR_{calc}(E^*, J^\pi) \tag{2}$$

with $P(\pi) = \frac{1}{2}$ and $P(J)$ following [10] :

$$P(J) \propto (2J + 1)\exp\left(-\frac{(J + 1/2)^2}{J_{rms}^2}\right) \tag{3}$$

with J_{rms} a free parameter also called spin cutoff. Its value is determined through a Bayesian comparison between experimental isomeric ratios and calculated ones. Finally we extracted the mean $\overline{J_{rms}}$ corresponding to the mean isomeric ratio $\overline{IR}$. The result is compared with a GEF calculation and the Madland-England modelling in Tab. 2. This work shows the limit of the current modelling of the isomeric ratio in the evaluation process. Indeed if all fission fragments have the same spin-cutoff, then in the framework of the Madland-England modelling, the isomeric ratio of ^{132}Sn should be equal to

Table 2. Extraction of a spin-cutoff from an isomeric ratio.

^{132}Sn	$\overline{IR}$	$\overline{J_{rms}}$
This work	0.054 ± 0.006	4.7 ± 0.2
GEF		7.1 ± 0.2
Madland-England a	0.67 ± 0.03	7.5 ± 0.5
Madland-England b	0.054 ± 0.006	2.6 ± 0.1

0.67 (Madl. a) which is incompatible with our measurement. On the contrary if the isomeric ratio is fixed to the experimental value, then the spin-cutoff of ^{132}Sn should be equal to 2.6 (Madl. b) which is incompatible with the extraction done with the FIFRELIN code. Finally, the spin-cutoff calculated by GEF is also in disagreement with the proposed value.

These limits show how necessary it is to build an IR database in order to test and validate the models involved in the evaluation process. This also questions on the mechanism of angular momentum generation [6, 14–17] which is the weakness of the nuclear data evaluation.

4 Conclusion

In this work, a new experimental method was developed in order to extract the isomeric ratio for nuclei where the identification is difficult. The beamcut method was validated on the test case of ^{136}I with a double analysis. The limits of the current modelling of isomeric ratios and angular momentum distribution involved in the evaluation process were pointed out. New measurements are planned in order to bring new information on the mechanism of angular momentum generation of fission fragments. A better understanding of the mechanism would permit to get more accurate nuclear data and satisfy the recommendation of the nuclear data evaluation community.

References

[1] O. Hahn and F. Strassmann, *Naturwissenschaften* **27** 11 (1939)
[2] L.Meitner and O.R Frisch, *Nature* **143** 239 (1939)
[3] G. Rimpault *et al.*, *Phys. Proc.* **31** 3 (2012)
[4] H. Faust, J. *Korean Phys. Soc.* **59** 879 (2011)
[5] R. Vandenbosch and J. Huizenga, Nuclear Fission (Academic Press, New York and London, 1973)
[6] F. Gönnenwein, I. Tsekhanovich and V. Rubchenya, *Int. J. Mod. Phys. E* **16** 410 (2007)
[7] P. Armbruster *et al.*, *Nucl. Instrum. Methods* **139** 213 (1976)
[8] G. Fioni *et al.*, *Nucl. Instrum. Meth. A* **332** 175 (1993)
[9] A. Chebboubi *et al.*, *14th NRM conference (CERN Proceedings)*, to be published.
[10] D.G. Madland and T.R. England, *Nucl. Sci. Eng.* **64** 859 (1977)
[11] K.-H Schmidt *et al.*, *Tech. Rep. NEA Data Bank* (2014)
[12] O. Litaize and O. Serot, *Phys. Rev. C* **82** 054616 (2010)
[13] D. Regnier, O. Litaize and O. Serot, *Comput. Phys. Comm.*, to be published.
[14] J.R. Nix and W.J. Swiatecki, *Nucl. Phys.* **71** 1 (1965)
[15] J. Rasmussen, W. Nörenberg and H. Mang, *Nucl. Phys. A* **136** 465 (1969
[16] M. Zielinska-Pfabé and K. Dietrich, *Phys. Lett. B* **49** 123 (1974)
[17] L. Bonneau, P. Quentin and I.N. Mokhailov, *Phys. Rev. C* **75** 064313 (2007)

EPJ Web of Conferences **111**, 08004 (2016)
DOI: 10.1051/epjconf/201611108004

Revision of the JENDL FP Fission Yield Data

Jun-ichi Katakura[1,a], Futoshi Minato[2], and Kazuya Ohgama[3]

[1] *Nagaoka University of Technology, 1603-1 Kamitomioka, Nagaoka, Niigata, Japan*
[2] *Japan Atomic Energy Agency, Nuclear Science and Engineering Center, 2-4, Shirakata, Tokai-mura, Ibaraki, Japan*
[3] *Japan Atomic Energy Agency, Advanced Fast Reactor Cycle System R&D Center, 4002, Narita, O-arai, Ibaraki, Japan*

Abstract. Some fission yields data of JENDL FP Fission Yields Data File 2011 (JENDL/FPY-2011) revealed inadequacies when applied to delayed neutron related subjects. The sensitivity analyses of decay heat summation calculations also showed some problems. From these results the fission yields of JENDL/FPY-2011 have been revised. The present report describes the revision of the yield data by emphasizing the sensitivity analyses.

1 Introduction

The JENDL/FPY-2011 file [1] was released in 2011. After the release of the file, several feedbacks from the users of the file were reported. One of them is from delayed neutron application [2] where it was shown that the yield of ^{86}Ge played significantly. The sensitivity analyses of decay heat summation calculation also showed inadequacies of some fission yield data.

The sensitivity analysis method of decay heat summation calculation has been developed [3] for the estimation of the uncertainties emerging from nuclear data used. In the course of the sensitivity analyses of fast neutron fission, the comparison of the uncertainties between fast and thermal neutron fissions was performed and it was found that the uncertainties of the calculated decay heat for thermal neutron fission were unexpectedly larger than those for fast neutron fission at short cooling times after a fission burst when the JENDL/FPY-2011 file was employed. During the search for the reason, it was found that the fission yields of the nuclides contributing to the large uncertainties of the decay heat calculations seemed to be too large. The yield data of the nuclides which showed unexpectedly large values were revised.

In the present paper, the comparison of the uncertainties between the thermal and the fast neutron fissions and the revision of the JENDL yield data are described.

2 Comparison of decay heat uncertainties for thermal and fast neutron fission

The uncertainties of decay heat summation calculations were derived for thermal and fast neutron fissions using JENDL FP Decay Data File 2011 (JENDL/FPD-2011) and JENDL/FPY-2011. The

[a] e-mail: j_katakura@vos.nagaokaut.ac.jp

total uncertainties of the calculated decay heat values at short cooling times after a fission burst are shown in Figure 1. In this figure, the results of ^{235}U and ^{239}Pu fissions are shown. In the case of ^{239}Pu

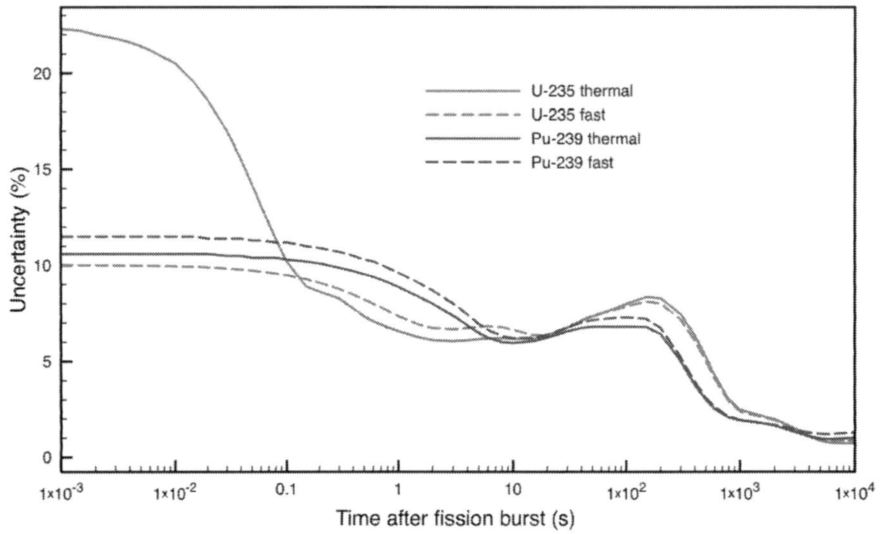

Figure 1. Uncertainties of decay heat summation calculations for thermal and fast neutron fissions using JENDL/FPD-2011 and JENDL/FPY-2011.

fission, the uncertainties of the fast neutron fission and the thermal neutron fission are not so different from each other. However, in the case of ^{235}U fission, the uncertainties of the thermal neutron fission are much larger than those of the fast neutron fission at shorter cooling times than 0.1 s. The nuclides contributing to the uncertainties at the short cooling times are shown in Figure 2 as a time depending sensitivity. On the left-hand side, the nuclides with large sensitivity for ^{235}U thermal neutron fission are indicated. On the right-hand side, those for ^{235}U fast neutron fission are indicated. As seen in the figures, the nuclide of ^{86}Ge has large contribution to the thermal fission but no significant contribution to the fast neutron fission. The contributions of ^{88}As and ^{100}Rb are not significant but sizable contribution to the uncertainties of the ^{235}U thermal neutron fission and no contribution to

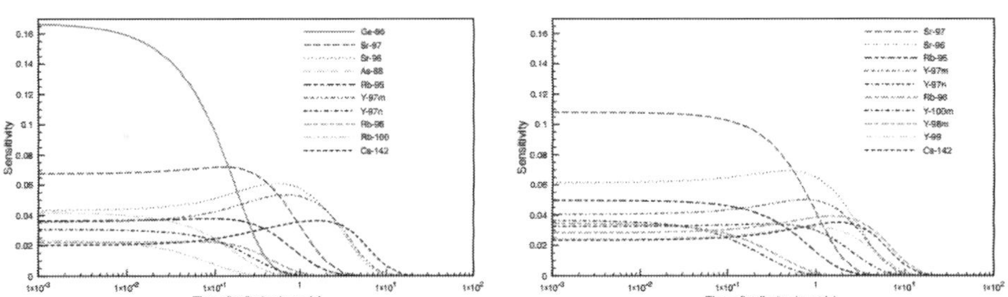

Figure 2. Important sensitivity coefficients at short cooling times after fission.

those of the fast neutron fission. The independent fission yields of those nuclides for the thermal neutron fission of ^{235}U in JENDL/FPY-2011 are listed in Table 1. As the fission yields depend on

Table 1. Comparison of independent fission yields of JENDL/FPY-2011.

Nuclide	Thermal	Fast	Ratio (Fast/Termal)
^{86}Ge	6.28×10^{-3}	5.41×10^{-6}	8.61×10^{-4}
^{88}As	1.24×10^{-3}	2.01×10^{-5}	1.62×10^{-2}
^{100}Rb	3.47×10^{-4}	1.55×10^{-7}	4.45×10^{-4}

incident neutron energy, it is true that the yields are different from each other for thermal and fast fission. However, the differences shown in Table 1 seem to be too large. The charge distribution of the same masses as those nuclides, that is, A = 86, 88 and 100 are shown in Figure 3. From Figure 3

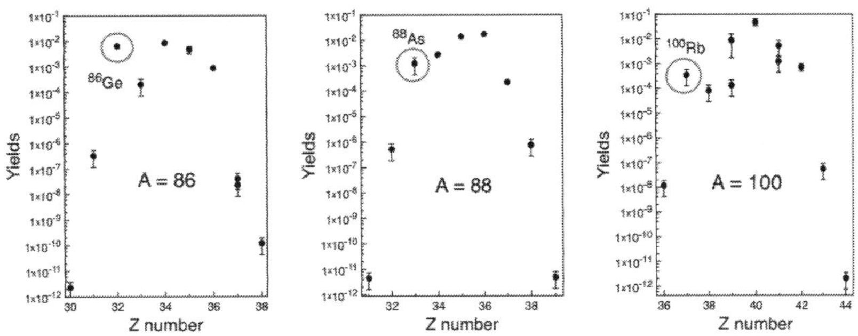

Figure 3. Charge distribution of fission yields for A = 86, 88 and 100.

the yields corresponding to ^{86}Ge (Z = 32 for A = 86), ^{88}As (Z = 33 for A = 88) and ^{100}Rb (Z = 37 for A = 100) look like abnormally large comparing with other yields.

3 Fission yields of JENDL/FPY

The fission yield data of JENEL/FPY-2011 are based on the evaluation of ENDF/B-VI [4]. Although some modifications were carried out for the compilation of the JENDL file because of the differences of decay chains and isomers adopted in the JENDL and ENDF files, basically the independent yields of the JENDL file were taken from those of ENDF/B-VI. As for the ^{86}Ge case, the problem of the ENDF yield data was claimed previously [5]. And recent analysis of delayed neutron also pointed out that the yield data of ^{86}Ge of JENDL/FPY-2011 severely affect the reactivity analysis of a reactor [2]. The independent fission yields of JENDL and ENDF are listed in Table 2. As pointed out above, the yield value of the JENDL file is similar to that of ENDF.

As the ENDF/B-VI yield data are based on the reported ones in the report of LA-UR-94-3106 [4] which contains detailed data sources used for the evaluation work. In the appendix A of the report, the measured cumulative yield of ^{86}Ge by ^{235}U thermal neutron fission is listed to be 0.87 ± 0.12 % as listed in Figure 4. The red line shows the yield of ^{86}Ge with the reference of 90rud1. The reference designated as 90rud1 is Radiochimica Acta 49, 155-191 (1990) [6]. However, the value of 0.87 ± 0.12 for ^{86}Ge is not found in the original paper. There is no yield data of ^{86}Ge in the paper. The isotope

Table 2. Independent yields of ^{86}Ge in ENDF/B-VI and JENDL/FPY.

ENDF/B-VI	JENDL/FPY-2011
6.28666×10^{-3}	6.27711×10^{-3}

```
86 as g   u235t   ms   fi   ul   0.0000000  0.0   1.4000000  1.2000000    86              u235t   100.0000000   75cle2
86 as g   u235t   ms   in        0.0200072 50.0   0.0200000  0.0100000     0                        0.0000000   79clel
86 as g   u235t   ms   fi   su   0.0000000  0.0   4.3000000  0.9000000    86              u235t   100.0000000   76woll
86 as g   u235t   sp   fi        0.0196587 50.0   1.0000000  0.5000000    86              u235t   100.0000000   80lanl
86 as g   u235t   cm   in        0.0000000  0.0   0.0390100 66.0000000%    0                        1.0000000   88wahl
86 as g   u235t   ms   fi   x4   0.0000000  0.0   4.8000000  1.0000000    86              u235t   100.0000000   76siel
86 ge g   u235t   sp   cu        0.8703130 13.8   0.8700000  0.1200000     0                        0.0000000   90rudl
86 kr g   u235f   ms   re   rp   1.8386961 11.2  53.5100000  0.0000000  no. relative values =    2                70ebel
86 kr g   u235f   ms   re   x4   0.0000000  0.0   2.2400000  3.0000000%  148 nd g        u235f     1.7500000   72kryl
86        u235f   es   cu   cm   1.9147334 30.0   1.9300000  0.0000000     0                        0.0000000   72sidl
86        u235f   cm   cu        0.0000000  0.0   1.9600000  4.5000000%    0                        0.0000000   77crol
86 kr g   u235f   ms   cu        1.9742588  2.0   1.9900000  0.0110000     0                        0.0000000   80mael
86 kr g   u235f   ms   cu        1.8418982  2.6   1.9300000  0.0500000  148 nd g        u235f     1.7500000   70lisl
```

Figure 4. Measured data extracted from Appendix A of LA-UR-94-3106.

whose measured yield is 0.87 ± 0.12 % is ^{86}Se, not ^{86}Ge. So the value of the appendix A of LA-UR-94-3106 seems to be incorrectly cited. As the cumulative yield of ^{86}Se seems to be mistaken to be that of ^{86}Ge, the independent yield of ^{86}Ge would be also adjusted to give the mistaken cumulative yield. As for other nuclide yields, similar mistakes were considered to have happened. Then the independent yields of the nuclides mentioned above were discarded and newly calculated to give new yield data of JENDL.

4 Calculation of independent yield

The independent fission yields were calculated following the equations in the report of JENDL/FPD-2011 [3]. The independent yield $y_i(A, Z)$ of a nuclide with charge Z and mass A is expressed by:

$$y_i(A, Z) = Y(A) \times f(A, Z) \times F_x(A, Z) \times r_i(A, Z), \tag{1}$$

where $Y(A)$ is the mass yield, $f(A, Z)$ the fractional independent yield, $F_x(A, Z)$ the pairing and odd-even correction factor and $r_i(A, Z)$ the isomeric yield ratio.

The fractional independent yield $f(A, Z)$ was calculated by:

$$f(A, Z) = \frac{N(A)}{\sigma_A \sqrt{2\pi}} \int_{Z-0.5}^{Z+0.5} \exp\left[-\frac{(Z - Z_p(A))^2}{2\sigma_A^2}\right] dZ, \tag{2}$$

where $Z_p(A)$ is the most probable charge of mass A, σ_A the Gaussian width of mass A and $N(A)$ the normalization factor. The normalization factor $N(A)$ is given as the summation of all elements of each mass becomes 1.0.

$$\sum_{Z=0}^{\infty} f(A, Z) \cdot F_x(A, Z) = 1.0. \tag{3}$$

The value of $Z_p(A)$ and σ_A were taken from LA-UR-94-3106.

The new charge distributions of A=86, 88 and 100 are shown in Figure 5. Comparing with Figure 3, the abnormal behaviors of ^{86}Ge, ^{88}As and ^{100}Rb seem to be disappeared. The new values were compared with those of the JEFF evaluated values [7], too. The new values and those of JEFF evaluation are listed in Table 3. It is found out that the new JENDL values are same order as those of JEFF.

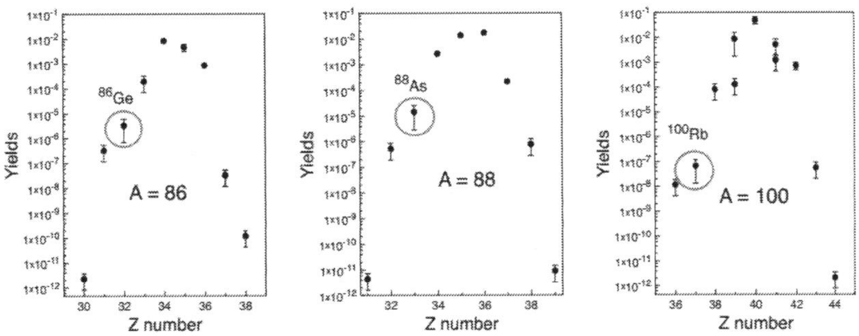

Figure 5. Charge distribution of new fission yields for A = 86, 88 and 100.

Table 3. New independent yields of JENDL and those of JEFF.

	^{86}Ge	^{88}As	^{100}Rb
New JENDL	3.4405×10^{-6}	1.4342×10^{-5}	6.5699×10^{-8}
JEFF 3.1.1	2.8497×10^{-6}	1.0704×10^{-5}	3.9483×10^{-8}

5 Summary

Sensitivity analysis of decay heat summation calculation showed some discrepancies between the uncertainties of thermal and fast neutron fissions for ^{235}U when the JENDL/FPY-2011 and the JENDL/FPD-2011 files. It was found that the several fission yields of the JENDL/FPY-2011 file had unexpectedly large and abnormal discontinuity of charge distribution. Those fission yields were updated. The updated yield data seem to remove the abnormal discontinuation of the charge distribution and to show rather smooth distribution. The sensitivity analysis is useful tool to analyze the uncertainty of decay heat calculation and to improve the data quality. The deficiencies of the fission yields were also suggested by the feedbacks from the users of the JENDL file. The user feedback is also important for improving the quality of the evaluated data.

References

[1] J. Katakura, "JENDL FP Decay Data File 2011 and Fission Yields Data File 2011", JAEA-Data/Code 2011-025, Japan Atomic Energy Agency (2011).

[2] G. Chiba, M. Tsuji, and T. Narabayashi, Nucl. Data Sheets, **118**, 401-404 (2014).

[3] J. Katakura, J. Nucl. Sci. Technol., **50**, 799-807 (2013).

[4] T.R. England and B.F. Rider, "Evaluation and Compilation of Fission Product Yields", LA-UR-94-3106, ENDF-349, Los Alamos National Laboratory (1994).

[5] K. Oyamatsu, J. Nucl. Sci. Technol. Supp. 2, 1109 (2002).

[6] G. Rudstam, P. Aagaard, B. Ekstrom, E. Lund, H. Gokturk, H.U. Zwicky, Radiochimica Acta **49**, 155-191 (1990).

[7] M.A. Kellet, O. Bersillon and R.W. Mills, "The JEFF-3.1/-3.1.1 radioactive decay data and fission yields sub-libraries", JEFF Report 20, OECD/NEA Data Bank (2009).

EPJ Web of Conferences **111**, 08005 (2016)
DOI: 10.1051/epjconf/201611108005

Microscopic predictions of fission yields based on the time dependent GCM formalism

D. Regnier[1,a], N. Dubray[1,b], N. Schunck[2], and M. Verrière[1]

[1] *CEA, DAM, DIF, 91297 Arpajon, France*
[2] *Nuclear and Chemical Science Division, LLNL, Livermore, CA 94551, USA*

Abstract. Accurate knowledge of fission fragment yields is an essential ingredient of numerous applications ranging from the formation of elements in the r-process to fuel cycle optimization in nuclear energy. The need for a predictive theory applicable where no data is available, together with the variety of potential applications, is an incentive to develop a fully microscopic approach to fission dynamics. One of the most promising theoretical frameworks is the time-dependent generator coordinate method (TDGCM) applied under the Gaussian overlap approximation (GOA). Previous studies reported promising results by numerically solving the TDGCM+GOA equation with a finite difference technique. However, the computational cost of this method makes it difficult to properly control numerical errors. In addition, it prevents one from performing calculations with more than two collective variables. To overcome these limitations, we developed the new code FELIX-1.0 that solves the TDGCM+GOA equation based on the Galerkin finite element method. In this article, we briefly illustrate the capabilities of the solver FELIX-1.0, in particular its validation for n+^{239}Pu low energy induced fission. This work is the result of a collaboration between CEA,DAM,DIF and LLNL on nuclear fission theory.

1 Introduction

Data on fission fragment yields play an essential role both in the conception of new generations of reactors as well as in fuel cycle technology. Other growing activities such as neutron interrogation would also benefit from a better knowledge of differential fission distributions involving the mass and charge but also the spin of the fragments. Over the last decades, part of the theoretical effort on fission has been focused on predicting induced fission yields based on the time-dependent generator coordinate method (TDGCM) associated with the Gaussian overlap approximation (GOA) [1, 2]. This approach provides a fully quantum mechanical description of the time evolution of the fissioning system. Recent applications provided the first fully-microscopic estimate of the mass and kinetic energy distributions of fission fragments for neutron-induced fission in the actinides region [3–5]. Note that these studies used the same solver based on the discretization of the TDGCM+GOA equations using finite differences in a regularly meshed hyper-cube. Because of the computational resources required by this numerical scheme, fission dynamics was only studied in 2-dimensional collective spaces. Yet,

[a] e-mail: david.regnier@cea.fr
[b] e-mail: noel.dubray@cea.fr

Article available at http://www.epj-conferences.org or http://dx.doi.org/10.1051/epjconf/201611108005

it is well-known from both semi-phenomenological and fully microscopic approaches that at least four or five collective variables play a role in the dynamics of fission [6–9]. Although possible in principle, extending the current scheme to $N > 2$ collective spaces would rapidly become prohibitive computationally. These reasons were the primary motivation to upgrade to a more flexible and more scalable discretization scheme. In this work, we give a quick overview of the new solver FELIX-1.0 [10] for the TDGCM+GOA equation. We briefly outline the numerical methods implemented before discussing some preliminary benchmark results on fission fragment yields obtained within this approach for the neutron-induced fission on ^{239}Pu.

2 TDGCM+GOA approach to compute fission yields

The time-dependent generator coordinate method (TDGCM) provides a fully quantum mechanical description of the time evolution of the fissioning system. It starts from the general observation that the time evolution of the many-body fissioning system is governed by the time-dependent Schrödinger equation. In our applications, the nuclear Hamiltonian $\hat{H}$ involved contains an effective two-body potential such as, e.g., the Skyrme or the Gogny interaction. In the TDGCM method, one chooses the following anszatz for the many-body state $|\Psi(t)\rangle$ associated with the fissioning nucleus [1, 2],

$$|\Psi(t)\rangle = \int d\boldsymbol{q}\; f(\boldsymbol{q}, t)|\Psi(\boldsymbol{q})\rangle. \tag{1}$$

In fission, the states $|\Psi(\boldsymbol{q})\rangle$ are solutions to the static Hartree-Fock-Bogoliubov (HFB) equations under a set of constraints $\boldsymbol{q}$ while the functions $f(\boldsymbol{q}, t)$ are to be determined. The constraints, also called collective variables, are in this work expectation values of the quadrupole and octupole moments. The quadrupole moment q_{20} characterizes the elongation of the nucleus whereas the octupole moment q_{30} is related to the asymmetry between the pre-fragments.

Inserting Eq.(1) in a variational form of the time-dependent Schrödinger equation yields a time-dependent integro-differential equation for the weight function $f(\boldsymbol{q}, t)$ known as the Hill-Wheeler equation. Numerically solving this equation has not yet been attempted in the case of fission as it would require a tremendous amount of computation. Instead, a widespread approach consists in assuming that the norm kernels $\langle\Psi(\boldsymbol{q})|\Psi(\boldsymbol{q}')\rangle$ can be approximated by a Gaussian form factor [11]. This Gaussian overlap approximation (GOA) yields a local, time-dependent Schrödinger-like equation for the dynamics of the fissioning system in the space of the collective variables $\boldsymbol{q}$,

$$i\hbar\frac{\partial}{\partial t}g(\boldsymbol{q}, t) = \left[-\frac{\hbar^2}{2}\sum_{kl}\frac{\partial}{\partial q_k}B_{kl}(\boldsymbol{q})\frac{\partial}{\partial q_l} + V(\boldsymbol{q})\right]g(\boldsymbol{q}, t), \tag{2}$$

where

- The unknown complex function $g(\boldsymbol{q}, t)$ is related to the weight function $f(\boldsymbol{q}, t)$ appearing in Eq.(1). It contains all the information about the dynamics of the system. In the limit of sharply peaked norm kernels, the quantity $|g(\boldsymbol{q}, t)|^2$ can be interpreted as the probability density for the system to have the collective characteristics $\boldsymbol{q}$ at time t;
- The potential field $V(\boldsymbol{q})$ and the inertia tensor $B_{kl}(\boldsymbol{q})$ are fully determined by the knowledge of the effective Hamiltonian $\hat{H}$ and the generator states $|\Psi(\boldsymbol{q})\rangle$.

The equation (2) is referred to as the TDGCM+GOA equation.

Our microscopic approach to compute the fission yields is a three-step process. First, we compute the generator states $|\Psi(\boldsymbol{q})\rangle$ for a range of values of the collective variables. This is the most time-consuming part, since any realistic two-dimensional fission calculation involves of the order of 50,000 points of the collective space. From these generator states, we can compute both the potential energy surface and the inertia tensor. A second step consists in numerically solving the TDGCM+GOA equation (2) starting from an initial function $g(\boldsymbol{q}, t = 0)$. This initial condition is arbitrarily built as a wave packet localized at low deformation, in the first well of the potential energy surface. The system begins its evolution with an average initial energy of 1 MeV above the inner barrier to simulate an induced fission. In the third and final step, we extract fission product yields from the time-dependent solution of the dynamic equation by defining a frontier in the collective space, which separates scissioned configurations from configurations where the system is still a whole nucleus. Computing the probability for the system to cross through any point of this line provides a first estimate of the fission yields. This raw data is then convoluted with a Gaussian to take into account effects such as quantum fluctuations of particle number. Figure 1 gives a schematic picture of the whole process.

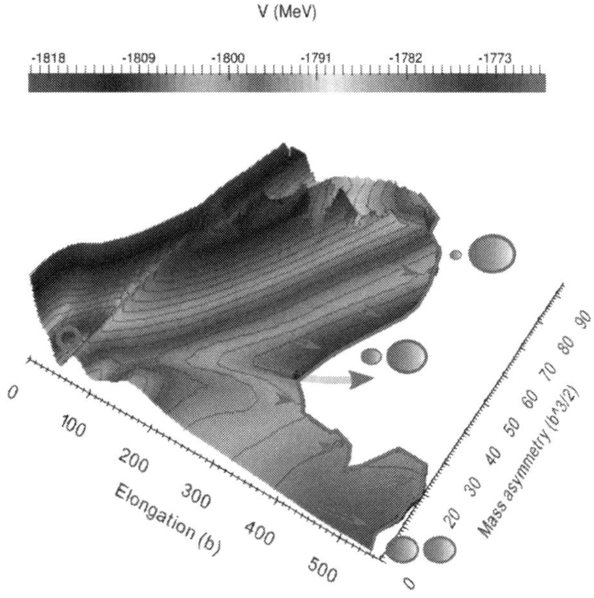

Figure 1. Schematic representation of the propagation of the wave packet $g(\boldsymbol{q}, t)$ solution of the TDGCM+GOA equation for a $n_{th}+^{239}Pu$ fission. The colored field represents the potential $V(\boldsymbol{q})$. The full red line accounts for the frontier between scissioned configurations and whole nucleus configurations. The wave packet evolves from the first potential well area to very deformed configurations. The potential landscape favors the diffusion of the wave packet through the so-called fission valley. Fission product yields are computed from the flux of the wave packet through each point of the frontier.

3 FELIX-1.0, a new solver for the TDGCM+GOA equation

As already mentioned, previous studies of induced fission within the TDGCM+GOA framework relied on the same code tdgcm2d developed and used at CEA and LLNL over the last decades. This code uses a finite difference scheme to discretize the function $g(\boldsymbol{q}, t)$ in the collective space. Because of the high computational cost of this numerical method, fission dynamics could only be studied with two collective variables. However, it is well-known from both semi-phenomenological and fully microscopic approaches that at least four or five collective variables play a role in the dynamics of fission [6–9]. Although possible in principle, extending the current scheme to $N > 2$ collective spaces would rapidly become prohibitive computationally: Each additional dimension multiplies both the pre-calculation time of the potential energy surface and the evolution time of the TDGCM equations by approximately a factor 100.

In order to fully exploit this microscopic technique, we have thus developed a new solver for the TDGCM+GOA equation called FELIX-1.0, with the following capabilities:

- handling of more than two collective variables for a better description of the fissioning system,

- full control on the numerical precision,
- discretization of the collective space on irregular meshes to focus the computational effort on physically-relevant regions.

FELIX-1.0 relies on the Galerkin finite element method for the $\boldsymbol{q}$ discretization of the TDGCM equation. This well-established method allows the use of irregularly-spaced meshes in the collective space, which results in turn in calculations scaling much more efficiently with the number of collective variables. In addition, standard p-refinement techniques, i.e., the use of higher-degree polynomial bases in each finite element, give a better control on the numerical precision of the calculations. A Cranck-Nicholson unitary scheme is used for the time integration of the wave packet. This implicit method requires solving a sparse linear system at each time step of the evolution. The numerous matrix inversion operations involved are performed with a quasi-minimal residual (QMR) algorithm. Finally, we note that there are virtually no restrictions in the number of collective variables used. The code FELIX-1.0 has been tested up to $N = 3$ but should also work for higher dimensions. All this work is summarized in [10]. The first version of the FELIX-1.0 solver was released under an open source GPL-2 license and will be available in the Computer Physics Communications program library.

4 Preliminary results on n+^{239}Pu

We performed a set of preliminary calculations for the pre-emission mass yields of the low energy induced fission on ^{239}Pu. Our first objective was to check the convergence of the calculations. Four parameters control the numerical precision of the results: the mesh spacing parameter h, the time step δt, the tolerance for matrix inversions, and the convergence criteria for the initial state calculation. Thanks to the efficiency of FELIX-1.0, we can perform multiple series of calculations to check the convergence associated with each one of these parameters. Figure 2 shows the calculated mass yields of the heavy fragment in $n_{th}+^{239}$Pu as a function of the mesh parameter h for the SkM* parametrization of the Skyrme force. Within a reasonable amount of computational time, the precision obtained reaches 2% in the peak region ($A_H \in [127;150]$) and is maintained below 6% in the wings of the distribution ($A_H < 127$ and $A_H > 150$).

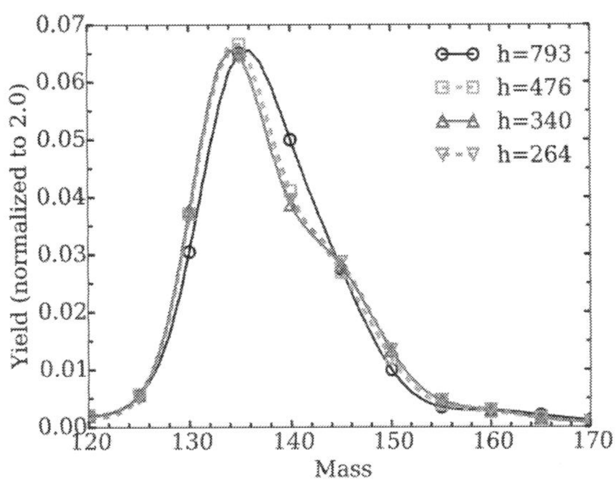

Figure 2. Convergence of the pre-neutron mass yields with respect to the mesh spacing parameter h (in fm^2 and fm^3 units for the q_{20} and q_{30} directions respectively) for the low energy induced fission n+^{239}Pu.

With this robust numerical scheme, we have solved the TDGCM+GOA equation for several potential energy surfaces based on both Skyrme and Gogny interactions. In a few cases, the computed mass yields show an encouraging 30% agreement with experiments in the mass range 127-150. On the other hand, the dispersion of our results with changes of potential energy surface may reach 100% at the peak. Although the interpretation of such discrepancy is still in progress, most of it seems to be related to the well-established presence of discontinuities [8] in the two dimensional collective space used. One possible way to overcome this drawback would be to increase the number of collective coordinates. Systematic, large-scale investigations of fission product yields in $n_{th}+^{239}Pu$ for a variety of microscopic inputs are currently on-going and will be presented elsewhere.

4.1 Conclusion

In this work we use a fully quantum mechanical approach to compute fission fragment yields for neutron-induced fission with thermal neutrons. We presented the new code FELIX-1.0 to solve the dynamics of fission within the TDGCM+GOA framework. FELIX-1.0 gives full control on the numerical precision of fission product yields in neutron-induced fission, and it scalability also enables series of dynamical calculations on several potential energy surfaces. Preliminary results suggest an important sensitivity of our two-dimensional approach to the input potential energy surface.

Acknowledgements

The research was carried out under the US-France International Agreement on Cooperation on Fundamental Research Supporting Stockpile Stewardship. This work was partly performed under the auspices of the U.S. Department of Energy by Lawrence Livermore National Laboratory under Contract DE-AC52-07NA27344.

References

[1] J.J. Griffin, J.A. Wheeler, Phys. Rev. **108**, 311 (1957)
[2] P. Reinhard, K. Goeke, Rep. Prog. Phys. **50**, 1 (1987)
[3] H. Goutte, J.F. Berger, P. Casoli, D. Gogny, Phys. Rev. C **71**, 024316 (2005)
[4] W. Younes, D. Gogny, Lawrence Livermore National Laboratory, Tech. Rep. LLNL-TR-586678 (2012)
[5] W. Younes, D. Gogny, Lawrence Livermore National Laboratory, Tech. Rep. LLNL-TR-586694, 2012 (2012)
[6] P. Möller, D. Madland, A.J. Sierk, A. Iwamoto, Nature **409**, 785 (2001)
[7] W. Younes, D. Gogny, Phys. Rev. C **80**, 054313 (2009)
[8] N. Dubray, D. Regnier, Comput. Phys. Commun. **183**, 2035 (2012)
[9] N. Schunck, D. Duke, H. Carr, A. Knoll, Physical Review C **90**, 054305 (2014)
[10] D. Regnier, M. Verriere, N. Dubray, N. Schunck, arXiv:1505.02704 (2015)
[11] J. Libert, M. Girod, J.P. Delaroche, Phys. Rev. C **60**, 054301 (1999)

EPJ Web of Conferences **111**, 08006 (2016)
DOI: 10.1051/epjconf/201611108006

Total Absorption Spectroscopy of Fission Fragments Relevant for Reactor Antineutrino Spectra and Decay Heat Calculations

A. Porta[1,a], A.-A. Zakari-Issoufou[1], M. Fallot[1], A. Algora[2,3], J.L. Tain[2], E. Valencia[2], S. Rice[4], V.M. Bui[1], S. Cormon[1], M. Estienne[1], J. Agramunt[2], J. Äystö[5], M. Bowry[4], J.A. Briz[1], R. Caballero-Folch[6], D. Cano-Ott[7], A. Cucouanes[1], V.-V. Elomaa[8], T. Eronen[8], E. Estévez[2], G.F. Farrelly[4], A.R. Garcia[7], W. Gelletly[2,4], M.B Gomez-Hornillos[6], V. Gorlychev[6], J. Hakala[8], A. Jokinen[8], M.D. Jordan[2], A. Kankainen[8], P. Karvonen[8], V.S. Kolhinen[8], F.G. Kondev[9], T. Martinez[7], E. Mendoza[7], F. Molina[2,10], I. Moore[8], A. B. Perez-Cerdán[2], Zs. Podolyák[4], H. Penttilä[8], P.H. Regan[4,11], M. Reponen[8,12], J. Rissanen[8], B. Rubio[2], T. Shiba[1], A.A. Sonzogni[13], C. Weber[8,14], and IGISOL Collaboration[8]

[1] *SUBATECH, CNRS/IN2P3, Université de Nantes, Ecole des Mines de Nantes, F-44307 Nantes, France*
[2] *Instituto de Fisica Corpuscular (CSIC-Universitat de Valencia), Apartado Correos 22085, E-46071 Valencia, Spain*
[3] *Institute of Nuclear Research, MTA ATOMKI, Debrecen, 4026 Hungary*
[4] *Department of Physics, University of Surrey, Guildford GU27XH, United Kingdom*
[5] *Helsinki Institute of Physics, University of Helsinki, FI-00014 Helsinki, Finland*
[6] *Universitat Politécnica de Catalunya (UPC), 08034 Barcelona, Spain*
[7] *Centro de Investigaciones Energéticas Medioambientales Y Tecnológicas, E-28040 Madrid, Spain*
[8] *Department of Physics, University of Jyväskylä, P.O. Box 35, FI-40014 Jyväskylä, Finland*
[9] *Argonne National Laboratory, Argonne, IL 60439, USA*
[10] *Present address: Comisión Chilena de Energía Nuclear, Post Office Box 188-D, Santiago, Chile*
[11] *National Physical Laboratory, Teddington, Middlesex, TW11 0LW, United Kingdom*
[12] *Present address: RIKEN, 2-1 Hirosawa, Wako, Saitama 351-0198, Japan*
[13] *National Nuclear Data Center, Brookhaven National Laboratory, Upton NY 11973-5000, USA*
[14] *Present address: Faculty of Physics, Ludwig-Maximilians University Munich, Am Coulombwall 1, D-85748 Garching, Germany*

Abstract. Beta decay of fission products is at the origin of decay heat and antineutrino emission in nuclear reactors. Decay heat represents about 7% of the reactor power during operation and strongly impacts reactor safety. Reactor antineutrino detection is used in several fundamental neutrino physics experiments and it can also be used for reactor monitoring and non-proliferation purposes. ^{92,93}Rb are two fission products of importance in reactor antineutrino spectra and decay heat, but their β-decay properties are not well known. New measurements of ^{92,93}Rb β-decay properties have been performed at the IGISOL facility (Jyväskylä, Finland) using Total Absorption Spectroscopy (TAS). TAS is complementary to techniques based on Germanium detectors. It implies the use of a calorimeter to measure the total gamma intensity de-exciting each level in the daughter nucleus providing a direct measurement of the beta feeding. In these proceedings we present preliminary results for ^{93}Rb, our measured beta feedings for ^{92}Rb and we show the impact of these results on reactor antineutrino spectra and decay heat calculations.

[a] e-mail: porta@subatech.in2p3.fr

1 Introduction

In reactor cores, beta decay of fission products is at the origin of the emission of a large amount of electron antineutrinos. About 10^{21} neutrinos per second are emitted from a 1 GW_{el} power reactor. This makes reactors a powerful source for neutrino oscillation experiments and suggests the possibility of using this massive flux for applied purposes such as fuel or reactor monitoring. Indeed the reactor antineutrino flux depends on fuel composition and reactor power [1, 2]. Motivated by an interest shown by the International Atomic Energy Agency in this idea [3], the scientific community started to actively explore this possibility with several dedicated experiments [4].
The beta decay of fission products is also at the origin of the decay heat which is about 7% of the reactor power during operation and the only source of energy when the reactor stops, generating strong safety constraints.
Both antineutrino spectra [5, 6]) and decay heat [7] can be calculated using the summation method which consists in summing the contribution to the desired quantity (decay heat or antineutrino energy) from each fission product present in the reactor core at a certain moment. In both fields, the advantage of this calculation respect to integral measurements is its usefulness in making predictions for future reactors. This is also an essential feature for safety and non-proliferation applications.
This method uses present nuclear data bases as input data for fission yields and decay data (energy spectra). Thus the precision of these data bases is extremely important for these calculations. One known problem affecting beta decay data bases is the so called "Pandemonium effect" [8], which arises from the difficulty that can occur in constructing nuclear level patterns for complex decays using Germanium detectors, especially when transitions are of high-energy or in regions of high level density, leading to a distortion in the reconstructed beta decay feeding.
The decays of ^{92,93}Rb are not well known in the data bases and they are candidates for the "Pandemonium effect". They have been selected [9] because they are major contributors to the antineutrino spectra. Moreover ^{92}Rb is indicated as priority 2 in decay heat calculations for the U/Pu cycle and priority 1 for the Th/U cycle [10, 11] .
In this paper we will present new preliminary results for the beta decay properties of ^{92,93}Rb obtained with TAS measurements and we will discuss the impact of these results on antineutrino spectra and decay heat calculations.

2 Total Absorption Spectroscopy (TAS) measurements of ^{92,93}Rb

A total absorption spectrometer should detect the de-excitation gammas emitted by the daughter nucleus after beta decay with an efficiency as close as possible to 100 %. The detection of the total gamma cascade allows for directly measuring the energy of the level fed and then the feeding probability can be deduced from the energy spectrum acquired during the experiment by solving the so called "inverse problem" $d = R \times f$, where d is the measured energy spectrum, R is the detector response and f is the feeding to the excited levels in the daughter nucleus [12, 13].
^{92,93}Rb have been measured at the IGISOL facility of the Jyväskylä University (Finland) [14] which is equipped with the double penning trap JYFLTRAP system [15], allowing for very precise selection of fission products. The TAS detector consisted of 12 BaF_2 crystals arranged in a compact geometry as shown in [16]. Each crystal is observed by a photomultiplier tube converting the scintillation light into an electrical signal proportional to the detected energy. The gamma detection efficiency is about 80 % at 5 MeV. A beta detector has been placed in the centre of this spectrometer, facing the source

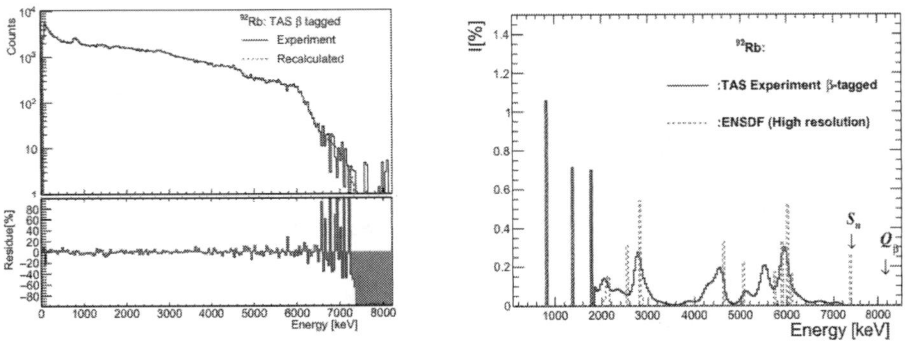

(a) (b)

Figure 1: (a) Upper panel: Comparison between cleaned raw data (black continuous line) and reconstructed spectrum (blue dotted line) from beta feeding after solving the inverse problem for ^{92}Rb. Lower panel: Residues between the two curves of the upper panel. (b) Comparison between the beta feeding from [9] in bleu continuous line and the feeding extracted from ENSDF data base [18] in red dotted line for ^{92}Rb. The g.s. feeding of ^{92}Rb from [9] is 87.5 ± 2.5 % instead of the 95.2 ± 0.7 % reported in ENSDF. Figures are from [9].

implantation zone. This means that the background can be reduced by making coincidences between the beta event and the following gamma emission from the levels in the daughter.

3 Data analysis

Raw data from the experiment have been cleaned of contaminants. Thanks to the purity of the beam provided by the Jyväskylä facility and to the beta coincidence arrangement described before, the only contaminants were produced by the daughter nucleus decay and the pile-up. These contaminants have been subtracted as described in [9] for ^{92}Rb. In the case of ^{93}Rb, its daughter nucleus, ^{93}Sr, has a neutron emission branch with 1.39 % probability. Both gamma and neutron contaminants have been reproduced and subtracted as described in [17].
The resolution of the inverse problem to obtain the feeding distribution from these cleaned data has been performed using a maximisation expectation algorithm based on the Bayes theorem and combined with a χ^2 minimisation [13]. For ^{92}Rb, the spectrum obtained from the feeding estimated with the inverse problem resolution is shown in figure 1a in blue dotted line and compared with the clean experimental spectrum in black continuous line (figure from [9]). The lower panel of this figure shows the residues between the two curves. The same analysis procedure has been applied to ^{93}Rb as shown in [17].

4 Results and impact on the antineutrino spectrum and decay heat calculations

Figure 1b shows the comparison between the beta feeding calculated as described in the previous section (blue continuous line) and the feeding extracted from the ENSDF data base [18] (red dotted

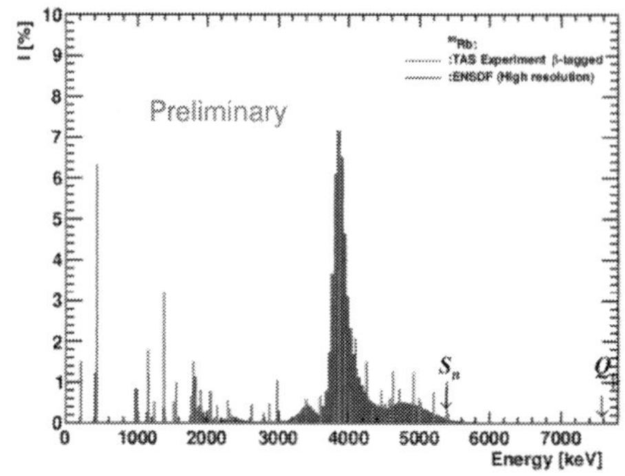

Figure 2: Comparison between the beta feeding from [17] in blue and the feeding extracted from ENSDF data base [18] in red for ^{93}Rb.

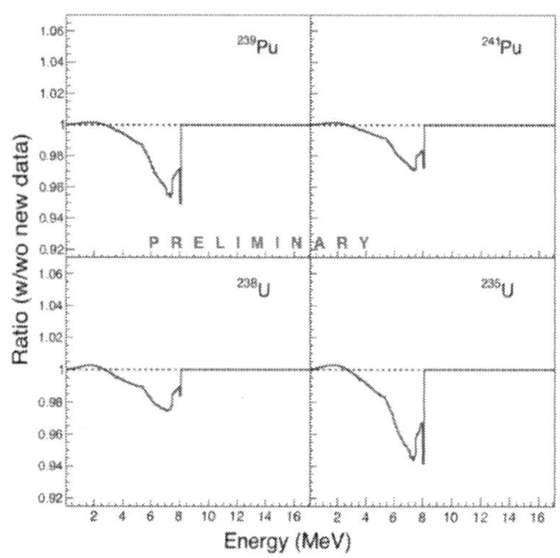

Figure 3: Ratio between antineutrino spectra for ^{235,238}U and ^{239,241}Pu calculated with the new TAS data for ^{92,93}Rb and with previous data from Rudstam et al. [20].

line) for ^{92}Rb (figure from [9]). The ground state (g.s.) feeding of ^{92}Rb from this work is 87.5 ± 2.5 % instead of the 95.2 ± 0.7 % reported in ENSDF.

Preliminary results for ^{93}Rb are shown in figure 2 and [17]. The results of this analysis (blue) are in quite good agreement with the data from ENSDF data base (red) that already contains results from previous TAS measurements for this nucleus from Greenwood et al. [19].

Figure 3 contains the ratio between the antineutrino spectra calculated with and without these new data and emitted by ^{235,238}U and ^{239,241}Pu isotopes. The spectra without these data are calculated as described in [6] using Rudstam's [20] results for ^{92,93}Rb. More details about the calculation of the spectra can be found in [9].

From figure 3 we can see that, as predicted in [9], the main impact of these data is in the 4-8 MeV region, with a maximum variation around 7-8 MeV of 6% for ^{235}U, 5% for ^{239}Pu, 3% for ^{241}Pu and

2.5% for ^{238}U.
The impact of these two nuclei shown in figure 3 underlines the importance of TAS measurements if we want to improve the calculation of reactor antineutrino spectra through the summation method. Regarding decay heat calculations, our measurements give a mean energy for the emitted gamma spectrum of 464.2 keV and for the emitted beta spectrum of 3.50 MeV. Mean values from ENDF-7.1.1 (g.s. to g.s. feeding 95.2 %) are 170 keV for the mean gamma energy and 3.63 MeV for mean beta energy [21]. This is mainly due to the change of g.s. to g.s. feeding from 95.2 % to 87.5 % and goes in the direction of increasing the decay heat component released by gammas and reducing the component due to beta particles. This nucleus is important for ^{235}U, for which summation calculations are less in agreement with Tobias' compilation [22] as shown in [23].

References

[1] A. Bernstein et al., J. Appl. Phys. **103**, 074905 (2008)
[2] Yu.V.Klimov et. al., Atomic Energy, **76**, 123 (1994)
[3] IAEA Report STR-361, Department of Safeguards. Proceedings of the first meeting of the Ad Hoc Working Group on Safeguards Applications utilising Antineutrino Detection and Monitoring, IAEA Report SG-EQGNRL-RP-0002 (2012).
[4] Applied Antineutrino Workshop 2015, presentations at: https://indico.in2p3.fr/event/10050/other-view?view=standard
[5] Th. A. Mueller et al., Phys.Rev. C **83**, 054615 (2011)
[6] M. Fallot et al., Phys. Rev. Lett. **109** , 202504 (2012)
[7] T. Yoshida et al., J. Nucl. Sci. Technol. **36**, 135(1999).
[8] J.C. Hardy et al., Phyics Letter B **71**, 307 (1977)
[9] A.-A. Zakari-Issoufou et al., Phys. Rev. Lett. **115**, 102503 (2015)
[10] Assessment of Fission Product Decay Data for Decay Heat Calculations , Nuclear Science NEA/WPEC-25 (2007).
[11] Decay Heat Calculation: Assessment of Fission Product Decay Data Requirements for Th/U Fuel, INDC(NDS)-0577
[12] D. Jordan, A. Algora, et al., Phys. Rev. C **87**, 044318 (2013).
[13] J. L. Tain et al., Nucl. Instrum. Methods. Phys. Res.Sect. A **571**, 728 (2012).
[14] I. D. Moore et al., Hyperfine Interact. **223**, 17 (2014).
[15] T. Eronen et al., Eur. Phys. J. A **48**, 46 (2012).
[16] A. Algora et al., Nucl. Data Sheets **120**, 12 (2014).
[17] A.-A. Zakari, Ph.D. thesis, Universit Nantes, France (2015).
[18] C. Baglin, Nucl. Data Sheets **113**, 2187 (2012)
[19] R. C. Greenwood, R. G. Helmer, M. H. Putnam, and K. Watt, Nucl. Instrum. and Methods Phys. Res. Sect. A **390**, 95 (1997).
[20] G. Rudstam et al., At. Data and Nucl. Data Tables **45**, 239 (1990).
[21] A. Sonzogni private communication.
[22] A. Tobias, Decay Heat, Prog. Nucl. Energy **5** 1 (1980).
[23] Total Absorption Gamma-ray Spectroscopy for Decay Heat Calculations and Other Applications, INDC(NDS)-0676

EPJ Web of Conferences **111**, 09001 (2016)
DOI: 10.1051/epjconf/201611109001

Differential Cross Sections and the Impact of Model Defects in Nuclear Data Evaluation

Georg Schnabel[1,2,a] and Helmut Leeb[1,b]

[1] *Atominstitut, TU Wien, Wiedner Hauptstrasse 8-10, 1040 Vienna, Austria*
[2] *CEA, Saclay, Irfu/SPhN, Gif-sur-Yvette, France*

Abstract. A statistically consistent method for the inclusion of the so-called model defects into Bayesian evaluation methods based on extensive modeling is presented. The method uses Gaussian processes to define a-priori probability distributions for possible model defects. The inclusion is of particular importance for the proper evaluation of differential data, i.e. angle-differential cross sections and spectra of ejectiles. The method is successfully applied to a simple realistic example which clearly shows the impact of model defects in a simultaneous evaluation of angle-differential and angle-integrated cross sections.

1 Introduction

Nuclear data evaluation yields consistent sets of reaction data, comprising cross sections, spectra and production cross sections for a given group of isotopes. Recent demands triggered by the development of novel nuclear technologies and applications include the extension of the energy range as well as the inclusion of uncertainty information. Because of the scarcity of neutron-induced reaction data beyond 20 MeV, corresponding evaluations depend strongly on nuclear models. Nowadays it is well established that Bayesian statistics [1] is the proper mean for a consistent combination of experimental data with values obtained from nuclear model calculations. Corresponding evaluation procedures yield mean values of the observables and associated covariance matrices containing uncertainty information. However, the unreflected inclusion of results from model calculations into a Bayesian evaluation may lead to unrealistically small uncertainties. This effect is particularly striking in angle-differential cross sections, which are strongly correlated and additionally include phase information not available in angle-integrated data.

In this contribution we consider this problem in more detail. Especially, we show at a schematic example in section 2 that this finding may be the consequence of the assumption of perfect models, frequently used in nuclear data evaluation. In section 3 we present a novel evaluation concept which provides within Bayesian statistics a consistent treatment of model defects on the basis of Gaussian processes. The advantages of this extended procedure are demonstrated at an example simultaneously evaluating simple, but realistic sets of angle-differential and angle-integrated cross section data. The results of this first application are very promising and therefore concluding remarks and an outlook on the potential of the method are given in the final section 4.

[a] e-mail: georg.schnabel@cea.fr

[b] Corresponding author: e-mail: leeb@kph.tuwien.ac.at

Article available at http://www.epj-conferences.org or http://dx.doi.org/10.1051/epjconf/201611109001

2 Standard Bayesian Evaluation Procedure

In a standard evaluation process, it is usually assumed that the model $M(\mathbf{p})$ is perfect and describes the experimental values of the observables $\boldsymbol{\sigma}_{\text{exp}}$ within the experimental uncertainties $\boldsymbol{\epsilon}_{\text{exp}}$, i.e. $\boldsymbol{\sigma}_{\text{exp}} = \boldsymbol{\sigma}_{\text{mod}}+\boldsymbol{\epsilon}_{\text{exp}}$. In most standard evaluation procedures [2], the combination of experimental and model values is governed by Bayes theorem [1],

$$\pi\left(\mathbf{p}\,|\,\boldsymbol{\sigma}_{\text{exp}}\right) = \left[\int d^d p\, \ell\left(\boldsymbol{\sigma}_{\text{exp}}\,|\,\mathbf{p}\right)\pi\left(\mathbf{p}\right)\right]^{-1} \ell\left(\boldsymbol{\sigma}_{\text{exp}}\,|\,\mathbf{p}\right)\,\pi\left(\mathbf{p}\right). \tag{1}$$

Here, $\pi(\mathbf{p}\,|\,\boldsymbol{\sigma}_{\text{exp}})$ is the a-posteriori distribution of the model parameters $\mathbf{p}$ conditioned on the measured values $\boldsymbol{\sigma}_{\text{exp}}$ of the observables, $\ell(\boldsymbol{\sigma}_{\text{exp}}\,|\,\mathbf{p})$ is the likelihood and $\pi(\mathbf{p})$ is the a-priori distribution which reflects the available *a-priori* knowledge. The integral in Eq. (1) acts as a normalization and is usually denoted as *evidence*. In the following discussion we substitute the parameter distributions $\pi(\mathbf{p})$, $\pi(\mathbf{p}\,|\,\boldsymbol{\sigma}_{\text{exp}})$ by cross section distributions $\pi(\boldsymbol{\sigma})$, $\pi(\boldsymbol{\sigma}\,|\,\boldsymbol{\sigma}_{\text{exp}})$. For details on this approach see e.g. [2, p. 37]. Bayes theorem is applicable for arbitrary probability distributions, but for the sake of simplicity we restrict the present discussion to multivariate normal distributions,

$$\rho(\mathbf{t}) = \frac{1}{\sqrt{(2\pi)^d|\mathcal{V}|}}\exp\left[-\frac{1}{2}\left(\mathbf{t}-\langle\mathbf{t}\rangle\right)^T \mathcal{V}^{-1}\left(\mathbf{t}-\langle\mathbf{t}\rangle\right)\right], \tag{2}$$

where the mean value $\langle\mathbf{t}\rangle$ and the covariance matrix $\mathcal{V}$ characterize the multivariate normal distribution denoted by $\mathcal{N}(\langle\mathbf{t}\rangle,\mathcal{V})$. Assuming a multivariate normal distribution for the experimental uncertainties with covariance matrix $\mathcal{B}$ gives rise to the likelihood $\ell(\boldsymbol{\sigma}_{\text{exp}}|\boldsymbol{\sigma}) \sim \mathcal{N}(\boldsymbol{\sigma},\mathcal{B})$. Using the further assumption of a multivariate normal a-priori distribution $\mathcal{N}(\boldsymbol{\sigma}_0,\mathcal{A}_0)$, Eq. (1) leads to a multivariate normal a-posteriori distribution $\pi(\boldsymbol{\sigma}|\boldsymbol{\sigma}_{\text{exp}}) \sim \mathcal{N}(\boldsymbol{\sigma}_1,\mathcal{A}_1)$. The evaluated mean values and the covariance matrix are available in closed-form,

$$\boldsymbol{\sigma}_1 = \boldsymbol{\sigma}_0 + \mathcal{A}_0\mathcal{S}^T\left(\mathcal{S}\mathcal{A}_0\mathcal{S}^T+\mathcal{B}\right)^{-1}\left(\boldsymbol{\sigma}_{\text{exp}}-\mathcal{S}\boldsymbol{\sigma}_0\right), \tag{3}$$

$$\mathcal{A}_1 = \mathcal{A}_0 - \mathcal{A}_0\mathcal{S}^T\left(\mathcal{S}\mathcal{A}_0\mathcal{S}^T+\mathcal{B}\right)^{-1}\mathcal{S}\mathcal{A}_0, \tag{4}$$

where $\mathcal{S}$ is the so-called sensitivity matrix providing the mapping from the angle/energy grid of the model vector $\boldsymbol{\sigma}$ to the grid of the experiment vector $\boldsymbol{\sigma}_{\text{exp}}$. Especially, we set up $\mathcal{S}$ in a way to implement bilinear interpolation for angle-differential and linear interpolation for angle-integrated cross sections.

In general, applications of the standard Bayesian evaluation procedures lead to reasonable mean values of the observables, while the uncertainties are frequently underestimated. In order to reveal the origin of this finding, we consider a schematic example of generated experimental data (Fig. 1a) and a linear model (Fig. 1b) as *a-priori* information. The result of the evaluation is displayed in Fig. 1c. The unrealistically small uncertainties associated with the evaluated observables are a direct consequence of the rigidity of the model. Moreover, in standard Bayesian evaluations the predictive power of the model has no influence on evaluated uncertainties – a questionable feature. The increase of the prior uncertainties, as frequently suggested, cannot cure the problem. The outcome of this schematic example indicates that one has to find either the *perfect model* or a proper method to account for model defects in the evaluation process.

3 Novel Evaluation Concept including Model Defects

In the past different approaches for the inclusion of model defects have been presented [3–6], but none of these is fully based on statistics and conserves sums. Therefore, we have formulated an extended

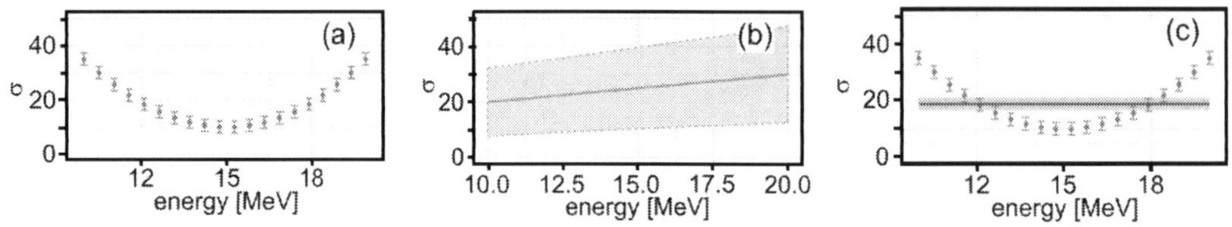

Figure 1. Evaluation based on schematic experimental data (red) and a model (blue), linear in the energy E. (a) generated experimental data set; (b) prior based on a linear model with given uncertainty; (c) Result of the standard Bayesian evaluation.

Bayesian evaluation procedure [2] assuming $\boldsymbol{\sigma}_{\text{exp}} = \boldsymbol{\sigma}_{\text{mod}} + \boldsymbol{\epsilon}_{\text{mod}} + \boldsymbol{\epsilon}_{\text{exp}}$. The inclusion of the model defects $\boldsymbol{\epsilon}_{\text{mod}}$ requires an extension of Eq. (1),

$$\pi(\mathbf{p}, \boldsymbol{\epsilon}_{\text{comb}} \,|\, \boldsymbol{\sigma}_{\text{exp}}) = \hat{C}\, \ell(\boldsymbol{\sigma}_{\text{exp}} \,|\, \mathbf{p}, \boldsymbol{\epsilon}_{\text{mod}})\, \pi(\mathbf{p})\, \pi(\boldsymbol{\epsilon}_{\text{comb}}) \tag{5}$$

with the extended evidence $\hat{C}$,

$$\boldsymbol{\epsilon}_{\text{comb}}^T = \left(\boldsymbol{\epsilon}_{\text{pred}}^T, \boldsymbol{\epsilon}_{\text{mod}}^T\right) \quad \text{and} \quad \pi(\boldsymbol{\epsilon}_{\text{comb}}) \propto \exp\left(-\frac{1}{2}\boldsymbol{\epsilon}_{\text{comb}}^T \mathcal{K}_0^{-1} \boldsymbol{\epsilon}_{\text{comb}}\right). \tag{6}$$

The factorized form of the prior in Eq. (5) is due to the assumption of vanishing a-priori correlation between the model parameters $\mathbf{p}$ and the model defects $\boldsymbol{\epsilon}_{\text{mod}}$. The covariance matrix $\mathcal{K}_0$ reflects the a-priori assumptions about possible model defects.

The determination of the model defect $\boldsymbol{\epsilon}_{\text{mod}}$ is the key issue of the procedure because it is beyond the theory underlying the model. Usually, there is no knowledge about the specific functional form of $\boldsymbol{\epsilon}_{\text{mod}}$, but frequently we have an idea about the smoothness of the cross section curve and the overall quality of the model. Therefore, we adopted the concept of Gaussian processes [7, 8], which allows us to specify a probability distribution for the shape of an unknown function using the expected smoothness of the function and the expected magnitude of the function values. More specifically, a probability distribution $\rho(.)$ on a function space $\mathcal{F}$ corresponds to a Gaussian process if for any finite set of $\{\mathbf{x}_i\}_{i=1,\ldots,N}$, $\forall \mathbf{x}_i \in \mathbb{R}^d$ the associated function values $\{f(\mathbf{x}_i)\}_{i=1,\ldots,N}$ follow a multivariate normal distribution. A Gaussian process is completely characterized by a mean function $m(\mathbf{x})$ and a covariance function $k(\mathbf{x}, \mathbf{x}')$. The covariance function is used to construct the covariance matrix $\mathcal{K}_0$ introduced in Eq. (6).

We adopt the assumption $m(\mathbf{x}) \equiv 0$ meaning that the model prediction is a-priori regarded as the most likely guess to represent the true cross section. Further, we use for the evaluation of the angle-differential cross sections the covariance function

$$\tilde{k}(E, \theta; E', \theta') = \frac{d\sigma}{d\Omega}(E, \theta)\frac{d\sigma}{d\Omega}(E', \theta') \sum_{i=1,2} \alpha_i(\theta)\alpha_i(\theta')g(E - E', \theta - \theta'; \delta_i, \lambda_i, \eta_i) \tag{7}$$

with $\alpha_2(\theta) = 1 - \alpha_1(\theta)$,

$$g(E, \theta; \delta, \lambda, \eta) = \delta^2 \exp\left(-\frac{E^2}{2\lambda^2}\right)\exp\left(-\frac{\theta^2}{2\eta^2}\right) \quad \text{and} \quad \alpha_1(\theta) = \left[1 + \exp\left(\frac{a - \theta}{b}\right)\right]^{-1}. \tag{8}$$

This covariance function contains two components: one component for small angles and another one for big angles. Each component i is characterized by the magnitude of the model defect δ_i and the smoothness hyperparameters λ_i and η_i. The transition between the two angle domains is achieved via the function $\alpha(\theta)$ depending on the parameters a and b. The covariance function in Eq. (7) also induces a covariance function for the model defects of the associated angle-integrated cross sections,

$$\tilde{k}(E, E') = (2\pi)^2 \int_0^{\pi} \int_0^{\pi} d\theta\, d\theta' \sin\theta \sin\theta' \, \tilde{k}(E, \theta; E', \theta') \,. \tag{9}$$

This relationship directly reflects the connection between angle-differential and associated angle-integrated cross sections.

If there is no idea about the values of the hyperparameters of the Gaussian process, one may determine these by maximizing the evidence $\hat{C}$. In general, this procedure leads to reasonable values for the hyperparameters (δ, λ, $\cdots$), but one must be aware that this method involves experimental values and is therefore not a completely independent *a-priori* information.

In principle, the extended Bayesian evaluation proceeds analogously to the standard procedure. Especially, one may use the linearized version for the Bayesian update (4). In addition to standard Bayesian evaluations, one also obtains mean values and covariance matrices for the model defects as well as cross correlations between observables and model defects.

The procedure outlined above can be applied simultaneously for angle-integrated and differential data. Especially for the evaluation of angle-differential cross sections, the inclusion of model defects is of great importance. Therefore, we present a test example of a simultaneous evaluation of a set of angle-differential cross sections at $E = 8.03$ MeV together with corresponding angle-integrated cross sections in the energy regime between 2 and 9 MeV. For the inclusion of model defects, we use the covariance functions (9) for angle-integrated cross sections and (7) for angle-differential cross sections. Assuming a model defect of $\delta_1 = 5\%$ in forward direction, we obtain the following values for the other hyperparameters by maximizing the evidence: $a = 23.014°$, $b = 4.127°$, $\lambda_1 = 1.121$ MeV, $\eta_1 = 60°$, $\delta_2 = 0.69$, $\lambda_2 = 0.523$ MeV, $\eta_2 = 2.944°$.

In Fig. 2 the evaluated differential cross sections at $E = 8.03$ MeV without and with model defects are shown. In both evaluations the experimental values of the differential cross sections are well reproduced. However, ignoring the model defects leads to an unrealistically small error band, while accounting for model defects leads to larger and more realistic uncertainties in the order of the experimental errors (see Fig. 3c). The method also yields the mean value of the model defect and its error band (Fig. 2c). Due to the conservation of sums, the inclusion of model deficiencies has also a direct impact on the evaluation of angle-integrated cross sections (Fig. 3). If model defects are ignored, the evaluated angle-integrated cross section deviates completely from the experimental value. Including model defects leads to a significantly better overlap between experimental and evaluated cross section. The remaining deviations reflect the fact that the values of the angle-integrated cross section are not compatible with available angle-differential data. Apart from mean values and associated uncertainties, the extended evaluation procedure yields also the complete set of covariance matrices.

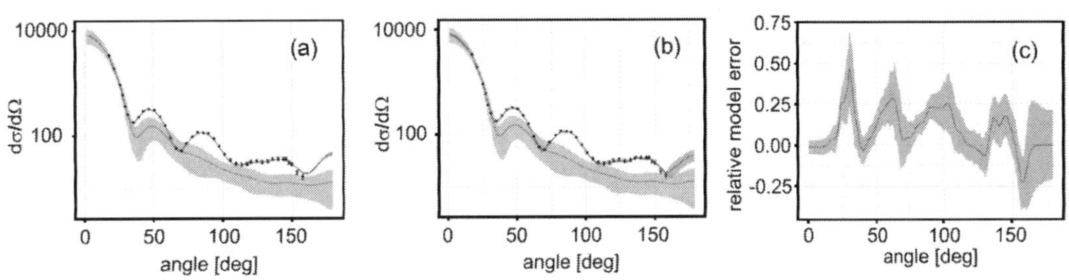

Figure 2. Evaluated (blue) and a-priori (red) angle-differential cross section data with the associated uncertainty band at $E = 8.03$ MeV. (a) Evaluation without model defects; (b) Evaluation accounting for model defects; (c) Mean value of the model defect and its uncertainty band.

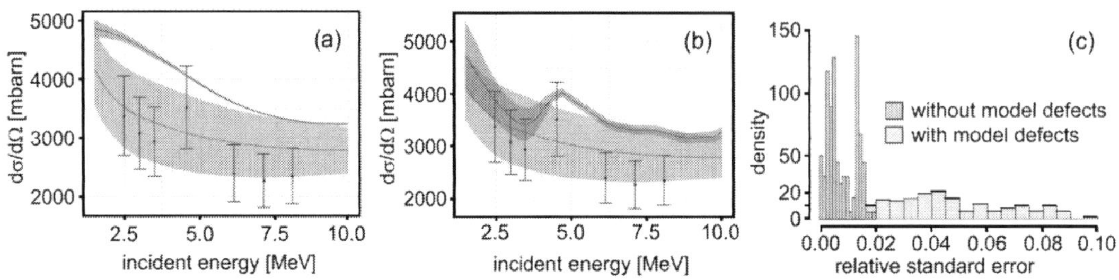

Figure 3. Evaluated (blue) and a-priori (red) angle-integrated cross section data with the associated uncertainty band. (a) Evaluation without model defects; (b) Evaluation accounting for model defects; (c) Histogram comparing the uncertainties without (yellow) and with (blue) model defects.

4 Conclusions and Outlook

We developed a statistically consistent procedure for the inclusion of model defects into nuclear data evaluations. The method is based on Gaussian processes, which allows to formulate the expected deviations from the nuclear model calculations in terms of reasonable assumptions for the covariance matrices. The method provides not only values for the evaluated observables and uncertainties, but also provides mean values and uncertainties of the model defects, which is an important information, especially for differential data. Applications of the method to evaluations of neutron-induced reaction data are currently in progress.

The presented method adresses the general problem of the assessement of the quality of models and is therefore not only of interest for nuclear data evaluation. Hence it has a high potential for applications in different fields of physics which strongly rely on modelling.

Acknowledgements

Work partly supported by the F4E Partnership Agreement grant F4E-FPA-168.02 and the associated KKKÖ project G. Schnabel of the Austrian Academy of Sciences as well as via the EURATOM project CHANDA. The views and opinions expressed herein do not reflect necessarily those of the European Commission.

References

[1] T. Bayes, *An Essay towards Solving a Problem in the Doctrine of Chances*, Philosophical Transactions **53**, 370 (1763)

[2] G. Schnabel, *Large Scale Bayesian Nuclear Data Evaluation with Consistent Model Defects*, PhD thesis, TU Wien (2015)

[3] M.T. Pigni, H. Leeb, *Uncertainty estimates of evaluated ^{56}Fe cross sections based on extensive modelling at energies beyond 20 MeV* in: Proc. Int. Workshop on Nuclear Data for the Transmutation of Nuclear Waste (GSI Darmstadt, Darmstadt, 2003) PO21

[4] H. Leeb, D. Neudecker, Th. Srdinko, Nucl. Data Sheets **109**, 2762 (2009)

[5] A. Trkov *et al.*, Nucl. Instr. Meth. A **212**, 3098 (2011)

[6] D. Neudecker, R. Capote, H. Leeb, Nucl. Instr. Meth. A **723**, 163 (2013)

[7] D.G. Krige, J. Chemical, Metallurgical and Mining Society **52**, 119 (1951)

[8] C.E. Rasmussen, Ch.K.I. Williams, *Gaussian Processes for Machine Learning* (MIT Prees, Cambridge, 2006), ISBN: 0-262-18253-X 978-0-262-18253-9

EPJ Web of Conferences **111**, 09002 (2016)
DOI: 10.1051/epjconf/201611109002

Comparison of deterministic and stochastic approaches for isotopic concentration and decay heat uncertainty quantification on elementary fission pulse

S. Lahaye[1,a], T. D. Huynh[1], and A. Tsilanizara[1]

[1] *CEA Saclay, DEN, DM2S, SERMA, LLPR F-91191 Gif-sur-Yvette, France*

Abstract. Uncertainty quantification of interest outputs in nuclear fuel cycle is an important issue for nuclear safety, from nuclear facilities to long term deposits. Most of those outputs are functions of the isotopic vector density which is estimated by fuel cycle codes, such as DARWIN/PEPIN2, MENDEL, ORIGEN or FISPACT. CEA code systems DARWIN/PEPIN2 and MENDEL propagate by two different methods the uncertainty from nuclear data inputs to isotopic concentrations and decay heat. This paper shows comparisons between those two codes on a Uranium-235 thermal fission pulse. Effects of nuclear data evaluation's choice (ENDF/B-VII.1, JEFF-3.1.1 and JENDL-2011) is inspected in this paper. All results show good agreement between both codes and methods, ensuring the reliability of both approaches for a given evaluation.

1 Introduction

These last years, most of depletion code systems have introduced uncertainty propagation functionalities. CEA fuel cycle code systems used in this study have implemented two different methods. DARWIN/PEPIN2 [1] propagates uncertainties using deterministic first order forward perturbations through the INCERD supervisor while MENDEL [2, 3] uses Monte Carlo sampling with samples created by CEA/DEN uncertainty platform URANIE [4].

Three nuclear data evaluations were used: ENDF/B-VII.1, JEFF-3.1.1 and JENDL-2011.

2 Depletion calculations

2.1 Concentrations and decay heat expressions

After an elementary fission pulse, the atom density $N_i(t)$ of the different fission products i is solution of the generalized Bateman equation when all neutronic reactions have been taken out:

$$\frac{\mathrm{d}N_i}{\mathrm{d}t}(t) \quad = \quad -\lambda_i N_i(t) + \textstyle\sum_{j=1}^{N} b_{j,i}\lambda_j N_j(t) \quad ; \quad N_i(0) \quad = \quad \gamma_i \tag{1}$$

where λ_i is the radioactive decay constant of fission product i, $b_{j,i}$ the radioactive decay branching ratio from j to i and γ_i the independent fission yield generating fission product i. Independent fission yields correspond to initial concentration of fission products after the elementary fission pulse.

[a] e-mail: sebastien.lahaye@cea.fr

For any cooling time t, total decay heat DH is computed as displayed in equation (2), where E_i is the total radioactive decay energy released by nuclide i disintegration (sum of α, β and γ components):

$$\mathrm{DH}(t) = \sum_i E_i \lambda_i N_i(t) \tag{2}$$

2.2 Uncertainty data

In this paper, we consider for radioactive decay constants, radioactive decay energies, radioactive decay branching ratios and independent fission yields all the uncertainty data available in each evaluations (ENDF/B-VII.1, JEFF-3.1.1 and JENDL-2011). If physical parameters are not associated to nuclear data uncertainties in the considered evaluation, its value is considered known without uncertainty. Table 1 summarizes the available uncertainty data in different data sets and evaluations.

Table 1. Number of physical parameters which uncertainty is not equal to 0, in different data sets.

	Decay constant	Total decay energy	Decay branching ratio	Ind. Fission yield (U235T)	Nb of isotopes in decay chain
JEFF-3.1.1	3204	1554	505	918	3849
ENDF/B-VII.1	2233	1285	866	998	3820
JENDL-2011	867	653	590	1067	1142

3 Propagation of uncertainty data

3.1 Correlations between parameters

Two uncertain nuclear data from different parameters types are always considered uncorrelated, as well as two radioactive decay constants or radioactive decay energies or independent fission yields, as no physical correlation are given in evaluation files. For independent fission yields, this assumption will stop as soon as independent fission yields covariance matrices will be available.

Radioactive decay branching ratios corresponding to the different decay processes from the same father are correlated to assure a sum equal to 1, as described in section 3.2.

3.2 Deterministic method

The deterministic method used in DARWIN/PEPIN2 to calculate the propagation of uncertainty is based on the first order Taylor series (linearity hypothesis). In the matrix form, the uncertainty of a variable Y due to the effect of uncertainties on variable X is given by the following formula (3):

$$\mathrm{Cov}(Y) = S_{Y/X}\mathrm{Cov}(X)S_{Y/X}^T \tag{3}$$

where $\mathrm{Cov}(Y)$ (resp. $\mathrm{Cov}(X)$) is the variance-covariance matrix of variable Y (resp. X). $S_{Y/X}$ and $S_{Y/X}^T$ are respectively the sensitivity matrix and its transposed matrix.

To take into account the correlation $r_{j,k}$ between radioactive decay ratio branches $b_{i,j}$ and $b_{i,k}$ from a given radioactive nuclide i, we introduce a unique correlation coefficient r_i between all radioactive decay branching ratio from the same father isotope i, with respect of the physical constraint $\sum_{j=1}^N b_{i,j} = 1$. It is deduced from equation (4):

$$\mathrm{Var}\left(\sum_{j=1}^{N} b_{i,j}\right) = 0 \tag{4}$$

Which leads to:

$$r_{jk} = \text{const} = \frac{-\sum_{j=1}^{N} \text{Var}(\text{b}_{\text{i,j}})}{2\sum_{1\leq j<k\leq N} \sqrt{\text{Var}(b_{i,j})\text{Var}(b_{i,k})}} \tag{5}$$

The uncertainty propagation in DARWIN/PEPIN2 involves the supervisor module INCERD. INCERD gathers all user's data and uncertainties data of physical parameters and launches depletion calculating in parallel mode (using MPICH2/OPENMPI) to determine the sensitivity coefficient values ($S_{Y/X}$). Then, INCERD performs the uncertainty propagation as described by equation (3).

3.3 Stochastic method

The stochastic method implemented in MENDEL propagates the uncertainty on output data using sampled input data. For each uncertain nuclear data parameter, sampling is done using the CEA/DEN uncertainty platform URANIE.

Decay branching ratios cannot be sampled using the correlation factor from equation (5), as incoherence in the evaluation data creates for ENDF/B-VII.1 and JENDL-2011 correlation coefficients outside $[-1, 1]$. MENDEL assumes all standard deviations are equal for a given father isotope, which leads to $r_i = -1/(N-1)$. All other parameters are sampled independently.

Radioactive decay energies, radioactive decay periods and radioactive decay branching ratios are considered to have positive Gaussian distribution (truncated) if their relative standard deviation is less than 50%, lognormal distribution if it is bigger. Due to large number of high relative standard deviation for fission yields, they are sampled by a lognormal distribution. Those assumptions enable to assure the positivity of the distributions and a good estimation of mean value and standard deviation.

All results shown in this paper result from a use of 2000 realisations.

4 Results and Discussion

4.1 Uncertainty quantification for decay heat

When propagating uncertainty data to total decay heat for the three nuclear evaluations, DARWIN/PEPIN2 and MENDEL results both show a very good agreement, but also huge differences between evaluations, as shown in figure 1.

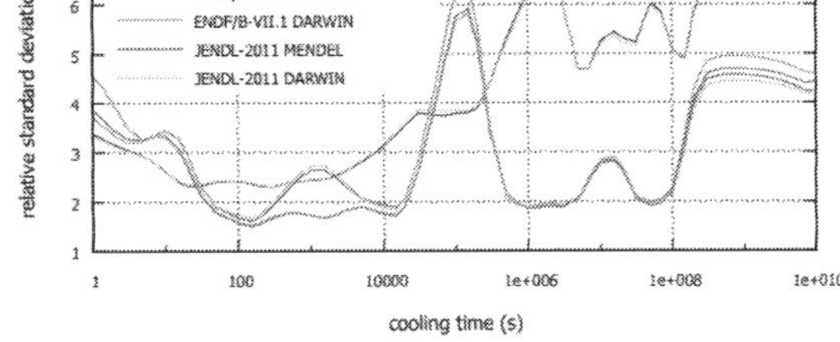

Figure 1. Uncertainty quantification of decay heat due to all parameters.

When observing the impact of each parameter separately, both codes for all evaluations show with the assumptions adopted here that fission yields is the most important contributor, followed by energies. Discrepancies are observed only for decay branching ratios, due to the differences in adopted correlations. This discrepancy can be important, as shown in the botom left of Figure 2.

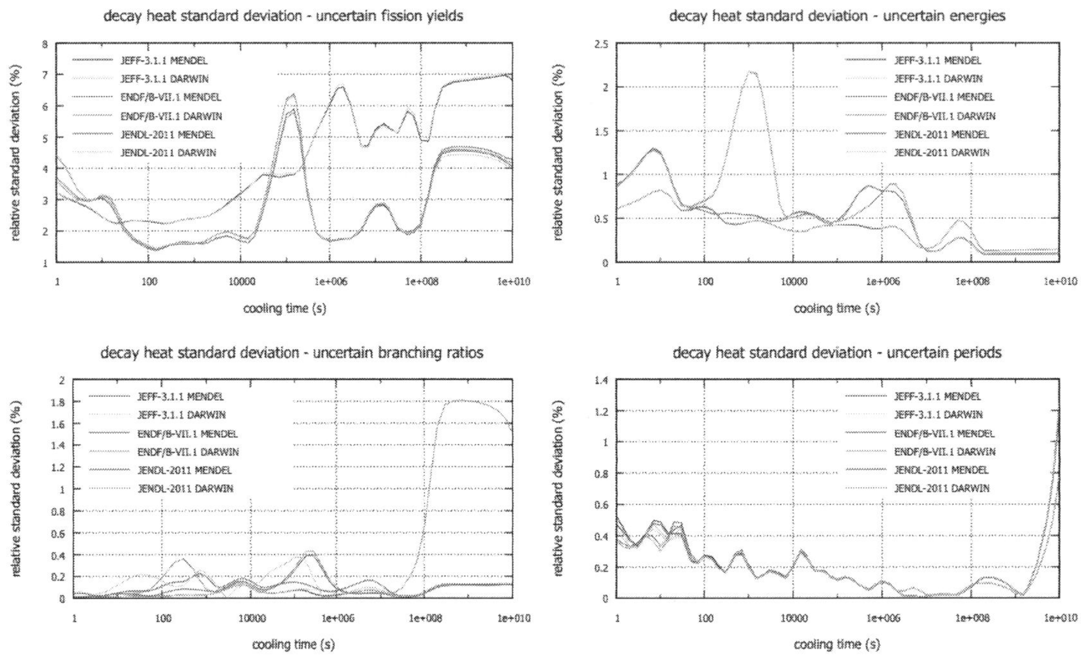

Figure 2. Uncertainty quantification of decay heat due to each parameter parameters.

4.2 Uncertainty quantification for nuclide density

DARWIN/PEPIN2 and MENDEL were used in this study to propagate uncertainty data to isotopic concentration of various fission products. Results are presented in this paper for Iodine-131 only but are representative of results for other nuclei. Uncertainty on Iodine-131 is due on independent fission yields (initial concentration), radioactive decay branching ratios and radioactive decay periods.

This work has been done assuming that uncertainty data not present in the evaluation corresponds to physical data known without uncertainty. This leads to no uncertainty of Iodine-131 concentration due to radioactive decay branching ratios for JEFF-3.1.1 in the bottom left of Figure 4.

Numerical precision effects can appear for when concentrations tends to 0 as MENDEL works in double precision and DARWIN/PEPIN2 in simple precision. MENDEL results where truncated to match DARWIN/PEPIN2 accuracy, but differences can still be seen for radioactive decay branching ratios around 10^8 s, when DARWIN/PEPIN2 goes down to zero before MENDEL.

5 Conclusion

This work validates DARWIN/PEPIN2 and MENDEL uncertainty quantification of isotopic concentrations for a Uranium-235 elementary thermal fission pulse.

The importance of differences between evaluations on uncertainty data leads several new needs from evaluation files. Correction on uncertainty levels incoherences like for ENDF/B-VII. and

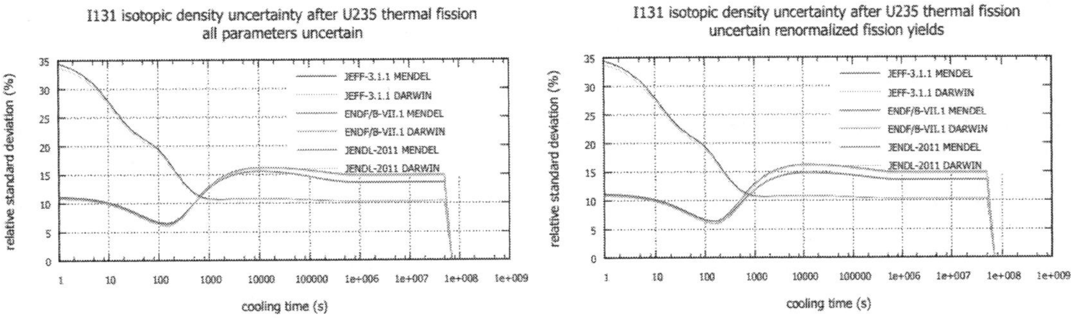

Figure 3. Uncertainty quantification of Iodine-131 concentration due to all parameters (left) and independent fission yields (left).

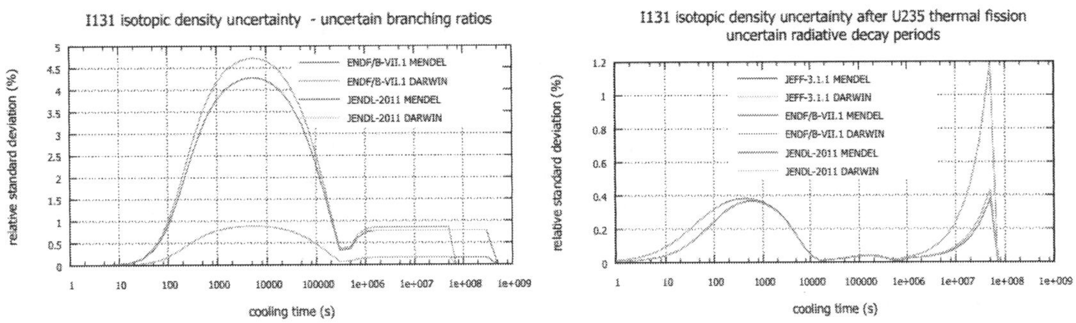

Figure 4. Uncertainty quantification of decay heat due to branching ratios (left) and periods (right).

JENDL-2011 radioactive decay branching ratios is needed to be able to use completely the data. Availability of covariances data on fission yields is expected, as well as the completion of uncertain data not present in the current evaluations.

Acknowledgement

The authors thank EDF and AREVA for their financial support to DARWIN/PEPIN2 and MENDEL codes development.

References

[1] A. Tsilanizara et al., Nuclear Science Technology Supplement **1**, 845 (2000)

[2] S. Lahaye et al., *First Verification and Validation Steps of MENDEL Release V1.0 Cycle Code System*, in *PHYSOR 2014* (2014)

[3] A. Tsilanizara et al., *Probabilistic approach for decay heat uncertainty estimation using URANIE platform and MENDEL depletion code*, in *SNA+MC 2013* (EDP Sciences, 2014)

[4] F. Gaudier, Procedia - Social and Behavioral Sciences, Sixth International Conference on Sensitivity Analysis of Model Output **2** (2010)

EPJ Web of Conferences **111**, 09003 (2016)
DOI: 10.1051/epjconf/201611109003

A Covariance Generation Methodology for Fission Product Yields

N. Terranova[1,2,a], O. Serot[2], P. Archier[2], V. Vallet[2], C. De Saint Jean[2], and M. Sumini[1]

[1] *Industrial Engineering Department, University of Bologna, Italy*
[2] *CEA-Cadarache, DEN, DER, SPRC, Saint-Paul-Lez-Durance, F-13108, France*

Abstract. Recent safety and economical concerns for modern nuclear reactor applications have fed an outstanding interest in basic nuclear data evaluation improvement and completion. It has been immediately clear that the accuracy of our predictive simulation models was strongly affected by our knowledge on input data. Therefore strong efforts have been made to improve nuclear data and to generate complete and reliable uncertainty information able to yield proper uncertainty propagation on integral reactor parameters. Since in modern nuclear data banks (such as JEFF-3.1.1 and ENDF/BVII.1) no correlations for fission yields are given, in the present work we propose a covariance generation methodology for fission product yields. The main goal is to reproduce the existing European library and to add covariance information to allow proper uncertainty propagation in depletion and decay heat calculations. To do so, we adopted the Generalized Least Square Method (GLSM) implemented in CONRAD (COde for Nuclear Reaction Analysis and Data assimilation), developed at CEA-Cadarache. Theoretical values employed in the Bayesian parameter adjustment are delivered thanks to a convolution of different models, representing several quantities in fission yield calculations: the Brosa fission modes for pre-neutron mass distribution, a simplified Gaussian model for prompt neutron emission probability, the Wahl systematics for charge distribution and the Madland-England model for the isomeric ratio. Some results will be presented for the thermal fission of U-235, Pu-239 and Pu-241.

1 Introduction

Nuclear data improvement gained an outstanding importance in the scientific community in the last decades, becoming a prior need in advanced nuclear system design. Future innovative nuclear systems are in fact imposing strict safety-by-design standards that require highly accurate engineering parameter uncertainty estimation. It has been quickly recognized that the value and the credibility of any uncertainty analysis was strictly dependent on the scientific quality of variances and correlations [1, 2] associated to the basic nuclear data. Supplying complete covariance information gives in fact the opportunity to estimate realistically the uncertainty on integral reactor parameters.
Fission product yields (FY) are fundamental nuclear data for burn-up and activation calculations, including those on decay heat, shielding, dosimetry, fuel handling, waste disposal and safety [3]. Nowadays full uncertainty information are not available in the present data banks (such as JEFF-3.1.1

[a] e-mail: nicholas.terranova@unibo.it

and ENDF/BVII.1) and the nuclear community expressed the need for full fission yield covariance matrices in core and fuel cycle analysis, to take into account complete uncertainty data [4]. Several efforts are being spent to generate reliable FY covariances by different organizations [7, 8]. In the present work we propose a possible alternative methodology to be applied in FY covariance generation. We chose to adopt the GLSM and the analytical marginalization tools available in CONRAD [5] (COde for Nuclear Reaction Analysis and Data assimilation), developed at CEA-Cadarache. In the following sections we will describe the main features of the procedure we adopted to generate preliminary results on the thermal fission of U-235, Pu-239 and Pu-241. Furthermore, covariance matrices produced through the present methodology have been tested in decay heat uncertainty evaluations using the CYRUS code [6], which allows to estimate decay heat sensitivity coefficients through direct perturbation calculations.

2 Description of the FY Covariance Generation Methodology

2.1 Main Definitions

Before starting to talk about the procedure that was actually used to generate FY covariances, let us recall some FY definitions which are useful hereinafter. In the ENDF format, two types of FY can be found

- Independent FY, $Y_{post}(A, Z, M)$, which are the probability to obtain a certain isomer of mass A, charge Z and isomeric state M as fission product after the prompt neutron emission, but before any radioactive decay. For this reason we also speak about post-neutron FY.
- Cumulative FY, $C(A, Z, M)$, which take into account radioactive decay. They are in fact the summation of all the contributions to a given isomer overall the entire decay time, since the fission process happened.

In formulas cumulative fission yields are given by

$$C_i = Y_i + \sum_j C_j b(j \rightarrow i), \tag{1}$$

where we called with C_i the cumulative yield for an isomer i identified by the triplet $(A, Z, M)_i$, with Y_i its corresponding independent yield Y_i and with $b(j \rightarrow i)$ the branching ratio which gives the probability that an isomer j, characterized by $(A, Z, M)_j$, decays into i.
Summing over the charge and the isomeric state we should obtain what is usually called independent mass yield, $Y(A) = \sum_{Z,M} Y_{post}(A, Z, M)$. Independent mass yields $Y(A)$ differ of few percent by chain yields, $Ch(A)$. The latters are equal to the sum of all stable long-lived cumulative yields for a given mass chain [3]. The main reason is that unstable neutron-rich fission isomers mostly decay through β-particle emission, without any mass change. In the following sections the semi-empirical models we used to reproduce the JEFF-3.1.1 library and to generate covariances will be discussed.

2.2 FY Semi-Empirical Models

In the present FY nuclear data library, such as JEFF-3.1.1, semi-empirical models are employed to calculate independent fission yields. Such models are used in a fitting procedure to interpolate-extrapolate missing data, especially for those fissioning systems which have few available experimental data. Independent fission yields can be written as the product of three factors

$$Y_{post}(A, Z, M) = Y_{post}(A) f(A, Z) R(A, Z, M) \tag{2}$$

where $f(A, Z)$ is the charge distribution giving the fractional independent yields and $R(A, Z, M)$ is the isomeric ratio. In JEFF-3.1.1 evaluation mass yields interpolation and extrapolation are deduced performing a multi-Gaussian systematics calculation to fit chain yields [3, 9]. Multiple Gaussians can in fact well represent the general trends of mass distributions, if the fluctuations due to prompt neutrons emission are ignored. Such systematics can be roughly justified thanks to the multi-channel fission theory proposed by Brosa [10]. In the Brosa model, however, only pre-neutron mass yields distributions can be represented by a summation of five Gaussian corresponding to three fission modes. The Standard I and Standard II modes are responsible for the asymmetric and the very asymmetric fission. The symmetric Super Long instead is centered in $A_f/2$, where A_f is the mass of the compound nucleus. In formulas we have

$$Y_{pre}(A) = \sum_{i=s1,s2} \sum_{\Delta=\pm D_i} \frac{N_i}{\sqrt{2\pi}\sigma_i} \cdot e^{-\frac{(A-A_f/2-\Delta)^2}{2\sigma_i^2}} + \frac{2N_{sl}}{\sqrt{2\pi}\sigma_{sl}} \cdot e^{-\frac{(A-A_f/2)^2}{2\sigma_{sl}^2}}, \quad \text{with} \sum_{i=s1,s2,sl} N_i = 1.0. \tag{3}$$

N_i are the weights of the several fission modes, σ_i the corresponding widths and D_i the deviations from the mass symmetry for the two asymmetric fission modes.

In JEFF-3.1.1 FY charge distributions $f(A, Z)$ are obtained using Wahl systematics for independent fission yields [11] and the Madland England [12] model is used instead to retrieve isomeric ratios.

In the present work, to represent JEFF-3.1.1 in a physical consistent way, we decided to adopt the Brosa model to represent pre-neutron mass distributions as it was meant to be. In order to finally obtain the independent mass yield distribution we seek, a prompt neutron emission probability model based on Gaussians distributions has been used. The Wahl systematics has been applied directly to the pre-neutron distributions, which have to be multiplied by prompt neutron emission probabilities $p_A(\nu)$ to give post-neutron yields. Such functions have been assumed as Gaussians centered in $\bar{\nu}$. The average number of prompt neutrons emitted $\bar{\nu}$ is a function of the pre-neutron mass A_{pre} of the primary fission fragments. Experimental measurements have shown a saw-tooth distribution for $\bar{\nu}(A_{pre})$, and such shape is roughly maintained between different fissioning systems. In formulas we get then

$$Y_{post}(A, Z, M) = \Big[\sum_{\nu_i=0}^{\infty} Y_{pre}(A+\nu_i) \cdot p_{A+\nu_i}(\nu_i) \cdot f(A+\nu_i, Z) \Big] \cdot R(A, Z, M). \tag{4}$$

Further explanation on the prompt neutron emission model can be found in [13]. Cumulative FY can be subsequently obtained through the **Q**-matrix formalism [3].

2.3 GLSM and Marginalization

The FY models presented in Sect. 2.2 have been employed in a GLSM to retrieve adjusted model parameters able to represent JEFF-3.1.1 FY library. Using CONRAD, we minimize a cost function [14]

$$\chi^2(\vec{x}) = (\vec{x} - \vec{x}_{prior})^T \mathbf{M}_{prior}^{-1} (\vec{x} - \vec{x}_{prior}) + [\vec{y} - \vec{f}(\vec{x})]^T \mathbf{M}_y^{-1} [\vec{y} - \vec{f}(\vec{x})], \tag{5}$$

using an iterative Newton-Raphson algorithm which gives

$$\mathbf{M}_x^{(n-1)} = \left[\mathbf{M}_{prior}^{-1} + \mathbf{G}^{(n-1)^T} \mathbf{M}_y^{-1} \mathbf{G}^{(n-1)} \right]^{-1}, \tag{6}$$

$$\vec{x}^{(n)} = \vec{x}^{(n-1)} - \mathbf{M}_x^{(n-1)} \cdot \left[\mathbf{G}^{(n-1)^T} \mathbf{M}_y^{-1} \cdot (\vec{f}(\vec{x}^{(n-1)}) - \vec{y}) + \mathbf{M}_{prior}^{-1} (\vec{x}^{(n-1)} - \vec{x}_{prior}) \right]. \tag{7}$$

Without entering into the details on how Eq. 6 and Eq. 7 are derived (see Ref. [15]), posterior model parameters $\vec{x}$ and their covariance matrix $\mathbf{M}_x$ can be found once priors $\vec{x}_{prior}$ and new experimental

values $\vec{y}$ are available, provided with corresponding covariance matrices $\mathbf{M}_{prior}$ and $\mathbf{M}_y$. $\vec{f}(\vec{x})$ are the FY models just presented and $\mathbf{G}^{(n)}$ their derivatives. We performed the adjustment using JEFF-3.1.1 evaluated FY data as pseudo-experimental values. Brosa fission mode characteristics, $\bar{\nu}$ values, a single width for all the prompt neutron Gaussian probabilities, Wahl model parameters and the Madland England cut-off were simultaneously adjusted to achieve a satisfactory JEFF-3.1.1 representation. Such procedure was performed including extra experimental values concerning miscellaneous fission quantities, as the total prompt neutron emission probability distribution and the total average number of prompt neutrons emitted by the fissioning system. The parameter covariance matrix $\mathbf{M}_x$ was finally deduced enabling a retroactive analysis fitting only Brosa, Wahl and Madland England parameters.
To better reproduce also the uncertainties provided by the European library, an analytical Marginalization technique [16] has been set up to introduce systematic uncertainties on the normalization of FY charge distribution. The main issue of this phase was to represent all the uncertainties of the hundreds of independent FY available for each fissioning system. For U-235 for example, we focused on the normalization of Xe and Cs distributions. Acceptable results have been obtained also for other fission products, as demonstrated by the decay heat uncertainty calculations.

3 Preliminary Results

In this section some results achieved using the methodology just described are presented. In particular we will show the capabilities of this FY model to represent JEFF-3.1.1 evaluated data, with the main goal to obtain consistent correlation matrices. Some results on the decay heat uncertainties calculation using the CYRUS code will be included.

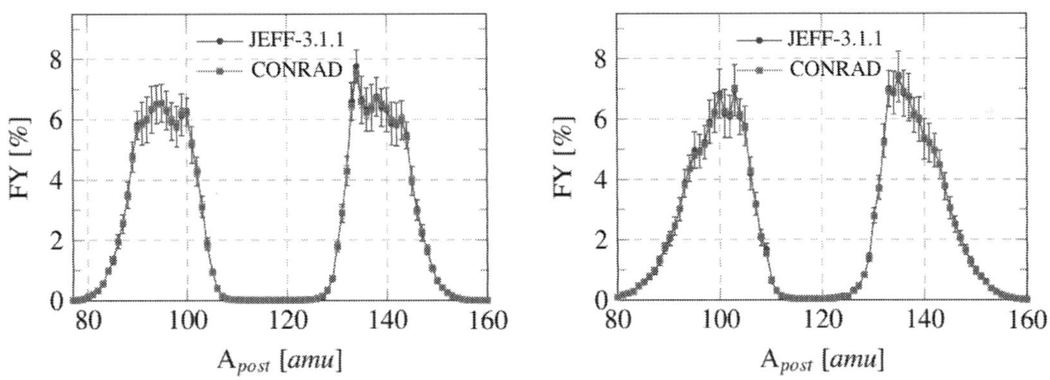

(a) Mass Independent FY Distribution U-235

(b) Mass Independent FY Distribution Pu-239

Figure 1: Comparison between JEFF-3.1.1 and CONRAD independent mass yields distribution for the thermal fissions of U-235 (left) and Pu-239 (right).

3.1 Covariance Matrix Generation

Satisfactory comparisons between CONRAD and FY evaluated in JEFF-3.1.1 have been performed for three thermal fissioning systems: U-235, Pu-239 and Pu-241. Hereinafter a few results will be presented just to point out the potential of the present methodology, without pretending to be exhaustive. Fig. 1(a) and Fig. 1(b) show how present FY models can nicely represent JEFF-3.1.1 mass independent fission yields for the thermal fission of U-235 and Pu-239. Similar results can be obtained also for the thermal fission of Pu-241. FY adjustment has been performed including miscellaneous

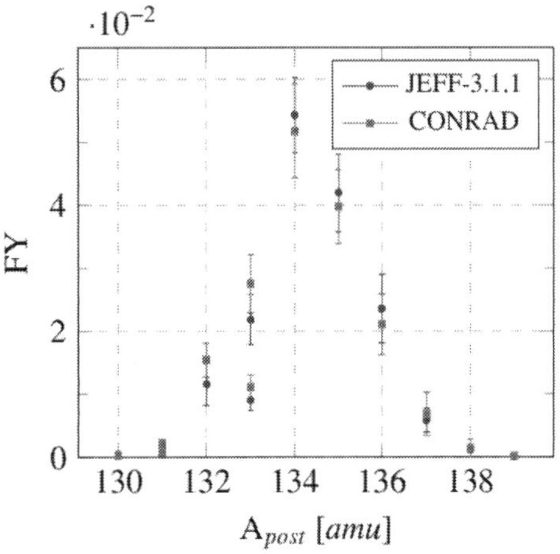

(a) Tellurium FY for Pu-241 fission

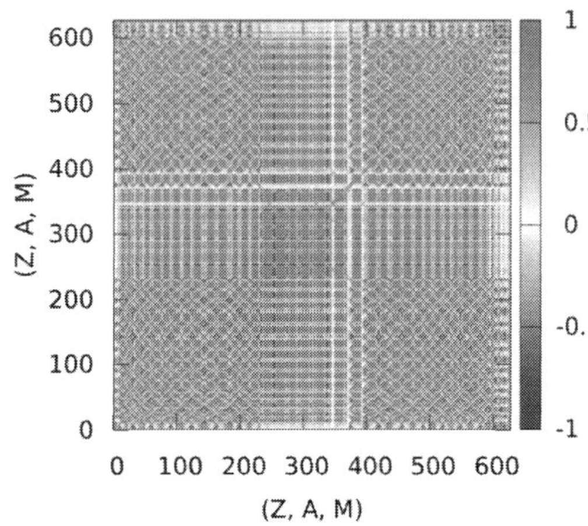

(b) Pu-241 Correlation Matrix

Figure 2: Pu-241 isomeric independent fission yields distribution (left, only Tellurium has been shown) and correlation matrix (right). Only FY>10^{-7} have been considered. Matrix coordinates are just cardinal numbers for a list of ~600 fission products, ordered for ascending charge, mass and isomeric state.

fission quantities such as the total prompt neutron emission probability $P(\nu)$ and the average number of prompt neutrons emitted by the fissioning system $\bar{\nu}_P$, which have been effectively reproduced [13]. In Fig. 2(a) and 2(b) Tellurium isotopic distribution and the full independent correlation matrix are presented for the thermal fission of Pu-241. Even isomeric fission yields are well described by the models and a correlation matrix can be found preserving agreement between CONRAD and JEFF results.

3.2 Decay Heat Uncertainty Estimation

To test fission yield covariance matrices, we used the CYRUS code, which performs decay heat uncertainty estimations. For the elementary thermal fission of U-235, the decay heat uncertainties due to U-235 fission yields provided by JEFF-3.1.1 and by CONRAD have been compared (see Fig. 3(a)). Results show reasonable agreement between the propagation of JEFF-3.1.1 and CONRAD uncertainties, without including correlations between data. Full CONRAD FY matrices propagation on the other hand provoked a dramatic decay heat uncertainty reduction, which has to be further investigated. The same exercise was repeated even for a PWR-UOX pin-cell (Fig. 3(b)), providing same conclusions and showing that FY uncertainties for fission products which are sensitive to such problems are well represented by the present methodology.

4 Conclusions

A methodology to generate FY covariance matrices has been presented. The main goal of such procedure is to reproduce the existing European FY library JEFF-3.1.1 in a physical consistent way and to generate covariances. Such covariances have been tested for applications in decay heat uncertainty calculations, showing how FY correlations can have an important impact on integral reactor parameter uncertainty reduction.

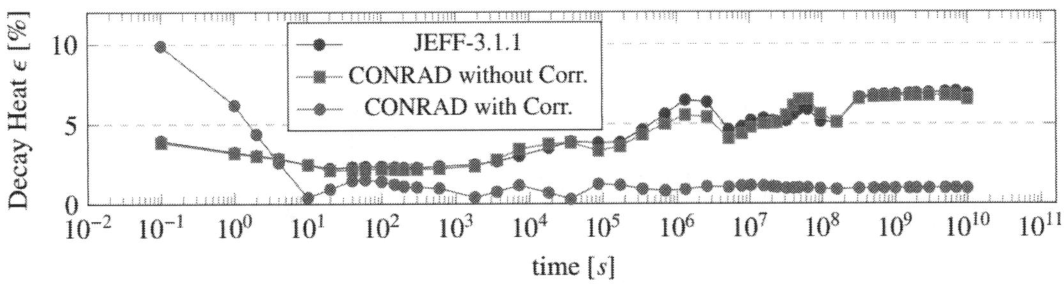

(a) Elementary Fission

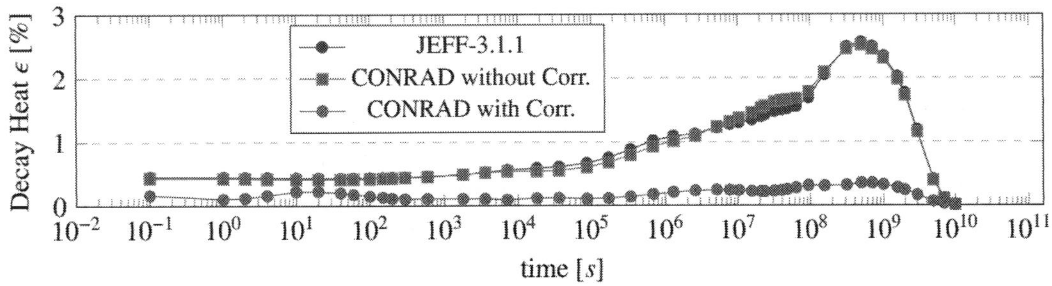

(b) PWR-UOX Pin-Cell

Figure 3: Decay heat uncertainty estimation due to U-235 FY propagation in elementary fission (top) and PWR-UOX pin-cell (bottom) calculations.

References

[1] G. Palmiotti et al., Nuclear Data Sheets 118, 596 (2014).
[2] G. Aliberti et al., Annals of Nuclear Energy 33, 700 (2006).
[3] R. Mills, "Fission Product Yields Evaluation", PhD Thesis, University of Birmingham, United Kingdom (1995).
[4] R. Mills, WPEC SG-37 Proposal, (2012).
[5] P. Archier et al., Nuclear Data Sheets 118, 488–490 (2014).
[6] V. Vallet et al., PHYSOR-2014, Kyoto (2014).
[7] L. Fiorito et al., Annals of Nuclear Energy 69, 331, (2014).
[8] M. T. Pigni et al., Nuclear Data Sheets 123, 231–236, (2015).
[9] M. A. Kellett et al., OECD-NEA JEFF Report 20, (2009).
[10] U. Brosa et al., Physics Reports 4, 167 (1990).
[11] A. C. Wahl, "Systematics of Fission-Product Yields", LA-13928, (2002).
[12] D. G. Madland, T. R. England, Nuclear Science and Engineering 64, 859 (1977).
[13] N. Terranova et al., Nuclear Data Sheets 123, 225–230 (2015).
[14] C. De Saint Jean et al., Journal Korean Physical Society 59, 1276 (2011).
[15] D. Smith, Probability, Statistics and Data Uncertainties in Nuclear Science and Technology, OECD-NEA Committee Series, (1990).
[16] B. Habert et al., Nuclear Science and Engineering 166, 276 (2010).

EPJ Web of Conferences **111**, 10001 (2016)
DOI: 10.1051/epjconf/201611110001

Dependence of Fission-Fragment Properties On Excitation Energy For Neutron-Rich Actinides

D. Ramos[1,a], C. Rodríguez-Tajes[1,2], M. Caamaño[1], F. Farget[2], L. Audouin[3], J. Benlliure[1], E. Casarejos[4], E. Clement[2], D. Cortina[1], O. Delaune[2,b], X. Derkx[5], A. Dijon[2], D. Doré[6], B. Fernández-Domínguez[1], G. de France[2], A. Heinz[7], B. Jacquot[2], A. Navin[2], C. Paradela[2,c], M. Rejmund[2], T. Roger[2], M.D. Salsac[6], and C. Schmitt[2]

[1]*GENP, USC, E-15784 Santiago de Compostela, Spain*
[2]*GANIL, CEA/DMS-CNRS/IN2P3, BP 55027, F-14076 Caen Cedex 05, France*
[3]*IPN Orsay, Université de paris-Sud, CNRS/IN2P3, F-91406 Orsay Cedex, France*
[4]*CIMA, UVigo, E-36310 Vigo, Spain*
[5]*LPC Caen, Université de Caen Basse-Normandie-ENSICAEN-CNRS/IN2P3, F-14050 Caen Cedex, France*
[6]*CEA Saclay, DMS/IRFU/SPhN, 91191 Gif-sur-Yvette Cedex, France*
[7]*Chalmes University of Technology, SE-41296 Göteborg, Sweden*

Abstract. Experimental access to full isotopic fragment distributions is very important to determine the features of the fission process. However, the isotopic identification of fission fragments has been, in the past, partial and scarce. A solution based on the use of inverse kinematics to study transfer-induced fission of exotic actinides was carried out at GANIL, resulting in the first experiment accessing the full identification of a collection of fissioning systems and their corresponding fission fragment distribution. In these experiments, a ^{238}U beam at 6.14 AMeV impinged on a carbon target to produce fissioning systems from U to Am by transfer reactions, and Cf by fusion reactions.
Isotopic fission yields of ^{250}Cf, ^{244}Cm, ^{240}Pu, ^{239}Np and ^{238}U are presented in this work. With this information, the average number of neutrons as a function of the atomic number of the fragments is calculated, which reflects the impact of nuclear structure around Z=50, N=80 on the production of fission fragments. The characteristics of the Super Long, Standard I, Standard II, and Standard III fission channels were extracted from fits of the fragment yields for different ranges of excitation energy. The position and contribution of the fission channels as function of excitation energy are presented.

1 Introduction

The fission of the nucleus is a complex process where intrinsic excitation energy transforms into a large deformation until the nucleus splits. During this collective motion, the single-particle shell structure of nucleons strongly influences the process if the excitation energy of the fissioning system remains moderate [1–3].

[a]e-mail: diego.ramos@usc.es
[b]Present address: CEA DAM DIF, F-91297 Arpajon, France
[c]Present address: EC-JRC-IRMM, Retieseweg 111, B-2440 Geel, Belgium

In general, experimental observations are limited to few observables, making it difficult to isolate the impact of specific nuclear properties on the whole process [4–6]. The present work, based on the use of inverse kinematics, provides a new set of measurements that relates the excitation energy of the fissioning system, produced either by fusion or transfer reactions, to the characteristics of the full isotopic fragment distribution. These new measurements for fissioning systems, some of them not accessible with any other technique, also provide nuclear data, important for nuclear energy applications, such as the development of next-generation nuclear reactors or the recycling of radioactive waste.

2 Experimental Setup

A ^{238}U beam, accelerated to 6.14A MeV of energy, impinges on a 100 $\mu g/cm^2$ ^{12}C target. Transfer reactions take place producing n-rich actinides with a certain excitation-energy distribution. The target-like nucleus is detected by using a telescope composed of two double-sided, annular, Si detectors. The two elements of the telescope (ΔE, E) provide the isotopic identification. The segmentation of the detector permits the scattering angle to be measured, which, together with the measurement of the energy, allows for the reconstruction of the excitation energy [7]. A fusion reaction is assumed for every fission event that is registered without any target-like recoil detected in the telescope.

The corresponding actinide may decay by fission. In such a case, one of the fragments passes through the VAMOS spectrometer and is fully identified in terms of charge state, atomic, and mass number at the focal plane. The detection setup includes a multiwire chamber for time-of-flight measurements, two drift chambers for position and angle determination, and an ionization chamber and a wall of Si detectors to perform dE-E measurements. The angles at the target position and the energy of the fission fragments are reconstructed by back-tracing methods based on the ionic optics of the spectrometer [8–10]. Finally, six clovers of the EXOGAM array of Ge detectors were surrounding the target for γ-ray measurements, yielding information on the sharing of the excitation energy between the fissioning system and the target-like recoil [7], and providing a cross-check of the fragment identification [11].

3 Fission Products

Data on five different fissioning systems are reported in this work: ^{250}Cf, produced in fusion reactions with a well-defined excitation energy of 46 MeV, ^{244}Cm ,^{240}Pu, ^{239}Np, and ^{238}U, produced in transfer reactions with distributions of excitation energy centered at 23 MeV, 10.7 MeV, 7.5 MeV, 7.4 MeV respectively, with a full width at half maximum of approximately 8 MeV [7]. Figure 1 shows the excitation energy distributions for each fissioning system produced by transfer reactions. In the case of ^{238}U, a 23% contamination from random coincidences between elastic scattering with fusion-fission events was subtracted. Figure 2 (a) shows the fission yields of these systems as a function of the fragment atomic number. Lighter fissioning systems at lower excitation energy show mostly asymmetric fission -this is the case of ^{238}U, ^{239}Np, and ^{240}Pu- while symmetric fission becomes more relevant for heavier fissioning systems at higher excitation energy, such as ^{244}Cm and ^{250}Cf. In the asymmetric fission, the heavier group of the distribution is observed to remain in the same Z-region for the different fissioning systems while the lighter group moves with the fissioning system. This behavior reflects the presence of structure effects at scission point, making certain configurations more favored than others, independently of the fissioning system. Figure 2 (b) shows that the stabilization of the heavy group is also present when displaying the fission yields as a function of the mass of the post-neutron evaporation fission fragments for the same fissioning systems. Figure 2 (c) shows the neutron excess, defined as the ratio of the average neutron number over the atomic number, of the post-neutron evaporation fission fragments. The observed sawtooth behavior is the signature of shell effects driven of <N>≈80. This behavior becomes smoother for the systems with higher excitation energy, which is coherent with the damping of shell effects. The sawtooth behavior of the neutron excess was already observed to evolve with the excitation energy for the case of ^{240}Pu [12].

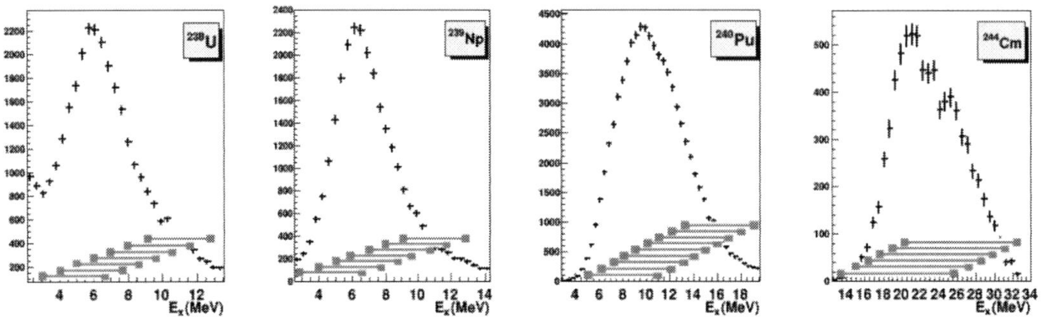

Figure 1. Distribution on excitation energy for each fissioning system produced by transfer reactions. Red lines indicates the gates on excitation energy considered for investigate the evolution of the fission-fragment distribution with the excitation energy.

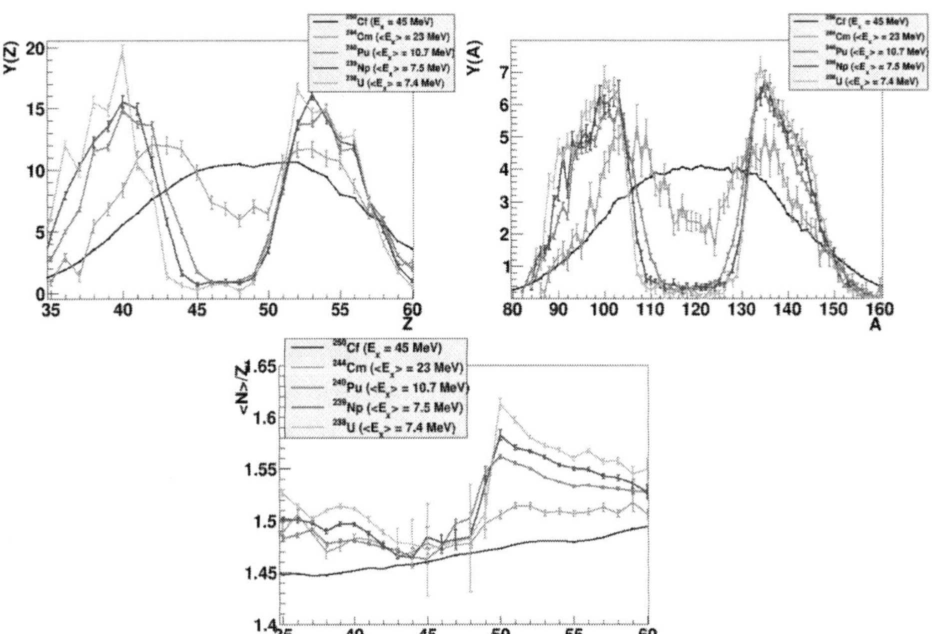

Figure 2. (a) Atomic number distributions of fission fragments for different fissioning systems. (b) Mass distributions of post-neutron evaporation fission fragments. (c) Neutron excess of post-neutron evaporation fission fragments as a function of their atomic number.

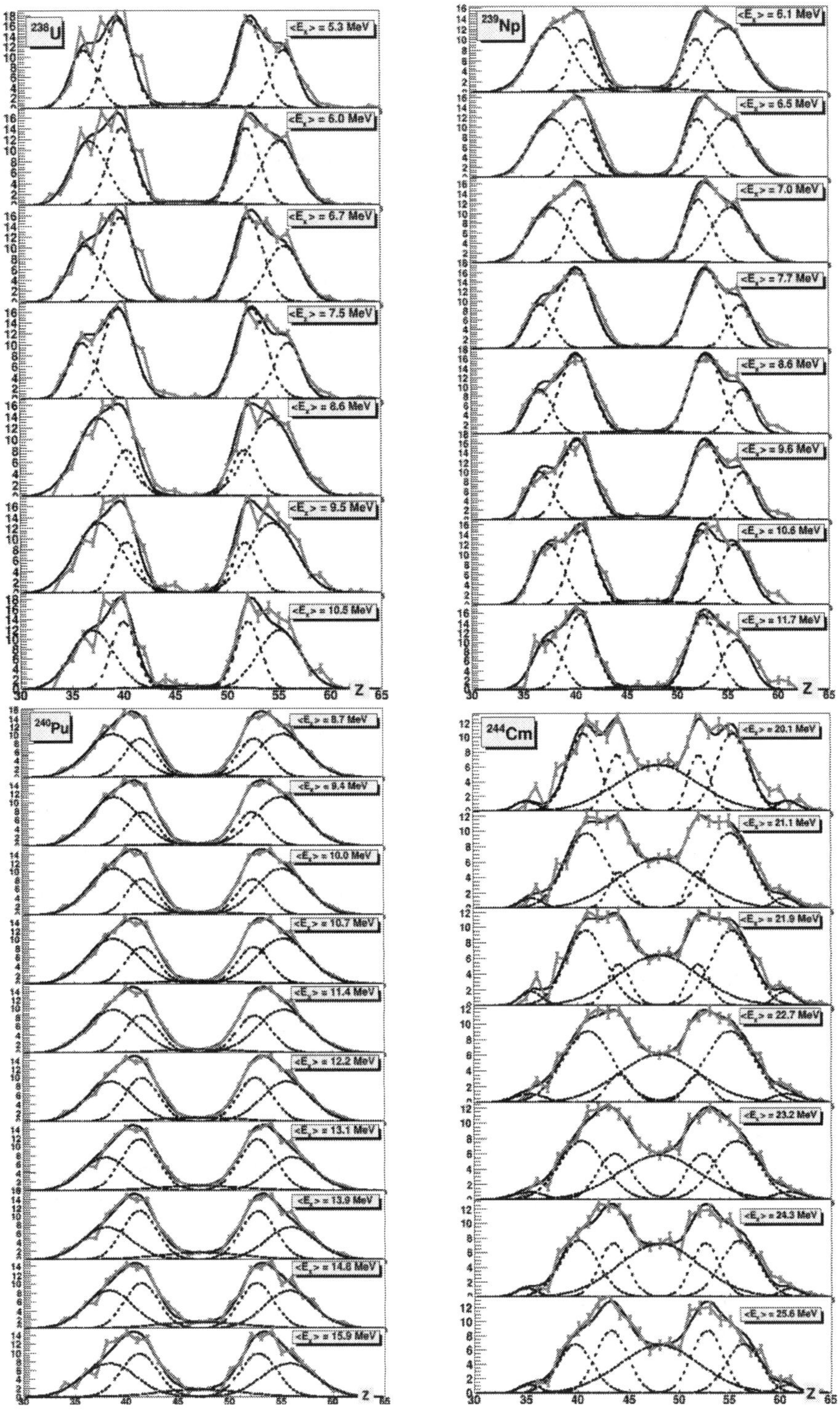

Figure 3. Atomic number distribution of fragments produced in the fission of ^{238}U, ^{239}Np, ^{240}Pu and ^{244}Cm. Different ranges of excitation energy were investigated. The Standard I and Standard II fission channels are fitted to the distributions of ^{238}U, ^{239}Np and ^{240}Pu. In the case of ^{244}Cm, the Standard III channel was needed to be included. The Super Long channel is fixed, the amplitude corresponds to the value at the distribution in the symmetry, and its width is fixed to 4, following the prescription [13].

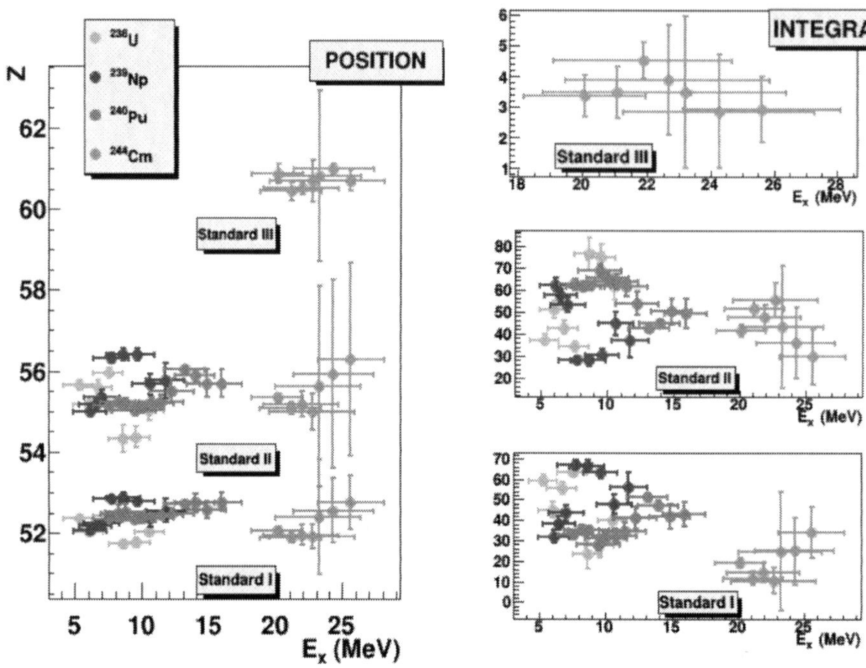

Figure 4. (left)Fitted positions and (right) integrals of the Standard I, Standard II and Standard III fission channels for ^{238}U, ^{239}Np, ^{240}Pu and ^{244}Cm as function of the excitation energy. A general trend in the position of the channels with the excitation energy is observed.

The fission fragment distribution is investigated also by scanning the excitation energy induced in the transfer reaction [7]. Overlapping gates are defined in the excitation energy distribution for the fissioning systems produced by transfer reactions; these gates are shown in Figure 1. The atomic number fragment distribution is studied for each gate and each fissioning system. Figure 3 shows the Z distributions of fission fragments for different ranges of excitation energy in the case of (a) ^{238}U, (b) ^{239}Np, (c) ^{240}Pu, and (d) ^{244}Cm. Also presented, is a fit of these distributions with different fission channels. The Super Long, Standard I, and Standard II fission channels are considered for the lightest systems, namely ^{238}U,^{239}Np, and ^{240}Pu. For the heavier system, ^{244}Cm, the Standard III channel has to be included to describe the distribution. The parameters of the Super Long channel are fixed: the amplitude corresponds to the value of the distribution at the symmetry position and the width is fixed to 4, following the prescription given in Ref. [13]. Figure 4 shows the positions (left) and the integrals (right) of the Standard I, Standard II, and Standard III fission channels obtained from the Z distribution fits as a function of the excitation energy. An evolution of the positions of the asymmetric fission channels is observed with the excitation energy. In general, there is an increase of the position with the excitation energy except in the case of ^{238}U, which shows a fast change with the excitation energy that can be explained by a remaining contamination from ^{250}Cf at low excitation energy. In Figure 5, the Super Long, Standard I, Standard II, and Standard III fission channels are fitted to the fragment Z distribution from fission of ^{250}Cf where only one well-defined excitation energy of 46 MeV is available. The positions of the S I, S II, and S III were fitted to 53.0, 56.1, and 59.7 respectively, the integrals were fitted to 5.2, 27.2, and 16.8 respectively.

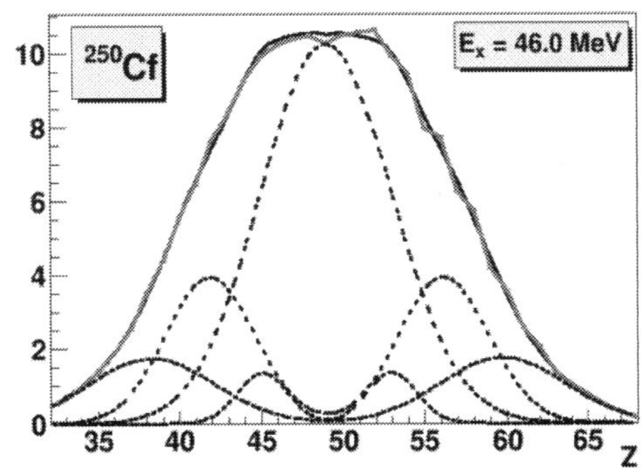

Figure 5. Atomic number distribution of fragments produced in the fission of ^{250}Cf with a well-defined excitation energy of 46 MeV. The Standard I, Standard II and Standard III fission channels are fitted to the distributions.The Super Long channel is fixed; the amplitude corresponds to the value of the distribution at symmetry, and its width is fixed to 4, following the K.-H. Schmidt prescription [13] . The positions of the S I, S II and S III were fitted to 53.0, 56.1 and 59.7, respectively, the integrals were fitted to 5.2, 27.2 and 16.8, respectively.

4 Conclusions and outlook

This experiment allowed the study of fission of five different systems, ^{238}U, ^{239}Np, ^{240}Pu, ^{244}Cm, and ^{250}Cf, most of them exotic nuclei. The reconstruction of the fission-induced reactions and the isotopic identification of their fission products show the evolution of the characteristics of the fragments as a function of the available excitation energy of the fissioning system.

The contribution of symmetric fission is observed to increase for heavier fissioning systems with higher excitation energy, as expected. The investigation of the neutron excess of the fragments reflects nuclear-structure effects driven of $<N> \approx 80$ in the region of $Z \approx 50$ that increase for lower excitation energies and lighter fissioning systems.

The study of the fission channels shows an evolution of the positions of the asymmetric channels, Standard I, Standard II, and Standard III with the excitation energy.

Finally, fragment velocities and total kinetic energies will become available as the analysis progresses and a more precise parametrization of the fission channels, taking into account the total kinetic energy distribution of the fission fragments, will be performed.

This work was partially supported by the Spanish Ministry of Research and Innovation under the Grants No. FPA2010-22174-C02-01 and RYC-2012-11585.

References

[1] J. L. Sida *et al.*, Nucl. Phys. A **502**, 233c (1989).

[2] S. Steinhäuser *et al.*, Nucl. Phys. A **634**, 89 (1998).

[3] C. Böckstiegel *et al.*, Nucl. Phys. A **802**, 12 (2008).

[4] A. Bail *et al.*, Phys. Rev. C **84**, 034605 (2011).

[5] J.-P. Bocquet and R. Brissot, Nucl. Phys. A **502**, 213 (1989).

[6] W. Lang *et al.*, Nucl. Phys. A **345**, 34 (1980).

[7] C. Rodríguez-Tajes *et al.*, Phys. Rev. C **89**, 24614 (2014).

[8] S. Pullanhiotan *et al.*, Nucl. Instr. and Methods A **593**, 343 (2008).

[9] M. Rejmund *et al.*, Nucl. Instr. and Methods A **646**, 184 (2011).
[10] M. Caamaño *et al.*, Phys. Rev. C **88**, 24605 (2013).
[11] A. Shrivastava *et al.*, Phys. Rev. C **80**, 051305 (2009).
[12] D. Ramos *et al.*, Acta Phys. Pol. B **46**, 443 (2015).
[13] K.-H. Schmidt *et al.*, Nucl. Phys. A **665**, 221 (2000).

EPJ Web of Conferences **111**, 10002 (2016)
DOI: 10.1051/epjconf/201611110002

Fission Fragment Angular Distribution measurements of ^{235}U and ^{238}U at CERN n_TOF facility

E. Leal-Cidoncha[1,a], I. Durán[1], C. Paradela[1,30], D. Tarrío[1,32], L.S. Leong[2,33], L.Tassan-Got[2], L. Audouin[2], S. Altstadt[3], J. Andrzejewski[4], M. Barbagallo[5], V. Bécares[6], F. Bečvář[7], F. Belloni[30], E. Berthoumieux[8], J. Billowes[10], V. Boccone[9], D. Bosnar[11], M. Brugger[9], M. Calviani[9], F. Calviño[12], D. Cano-Ott[6], C. Carrapiço[13], F. Cerutti[9], E. Chiaveri[9], M. Chin[9], N. Colonna[5], G. Cortés[12], M.A. Cortés-Giraldo[14], M. Diakaki[8], C. Domingo-Pardo[16], R. Dressler[17], N. Dzysiuk[18], C. Eleftheriadis[19], A. Ferrari[9], K. Fraval[8], S. Ganesan[20], A.R. García[6], G. Giubrone[16], M.B. Gómez-Hornillos[12], I.F. Gonçalves[13], E. González-Romero[6], E. Griesmayer[21], C. Guerrero[14], F. Gunsing[8,9], P. Gurusamy[20], A. Hernández-Prieto[9], D.G. Jenkins[22], E. Jericha[21], Y. Kadi[9], F. Käppeler[23], D. Karadimos[15], N. Kivel[17], P. Koehler[24], M. Kokkoris[15], M. Krtička[7], J. Kroll[7], C. Lampoudis[19], C. Langer[3], C. Lederer[3,25], H. Leeb[21], S. Lo Meo[27], R. Losito[9], A. Mallick[20], A. Manousos[19], J. Marganiec[4], T. Martínez[6], C. Massimi[26], P.F. Mastinu[18], M. Mastromarco[5], M. Meaze[5], E. Mendoza[6], A. Mengoni[27], P.M. Milazzo[28], F. Mingrone[26], M. Mirea[29], W. Mondelaers[30], A. Pavlik[25], J. Perkowski[4], A. Plompen[30], J. Praena[14], J.M. Quesada[14], T. Rauscher[34,31], R. Reifarth[3], A. Riego[12], M.S. Robles[1], F. Roman[9], C. Rubbia[9], M. Sabaté-Gilarte[9,14], R. Sarmento[13], A. Saxena[20], P. Schillebeeckx[30], S. Schmidt[3], D. Schumann[17], G. Tagliente[5], J.L. Tain[16], A. Tsinganis[9], S. Valenta[7], G. Vannini[26], V. Variale[5], P. Vaz[13], A. Ventura[27], R. Versaci[9], M.J. Vermeulen[22], V. Vlachoudis[9], R. Vlastou[15], A. Wallner[35], T. Ware[10], M. Weigand[3], C. Weiß[9], T. Wright[10], and P. Žugec[11]

[1] *Universidade de Santiago de Compostela (USC), Spain*
[2] *Centre National de la Recherche Scientifique/IN2P3 - IPN, Orsay, France*
[3] *Johann-Wolfgang-Goethe Universität, Frankfurt, Germany*
[4] *Uniwersytet Lódzki, Lodz, Poland*
[5] *Istituto Nazionale di Fisica Nucleare, Bari, Italy*
[6] *Centro de Investigaciones Energeticas Medioambientales y Tecnológicas (CIEMAT), Madrid, Spain*
[7] *Charles University, Prague, Czech Republic*
[8] *Commissariat à l'Énergie Atomique (CEA) Saclay - Irfu, Gif-sur-Yvette, France*
[9] *European Organization for Nuclear Research (CERN), Geneva, Switzerland*
[10] *University of Manchester, Oxford Road, Manchester, UK*
[11] *Department of Physics, Faculty of Science, University of Zagreb, Croatia*
[12] *Universitat Politecnica de Catalunya, Barcelona, Spain*
[13] *CTN, Instituto Superior Técnico, Universidade Técnica de Lisboa, Lisboa, Portugal*
[14] *Universidad de Sevilla, Spain*
[15] *National Technical University of Athens (NTUA), Greece*
[16] *Instituto de Física Corpuscular, CSIC-Universidad de Valencia, Spain*
[17] *Paul Scherrer Institut, Villigen PSI, Switzerland*
[18] *Istituto Nazionale di Fisica Nucleare, Laboratori Nazionali di Legnaro, Italy*
[19] *Aristotle University of Thessaloniki, Thessaloniki, Greece*

[a] e-mail: esther.leal@usc.es

[20] Bhabha Atomic Research Centre (BARC), Mumbai, India
[21] Atominstitut, Technische Universität Wien, Austria
[22] University of York, Heslington, York, UK
[23] Karlsruhe Institute of Technology, Campus Nord, Institut für Kernphysik, Karlsruhe, Germany
[24] Department of Physics, University of Oslo, N-0316 Oslo, Norway
[25] University of Vienna, Faculty of Physics, Austria
[26] Dipartimento di Fisica, Università di Bologna, and Sezione INFN di Bologna, Italy
[27] Agenzia nazionale per le nuove tecnologie, l'energia e lo sviluppo economico sostenibile (ENEA), Bologna, Italy
[28] Istituto Nazionale di Fisica Nucleare, Trieste, Italy
[29] Horia Hulubei National Institute of Physics and Nuclear Engineering (IFIN-HH) Bucharest-Magurele, Romania
[30] IRMM, European Commission JRC, Retieseweg 111, B-2440 Geel, Belgium
[31] Department of Physics and Astronomy - University of Basel, Basel, Switzerland
[32] Department of Physics and Astronomy, Uppsala University, Sweden
[33] JAEA, Japan
[34] CAR, School of Physics, Astronomy and Mathematics, University of Hertfordshire, UK
[35] Research School of Physics and Engineering, Australian National University, Australia

Abstract. Neutron-induced fission cross sections of ^{238}U and ^{235}U are used as standards in the fast neutron region up to 200 MeV. A high accuracy of the standards is relevant to experimentally determine other neutron reaction cross sections. Therefore, the detection efficiency should be corrected by using the angular distribution of the fission fragments (FFAD), which are barely known above 20 MeV. In addition, the angular distribution of the fragments produced in the fission of highly excited and deformed nuclei is an important observable to investigate the nuclear fission process.

In order to measure the FFAD of neutron-induced reactions, a fission detection setup based on parallel-plate avalanche counters (PPACs) has been developed and successfully used at the CERN-n_TOF facility. In this work, we present the preliminary results on the analysis of new ^{235}U(n,f) and ^{238}U(n,f) data in the extended energy range up to 200 MeV compared to the existing experimental data.

1 Introduction

The angular distribution of the fragments produced in the fission of an excited nucleus (FFAD) provides relevant information of the fission barrier structure and the transition levels. A good knowledge on the FFAD is also required in the calculation of the detection efficiency for fission fragments (FF) to provide accurate measurements of fission cross sections. The fission cross sections of ^{235}U and ^{238}U are used as standards, the first in the neutron energy from 150 keV to 200 MeV and the second from 2 MeV up to 200 MeV. The recently published n_TOF data of the ^{238}U/^{235}U fission cross section ratio [1] were obtained fitting the anisotropy parameter of the FFAD data available in the EXFOR database. The FFAD experimental data for the (n,f) reactions of ^{235}U and ^{238}U published in the EXFOR [2] database reach 20 MeV, the only ^{238}U data exceeding this energy up to 100 MeV are those of Ryzhov et al. [3]. The recently published data of ^{238}U and ^{235}U by Vorobyev et al. [4] and the data of Leong [5] reach 200 MeV. Measurements at intermediate neutron energies would be of interest in the next calculations of the ^{235}U(n,f) and ^{238}U(n,f) cross sections. In this work we present the preliminary results of the FFAD of ^{235}U and ^{238}U up to 200 MeV measured at the n_TOF facility (CERN).

2 Experimental setup

The experiment was accomplished in the CERN Neutron Time-of-Flight (n_TOF) facility [6]. The white neutron beam is produced through spallation induced by a 20 GeV/c bunched proton beam on a lead target and it is moderated in a 4 cm thick layer of borated water. The neutron beam covers the energy range from thermal up to 1 GeV travelling along approximately 185 m from the spallation target to the chamber containing the detectors and targets, which is located in the experimental area EAR 1. A sweeping magnet is located in the neutron path to remove charged particles from the beam and two collimators are used to give shape to the beam, where the second determines the beam diameter of 8 cm for fission measurements.

Nine targets, produced at IPN-Orsay, were used in this experiment: three of ^{234}U, one of ^{237}Np, two of ^{235}U and three of ^{238}U. They consisted of a thin radioactive layer 8 cm in diameter deposited on an Aluminium foil of different thickness: 2.5 μm for the first six targets in the beam direction and 0.7 μm for the last three targets (two of them of ^{238}U and one of ^{235}U).

Ten Parallel Plate Avalanche Counters (PPACs), with the targets alternated, were used to measure the (n,f) reactions. Those are gas detectors constituted by one central anode and two cathodes. The gaps between the electrodes are filled with C_3F_8, which is a non-flammable gas, at 4 mbar pressure. The anode signal has a very fast response, with a time resolution of about 500 ps and the cathodes are segmented with parallel strips in the X and Y directions in order to provide the impact position of each fission fragment in the PPAC plane.

Detectors and targets were located inside a stainless steel chamber filled with the gas, tilted 45° with respect the neutron beam in order to enlarge the angular range from 0° to 90° relative to the non-tilted setup in which the angular acceptance was limited between 0° and 60°; a discussion on the angular acceptance can be found in Ref. [8].

3 Data analysis and preliminary results

The principle of the fission identification is based on the simultaneous detection of the signals produced by the back-to-back emitted FFs in the two PPACs flanking the target. For this purpose the anode signals are used due to their fast response. The coincidence method allows us to discard most of the α background and of the products of spallation reactions. The cathode signals are used to reconstruct the detection position in both PPACs. Assuming that both FFs are back-to-back emitted in the laboratory reference system, their trajectory can be reconstructed and the emission angle determined. This assumption is valid because the error introduced in the angle measurement by the linear momentum transfer even at large neutron energies is negligible, as explained in Ref. [7].

In the tilted PPACs setup used in this work, the emission angle is calculated as the scalar product of two vectors, one for the FF trajectory ($\vec{V}_{FF}$), and another for the beam direction ($\vec{V}_{beam}$). As shown in Fig. 1, the cosine of the emission angle is given by (1).

$$\cos\theta = \frac{\vec{V}_{FF} \cdot \vec{V}_{beam}}{|\vec{V}_{FF}| \cdot |\vec{V}_{beam}|} \quad (1)$$

Making the assumption that the angular dependence of the FFs detection efficiency is constant along the whole energy range, we can deduce it from the actual FFAD obtained for the ^{235}U targets below 1 keV, where it is known to be isotropic [8]. The thickness of the respective targets backings must be taken into account.

The experimental FFAD ($W(\theta)$) can be fitted by the sum of Legendre polynomials:

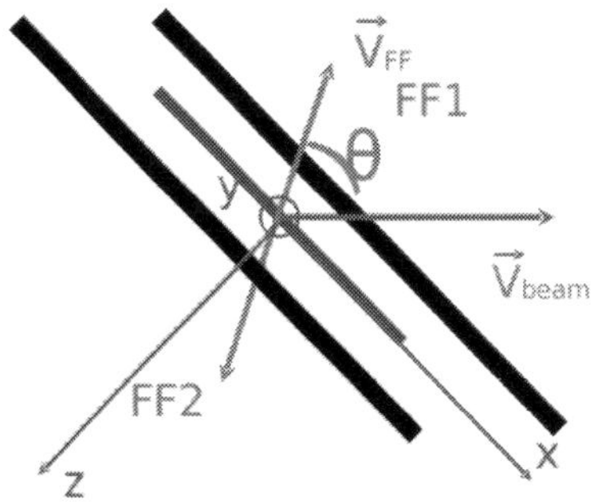

Figure 1. Scheme of the reference frame used to reconstruct the trajectories of the FFs, as is explained in [7, 8].

$$W(\theta) = A_0 \cdot \left[1 + \sum_{L_2}^{L_{max}} A_L \cdot P_L(\cos\theta) \right] \tag{2}$$

where A_L are the coefficients and L is the order of the polynomial. Only even terms must be included due to the symmetry of the fission fragments. In this work we have fitted ($W(\theta)$) by the sum of Legendre polynomials up to 4th order because the 2nd order polynomial alone is not sufficient to accurately reproduce the angular distribution behaviour, see Fig. 2.

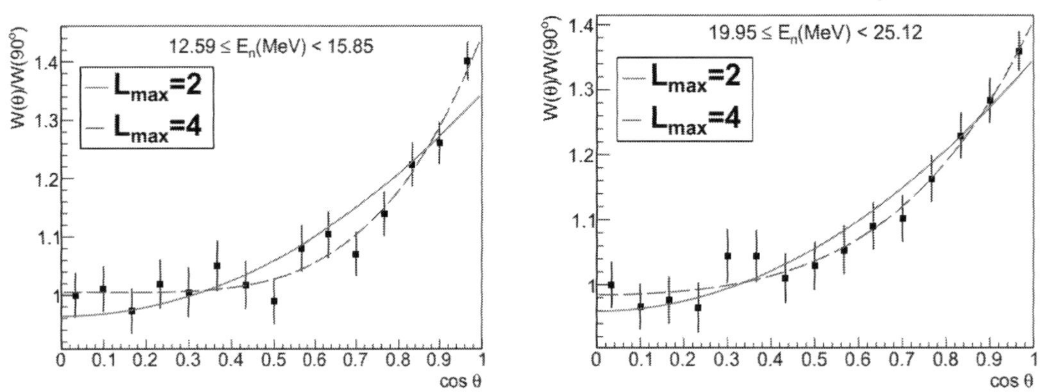

Figure 2. Angular distribution measured with one of the ^{235}U targets from 12.59 MeV to 15.85 MeV (left panel) and with one of the ^{238}U targets from 19.95 MeV to 25.12 MeV(right panel) fitted up to the 2nd and 4th order Legendre polynomials.

The FFAD dependence on the neutron energy is commonly described in a simple way using the anisotropy parameter, defined as ($W(0°)/W(90°)$). It can be expressed as:

$$W(0°)/W(90°) = \frac{1 + A_2 + A_4}{1 - \frac{1}{2} \cdot A_2 + \frac{3}{8} \cdot A_4} \tag{3}$$

where the coefficients are obtained from the fit to the experimental FFAD data.

Our preliminary results on the anisotropy parameter in function of the neutron energy are shown in Fig. 3, where the mean values have been calculated for both ^{235}U targets and for the three ^{238}U targets. The error bars include only the propagation of the statistical uncertainties.

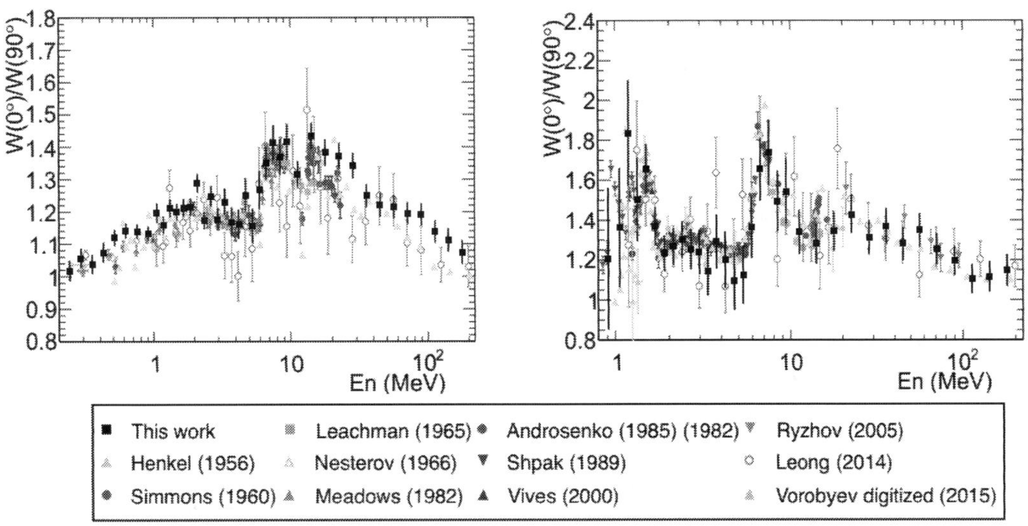

Figure 3. The anisotropy parameter measured in this work with the two ^{235}U targets (left panel) and with the three ^{238}U targets (right panel) compared with the available experimental datasets.

When compared with the available ^{235}U experimental data up to 20 MeV, the anisotropy parameter data here presented agree in general within their uncertainties. Above 20 MeV and up to 200 MeV there are only two datasets, Vorobyev et al. [4] and Leong [5] giving values 8% below the present data at 100 MeV.

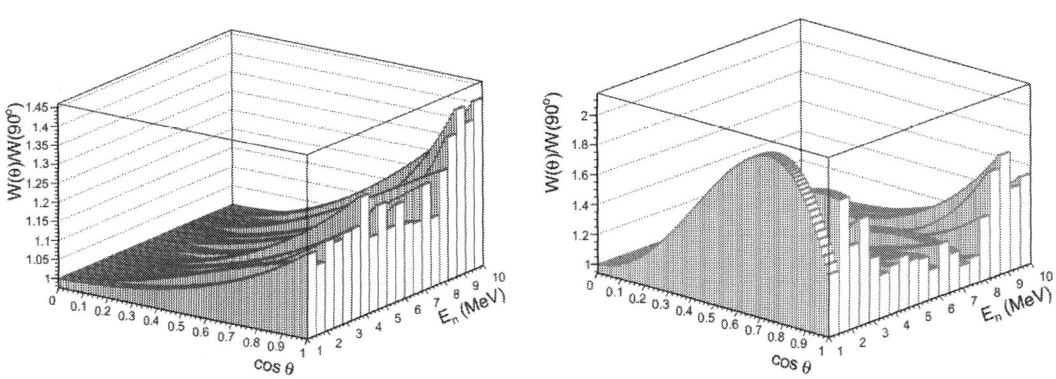

Figure 4. Evolution of the FFAD found in this work for ^{235}U (left panel) and ^{238}U (right panel) in the energy range from 1 MeV to 10 MeV.

Concerning ^{238}U, the anisotropy parameter data presented in this work are in general in good agreement with previous works in the entire energy range up to 200 MeV. It is worth to note that the statistics is not enough to describe with accuracy the behaviour of the ^{238}U angular distributions at the thresholds of the different fission chances where the anisotropy shows large variations for small changes in the neutron energy. This is better seen in Fig. 4, where the evolution of the FFAD is shown in the energy range from 1 to 10 MeV. Conversely to ^{235}U, in the case of ^{238}U the A_4 parameter in the Legendre expansion changes very suddenly from positive to negative giving as a result an angular distribution that can not be described by a simple anisotropy parameter defined as $(W(0°)/W(90°))$.

4 Conclusions

The angular distributions of the ^{235}U(n,f) and ^{238}U(n,f) reactions have been measured at the n_TOF facility (CERN) using parallel-plate avalanche counters (PPACs). In this work the preliminary results of the anisotropy parameter up to 200 MeV are presented and compared with the datasets available in EXFOR and the more recent data reaching 200 MeV [4, 5].

In general, the results for both ^{235}U and ^{238}U are in good agreement with previous measurements up to 20 MeV. From 20 MeV to 200 MeV, the results for ^{235}U are higher than those of the two datasets recently published [4, 5] (around 8% at 100 MeV); and for ^{238}U the present data are within the uncertainties of previous ones [3–5].

The results on ^{238}U show the high complexity of its fission mechanism, already seen for ^{232}Th [7] and ^{234}U [9]. A better understanding of this behaviour around the thresholds for first and third chance fission for even-even isotopes requires much higher statistics in order to resolve the sharp changes of the anisotropy parameter near the fission chance thresholds. Accordingly, new measurements are planned in the forthcoming n_TOF campaigns, taking advantage of the high resolution of the PPACs setup at the n_TOF facility.

References

[1] C. Paradela et al., Phys. Rev. C **91**, 024602 (2015).

[2] EXFOR (Experimental Nuclear Reaction Data) Nucl. Data Sheets **120**, 272 (2014).

[3] I. V. Ryzhov et al., Nucl. Phys. A **760**, 19 (2005).

[4] A. S. Vorobyev et al., JETP Letters **102**, 231 (2015).

[5] L. S. Leong, *Ph.D. thesis*, (Universitè Paris Sud) (2013).

[6] C. Guerrero et al., Eur. Phys. J. A **49**, 27 (2013).

[7] D. Tarrío, *Ph.D. thesis*, (Universidade de Santiago de Compostela) (2012).

[8] D. Tarrío et al., Nucl. Instrum. Meth. A **743**, 79 (2014).

[9] C. Paradela et al., Phys. Rev. C **82**, 034601 (2010).

EPJ Web of Conferences **111**, 10003 (2016)
DOI: 10.1051/epjconf/201611110003

Sensitivity studies of spin cut-off models on fission fragment observables

L. Thulliez[1,2,a], O. Litaize[1], and O. Serot[1]

[1] CEA, DEN, DER, SPRC, F-13108 Saint Paul Lez Durance, France
[2] CEA, DSM, Irfu, SPhN F-91191 Gif-sur-Yvette, France

Abstract. A fission fragment de-excitation code, FIFRELIN, is being developed at CEA Cadarache. It allows probing the characteristics of the prompt emitted particles, neutrons and gammas, during the de-excitation process of fully accelerated fission fragments. The knowledge of the initial states of the fragments is important to accurately reproduce the fission fragment observables. In this paper a sensitivity study of various spin cut-off models, completely defining the initial fission fragment angular momentum distribution has been performed. This study shows that the choice of the model has a significant impact on gamma observables such as spectrum and multiplicity and almost none on the neutron observables.

1 Introduction

During the fission fragment de-excitation, nuclei emit neutrons and gammas to loose their excitation energy and angular momentum in order to reach their ground state. The properties of those prompt emitted particles is of particular interest for nuclear reactors, for example to quantify the damage they can cause on the nuclear reactor vessel which has a direct impact on its lifetime. A Monte-Carlo computer code named FIFRELIN [1] [2] developed at CEA-Cadarache allows studying nuclear fission through the nuclear fission fragment de-excitation process. This is performed through a Monte-Carlo implementation of the so-called Hauser-Feshbach algorithm which takes into account the n/γ competition [2]. Practically the notion of nuclear realization is used as proposed for capture reactions in [3]. In this work the impact of nuclear models on fission observables, i.e. the outputs of the code, is investigated. By comparing the observables estimated by the code with the available experimental data, the relevance of the models can be inferred. If results related to major observables are in good agreement with available data, other calculated observables, which could not be measured, could be legitimately considered as pertinent. This confers to FIFRELIN a predictive power.
This paper will focus on the impact of the spin cut-off parameter models which defined the total angular momentum distribution from which the initial spins of the fragments are sampled. This entity mainly drives the characteristics of gammas such as their energy, their multiplicity and their spectrum. In order to perform this study, only the spin cut-off parameters of the simulation are tuned, the other parameters are unchanged. The spin cut-off models are tested in the case of the ^{252}Cf spontaneous fission.

[a] e-mail: loic.thulliez@cea.fr

2 Spin cut-off models involved in the initial spin distribution

In order to perform the de-excitation of each fragment, their excitation energy E^* and spin parity J^π have to be assigned. The procedure to determine E* is described in detail in [1]. An equal probability for parity is considered: p(π = +)=p(π = −)=$\frac{1}{2}$ The initial total angular momentum J, of the light (L) and heavy (H) fragments, is sampled in the distribution given by equation (1).

$$P_{L,H}(J) = \frac{(2J+1)}{2\sigma^2_{L,H}} exp\left(-\frac{(J+1/2)^2}{2\sigma^2_{L,H}}\right) \tag{1}$$

where $\sigma^2_{L,H}$ is the spin cut-off parameter, related to the mean and the width of the distribution.

Four different models have been implemented in FIFRELIN for the spin cut-off determination. Each model has free parameters which are adjusted to reproduce the ^{252}Cf total prompt neutron multiplicity $\bar{\nu} = 3.76$. The tested models are the following :

- **Constant** model, in which the light and the heavy fragment groups are described by a constant spin cut-off:

$$\sigma_{L,H} = K_{L,H} \tag{2}$$

  K_L and K_H are two free parameters .

- **Linear** model, based on Armbruster experimental data [4], in which the spin cut-off of each mass group is linear as a function of the mass :

$$\sigma_{L,H} = a_{L,H}A + b_{L,H} \tag{3}$$

  The coefficients a_L and a_H are the one from Armbruster while the b_L and b_H are simulation free parameters. This prescription is used to restrict the number of free parameters in the simulation.

- **Inertia** model, which was formulated by Bethe [5] depends on the mass and on the temperature of the nucleus. It takes into account shell effects through the Ignatyuk level density a parameter [6] used to calculate the nucleus temperature T.

$$\sigma_{Inertia} = \alpha_{Inertia}\sqrt{I_{rig}T} \tag{4}$$

  with $\alpha_{Inertia}$ a free parameter, I_{rig} the moment of inertia of the nucleus assuming it as a pure rigid rotor and T the nucleus temperature. The temperature is calculated considering the nucleus as a Fermi gas and in taking the Ignatyuk prescription for the level density parameter a calculation. Thus, the "Inertia" model lowers the number of free parameters by one contrary to the "Constant" and "Linear" models.

- **Inertia+Shell** model, which comes from observations following microscopic calculations [7] [8] and whom pointed out that the ratio $\frac{\sigma^2}{T}$ is not a constant but shows some shell effects, as already obtained from semi-empirical approaches [9] [10]. Thus the following spin cut-off can be formulated:

$$\sigma_{Inertia+Shell} = \alpha_{Inertia+Shell}\sqrt{\frac{a}{\tilde{a}}I_{rig}T} \tag{5}$$

  with $\alpha_{Inertia+Shell}$ a free parameter, a the Ignatyuk level density parameter [6] and $\tilde{a}$ its asymptotic parameter. As in the previous model, the number of free parameters is one.

The mean total angular momentum as a function of the pre-neutron fission fragment mass A obtained with the four models are plotted on Figure 1. In the region of high mass yields, i.e. around A=108 and A=145, the Constant model is a mean of the Inertia+Shell model. Consequently, one expects that those two models give approximately the same results. Figure 1 shows that Inertia and Inertia+Shell models do not have pronounced shell effects. This is due to the high initial energy given to primary fragments, which almost vanishes the shell effects in the level density parameter.

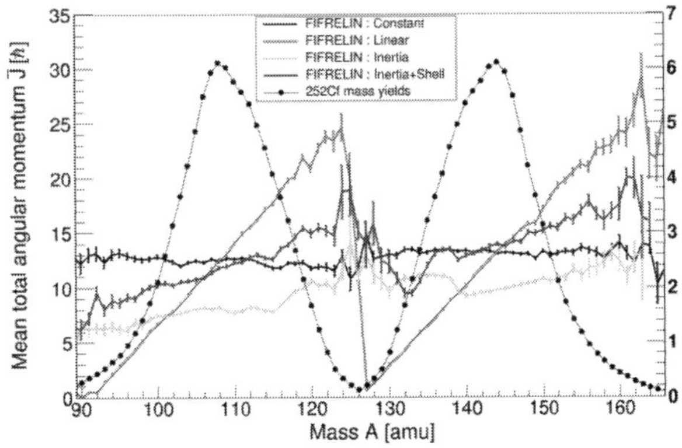

Figure 1. Mean Total angular momentum as a function of the pre-neutron fission fragment mass for the four tested models (left Y-axis) with the ^{252}Cf mass yields (right Y-axis).

3 Impact of spin cut-off models on fission observables

In this section, the four models are applied to the spontaneous fission of ^{252}Cf, compared to neutron and gamma experimental data and discussed.

3.1 Gamma observables

Figure 2 shows the prompt fission gamma spectra (PFGS) obtained with different initial spin distributions and compared with experimental data [11] [12]. Below 1.2 MeV, the spectrum obtained with the Inertia model underestimates experimental data contrary to the three others, which slightly overestimate the data of about one gamma per fission per MeV. In the gamma energy range between 1.2 MeV and 8.0 MeV (right side of Figure 2) all models agree with experimental data.
The discrepancies below 1.2 MeV with the Inertia model are clearly explained from the averaged values of the sampled total angular momenta, $\bar{J}_L$ = 8.00 $\hbar$ and $\bar{J}_H$ = 10.45 $\hbar$, which are significantly lower than the ones obtained with the three other models (see Table 1). The relation between a high initial total angular momentum and a high emission of low energy gamma can be understood in term of n/γ competition. The higher the initial total angular momentum is, the higher in excitation energy the competition will occur in order to decrease the angular momentum of the fragment [2]. For example the competition starts around the neutron separation energy S_n for low angular momenta ($J_i \approx 3\ \hbar$) and around S_n + 3 MeV for high angular momenta ($J_i \approx 16\ \hbar$). The competition starts higher in energy in order to give the nucleus the right spins to allow a gamma transition from the continuum to the first discrete levels. Consequently, in the n/γ competition zone, the excess of angular momentum is expected to be mainly evacuated by a cascade of low energy gammas.

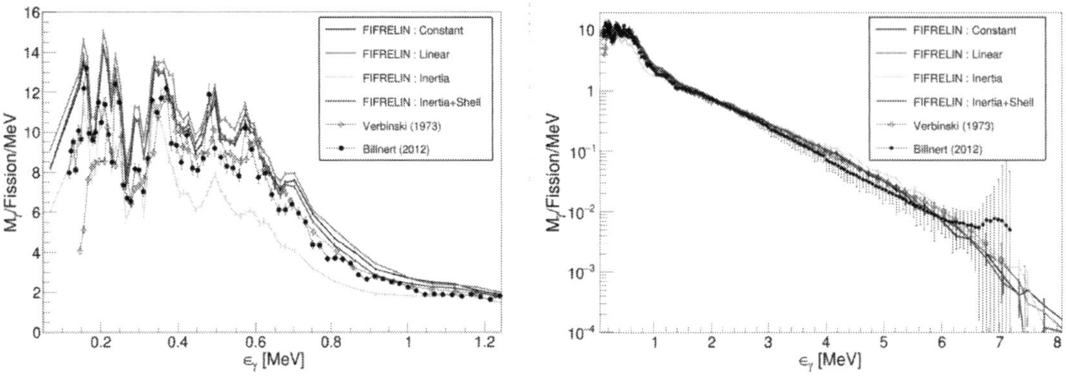

Figure 2. Prompt fission gamma spectra at low energy (left side) and up to 8 MeV (right side) compared with experimental data [11] [12].

Table 1 presents the main relevant averaged gamma quantities obtained experimentally and with FIFRELIN calculations. When comparing experimental and simulated averaged gamma quantities, the experimental setup such as energy threshold or coincidence time have to be taken into account in FIFRELIN calculations. Practically, when the threshold increases, the mean gamma energy increases (due to a cut in the low energy part of the spectrum) and the average multiplicity decreases (preserving a constant total gamma energy). Otherwise the comparison is fortuitous as Albinsson shown on the gamma multiplicity as a function of fragment mass [13] and reported by Brunson [14]. In our case a 110 keV threshold and 3 ns coincidence time are used, comparable with Billnert data. Table 1 shows that on average the Constant, Linear and Inertia+Shell spin cut-off models predict around one more gamma than experimental values and deposit an additional amount of 1 MeV in the gamma channel.

Table 1. Comparison between experimental data citeVerbinski [12] and FIFRELIN calculations obtained with Billnert experimental parameters ($E_\gamma^{threshold}$ = 110 keV; T_{coinc} = 3 ns). In Verbinski experiment, parameters were $E_\gamma^{threshold}$ = 140 keV and T_{coinc} = 10 ns, leading to a lower multiplicity $\bar{M}_\gamma$ and a higher average energy per quantum $<\epsilon_\gamma>$.

^{252}Cf	$\bar{J}_L$	$\bar{J}_H$	$<\epsilon_\gamma>$	$\bar{M}_\gamma$	$<E_\gamma^{tot}>$
Verbinski (1973)	/	/	0.88 ± 0.04	7.80 ± 0.3	6.84 ± 0.3
Billnert (2012)	/	/	0.80 ± 0.01	8.30 ± 0.08	6.64 ± 0.08
Constant	12.39	13.18	0.81	9.40 ± 0.02	7.74 ± 0.01
Linear	13.17	12.54	0.80	9.99 ± 0.02	8.07 ± 0.01
Inertia	8.00	10.45	0.90	6.89 ± 0.03	6.29 ± 0.02
Inertia+Shell	12.22	13.96	0.80	9.46 ± 0.03	7.67 ± 0.02

Figure 3 represents the average gamma multiplicity as a function of fission fragment mass. It is not well established that a saw-tooth shape can be expected. A detailed study of Albinsson for ^{235}U(n_{th},f) reaction [13] has shown that a flat or a saw-tooth behaviour can be obtained, depending on the experimental setup and gamma energy range. Even an inverse saw-tooth shape can be obtained. Here we simply show that a saw-tooth behaviour is directly obtained if the J_{RMS} of the initial spin distribution has a saw-tooth shape itself (Linear model). However an increase of the multiplicity is observed for heavy fragments as a function of mass whatever the used model. The expected saw-tooth behaviour of the Linear model is found but does not reproduce the Schmidt-Fabian data [15]. The three other models give results which have a less pronounced behaviour. However the Inertia model always underestimates the data, while the Constant and Inertia+Shell models are closer.

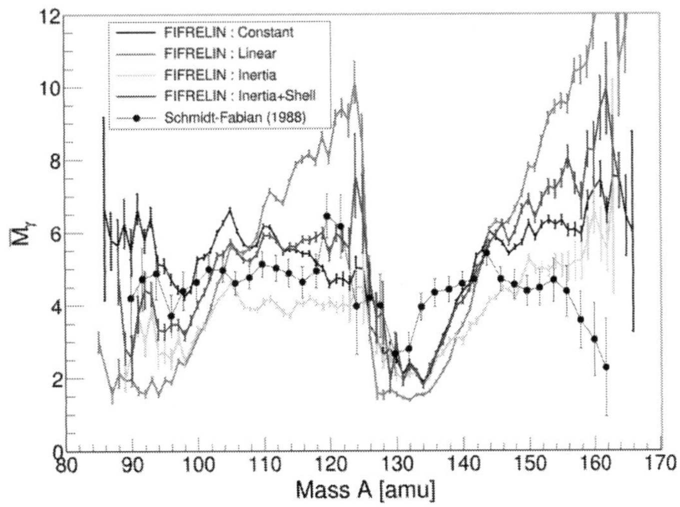

Figure 3. Comparison of the average gamma multiplicity as a function of the pre-neutron mass between Schmidt-Fabian data [15] and FIFRELIN calculations with different spin cut-off models.

In conclusion, the choice of the spin cut-off models for the initial total angular momenta has a great impact on gamma observables as expected. This study shows on one hand that the experimental parameters involved during the measurements must be taken into account in any calculations and on the other hand that for the moment no spin cut-off model allows reproducing the whole gamma observables. More experimental data are needed in order to confirm the previous one and to perform other verifications of the models on other fissioning nuclei.

3.2 Neutron observables

Figure 4 represents the neutron multiplicity as a function of mass and the prompt neutron spectra ratio to a Maxwellian (T=1.42 MeV), showing that there are no significant differences between the four tested models for those neutron observables, contrary to what could be expected due to a smaller n/γ competition zone in the case of the Inertia model.
In the left side of Figure 4, all models are in good agreement with the different experimental data sets. Nevertheless the Constant and Inertia+Shell models seem to better reproduce the data, especially for the very asymmetric fission. The neutron spectra represented on figure 4 are not different from one model to another. However at energy below 0.2 MeV the Inertia and Inertia+Shell models seem to have some structures contrary to the two others. This could be a hint for further investigations.

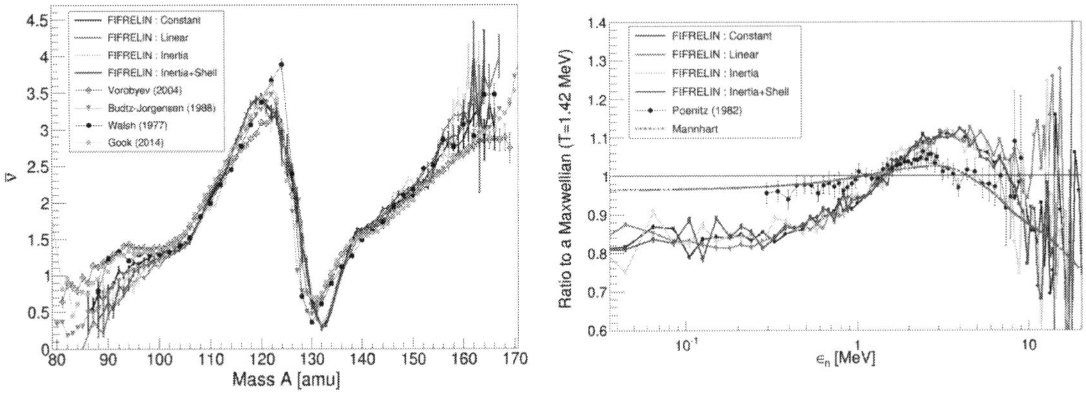

Figure 4. Comparison of neutron multiplicity as a function of mass (left side) and prompt fission neutron spectra ratio to a Maxwellian at T=1.42 MeV (right side) between experimental data and FIFRELIN calculations performed with different spin cut-off models.

4 Conclusion

In this work the impact of spin cut-off models for initial spin distribution on fission observables was studied through the spontaneous fission of ^{252}Cf. It has been shown that for the moment no model allows reproducing gamma observables accurately. However the Constant and the Inertia+Shell models seem to better reproduce experimental data than the others. The choice of the spin cut-off model has been shown to have no impact on neutron observables, and consequently does not softer the hard PFNS obtained with FIFRELIN. In order to try to improve the calculation of this observable, new energy sharing models are under investigation.
In perspective works, the presented spin cut-off models will be tested on other fissioning nuclei to validate these tendencies.

References

[1] O. Litaize *and* O. Serot, Phys. Rev. C **82**, 054616 (2010)
[2] D. Regnier, Ph.D. thesis, Univeristy of Grenoble (2013)
[3] F. Becvar, Nucl Instrum. Methods Phys. Res., Sect. A **417(2-3)**, 434 (1998)
[4] P. Armbruster et al., Z. Naturforsch. **26 a**, 512 (1971)
[5] H. A. Bethe, Phys. Rev. **vol 50**, 332 (1936)
[6] A. V. Ignatyuk et al., Sov. J. Nucl. Phys. **Vol 21**, 255 (1975)
[7] A. J. Koning, S. Hilaire *and* S. Goriely, Nucl. Phys. A **810**, 13 (2008)
[8] S. Goriely, Nucl. Phys. A **605**, 28 (1996)
[9] S. K. Kataria et al., Phys. Rev. C **18**, 549 (1978)
[10] H. Vonach et al., Phys. Rev. C **38**, 2541 (1988)
[11] V. V. Verbinski et al., Phys. Rev. C **7(3)**, 1173 (1973)
[12] R. Billnert et al., Phys. Rev. C **87(2)**, 246 (2013)
[13] H. Albinsson, Aktiebolagand Atomenergi Report **AE-420** (1971)
[14] G. S. Brunson, Los alamos Report **LA-9408-T** (1982)
[15] C. Wagemans, The Nuclear Fission Process (CRC Press 1 edition, September 1991)

EPJ Web of Conferences **111**, 11001 (2016)
DOI: 10.1051/epjconf/201611111001

Development of a gaseous proton-recoil detector for fission cross section measurements below 1 MeV neutron energy

P. Marini[1], L. Mathieu[1a], M. Aïche[1], S. Czajkowski[1], B. Jurado[1], I. Tsekhanovich[1]

[1]*Centre d'Etudes Nucléaires de Bordeaux Gradignan, 19, Chemin du Solarium, 33175 GRADIGNAN, France*

Abstract. The elastic H(n,p) reaction is sometimes used to measure neutron flux, in order to produce high precision measurements. The use of this technique is not straightforward to use below incident neutron energy of 1 MeV, due to a high background in the detected proton spectrum. Experiments have been carried out at the AIFIRA facility to investigate such background and determine its origin and components. Based on these investigations, a gaseous proton-recoil detector has been designed, with a reduced low energy background.

1 Introduction

The development of Generation IV nuclear system relies on accurate system modelling, to determine their behaviour and its evolution with time, irradiation and/or in incidental/accidental situations. These simulations require high quality physical models and nuclear data to meet the requested precision [1]. Despite having been studied for decades, large discrepancies and uncertainties exist on nuclear data in the actinide region [2]. In particular, the knowledge of fission cross sections is of primary importance for determining the nuclear core behaviour or waste incineration capabilities.

The neutron flux maximum in a fast reactor, such as the Sodium Fast Reactor, is located around 100 keV [3]. At these energies, cross sections of non-fissile isotopes are quite low. Nevertheless, some of them are not negligible (for instance $^{240,\,242}$Pu(n,f)) and may have an impact on the core behaviour. Strong requests are then addressed to improve the fission cross section in this energy range.

Cross section measurements are often normalized to standard reactions such as ^{235}U(n,f), ^{238}U(n,f) or ^{237}Np(n,f). Since the systematic normalization of the measurements to the mentioned cross sections introduces a strong correlation between the different results, independent measurements are carried out relative to the H(n,p) scattering. In this method, the neutron flux is converted into a proton flux by an H-rich foil. As a charged particle, proton can be easily detected, for instance in a silicon detector. Different types of proton-recoil detector exist, depending on the type of experiment, the facility environment (neutron beam vs. isotropic neutron source), the neutron energy range or the target accuracy [4-13].

[a] Corresponding author: mathieu@cenbg.in2p3.fr

2 Proton-recoil usual technique

The ACEN group of the Centre d'Etudes Nucléaires de Bordeaux Gradignan (CENBG) has developed in the 2000's its own proton-recoil setup to perform accurate measurements in the MeV energy range [4]. It consists of a polypropylene foil with a thickness ranging from 20 μm to 50 μm, for neutron energy of 2 MeV to 5 MeV respectively. A Silicon telescope is placed at 7.5 cm downstream on the beam axis to detect protons scattered at very low scattering angle. These protons have kinetic energies close to the incident neutron energy. Nearly mono-energetic incident neutrons produce a clear proton peak in the proton spectrum, with a very good signal-to-background ratio. A collimator placed in front of the Si detector is used to accurately define the detection efficiency. Fission cross sections were measured with this experimental setup, with final uncertainties as low as 3-4% [4].

In 2012, the setup was adapted to neutron energy around 1 MeV for measurement at the Bruyères-Le-Chatel 4MV Van-de-Graaff accelerator. The silicon telescope was replaced by a single layer silicon detector and the polypropylene foil thickness was reduced to 4 μm. At these energies, the signal-to-background ratio at the proton peak is strongly reduced, due to a strong background at lower energies. Such background prevents accurate neutron flux measurements at neutron energy of few hundreds keV.

Measurement was performed in 2014 at the AIFIRA facility to investigate the background and how to reduce it. Several runs were performed with a 50 μm-thick silicon detector, for incident neutron energies ranging from 1.6 MeV down to 500 keV. The proton spectra obtained are shown on Figure 1. As the incident energy decreases, the proton peak shifts toward lower energies and is progressively hidden by a strong background. At E_n = 700 keV, the proton peak is still visible but its integral (used to infer the neutron flux) cannot be determined with an accuracy better than 30%. At E_n = 500 keV, the proton peak is barely visible.

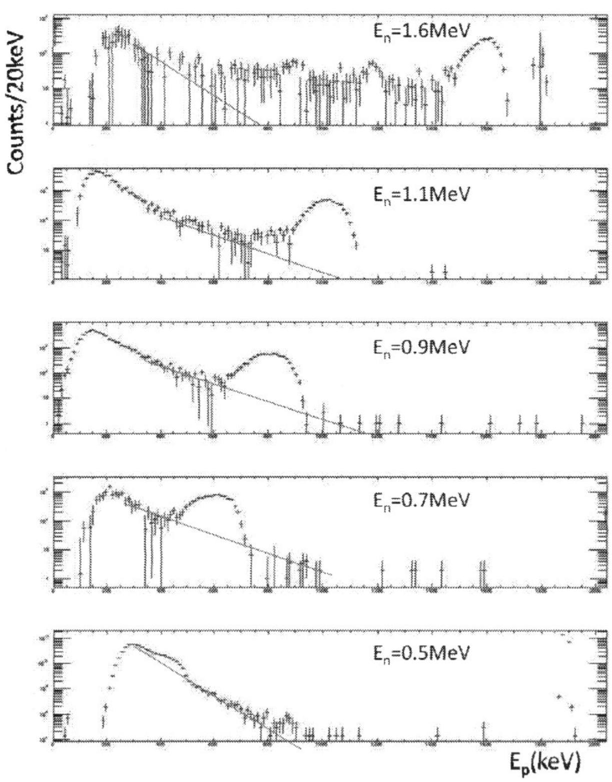

Figure 1. Proton spectra for different incident neutron energies (from 1.6 MeV to 0.5 MeV). The proton peak can be easily distinguished from the background for E_n greater than 1 MeV. For lower energies, the signal-to-background ratio is strongly reduced, preventing any accurate determination of the neutron flux.

Simulations were performed with the MCNP code to reproduce these results. Several hypothesises were tested to reproduce the exponential shaped background. The simulations show that low energy γ-rays (few hundreds of keV) can produce electrons by Compton scattering, which deposit their energy in the silicon detector. Contrarily to protons, which interact in the first few microns of the detectors (range of 18, 9 and 2 μm at 1.1, 0.7 and 0.2 MeV respectively), the electrons interact with the whole detector thickness. Even if they deposit only a small fraction of their energy, a 50 μm-thick silicon detector is then not suitable to discriminate low energy protons in a not negligible electron flux.

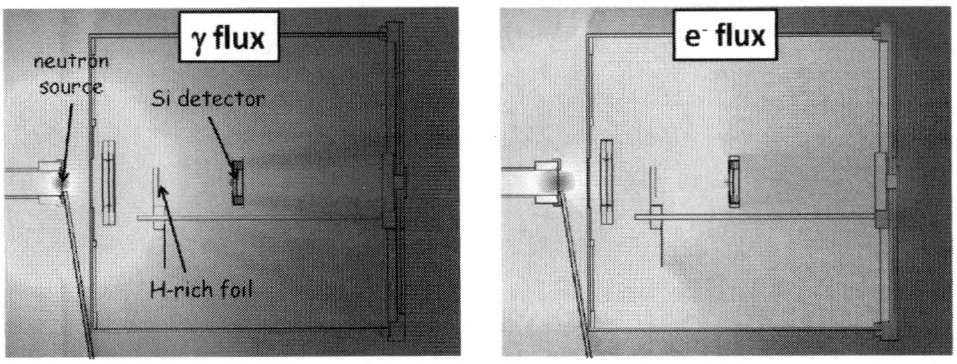

Figure 2. Simulations of γ-ray and electron flux in the experimental setup, with a γ-ray source of 200 keV located at the neutron production target. Electrons produced by Compton scattering are present in the whole setup and some deposit their energy in the Si detector.

3 DPR Design

Thinner Si detectors, with thickness down to few μm, would reduce the electron noise contribution. However this solution is not straightforward. The ACEN group decided to design a gaseous proton-recoil detector: the protons emitted by an H-rich foil are detected in an ionization chamber, filled with a CF_4 gas and segmented into two chambers ΔE and E separated by a collimator. The gas pressure is adjusted to the proton energy range in order to fit with the detector length. Electron energy deposition in such a low density material is very low, leading to a strongly reduced background. The coincidence between the two ionization chambers ΔE and E also significantly reduces parasitic recoil protons coming from other sources than the H-rich foil. A scheme of this detector is presented in Figure 3.

To perform high precision measurements, the detector efficiency has to be very well defined. The collimator between the two ionization chambers and a second collimator after the H-rich foil allow us to determine a precise geometrical efficiency. The CF_4 gas and ceramic materials used for the structures were chosen for very low or absent hydrogen content, so that no recoil protons can be emitted from the gas or the frame which would induce systematic errors or background.

For low incident neutron energies, the deposited energy in the gas is quite low. The generated primary electrons are collected and amplified by a Micromegas detector. To reduce electronic background, the detector is segmented into 64 pads. This will also make a crude track analysis possible, to reject false coincidence with abnormal trajectories (for instance not starting from the H-rich foil). Moreover, the choice of the fast CF_4 gas (i.e high electron mobility) reduces the pile-up of the signal with a possible background noise.

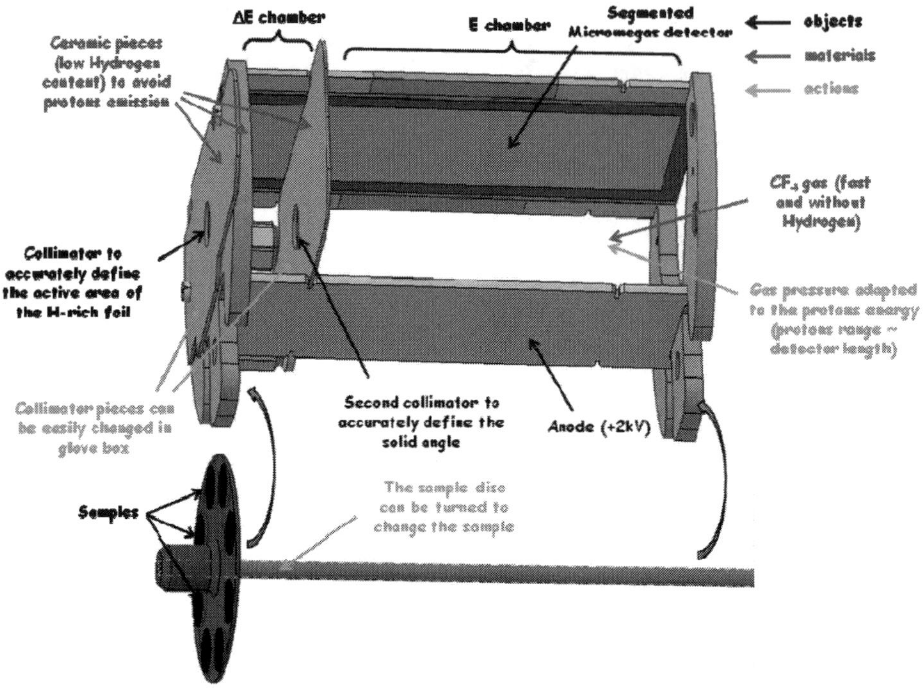

Figure 3. Scheme of the gaseous proton-recoil detector. For representation purpose, the sample disc has been separated from the proton-recoil detector.

To reduce the proton energy loss in the H-foil itself, several samples of different thicknesses (from 1 μm to 10 μm) are mounted on a disc (on the bottom of Figure 3). Thinnest samples are evaporated tristearine, while the thickest ones are polypropylene films. An empty hole is also available. The disc can be turned to select the sample thickness adapted to the incident neutron energy, or to perform background measurement without sample (empty hole).

To gaseous proton-recoil detector has been designed by the CENBG and will be manufactured in late 2015. The electronic connections will be set in early 2016, for an operational release planned in the first half of 2016.

4 Conclusions

Neutron flux measurements via the elastic scattering H(n,p) reaction is challenging for neutron energies below 1 MeV. Gamma rays produced by the neutron production target generate electrons which induce a strong background in the proton detector, preventing an accurate determination of the number of recoil protons.
A new detector based on the proton detection in a gas is being designed at the CENBG. Its low electron sensitivity is suited for neutron energies below 1 MeV in a gammas and electrons dense environment (i.e Van-de-Graaff type accelerator). The commissioning will take place in the first half of 2016.

References

1. G. Aliberti, G. Palmiotti, M. Salvatores, T. Kim, T. Taiwo, M. Anitescu, I. Kodeli, E. Sartori, J. Bosq, J. Tommasi, Ann. Nucl. Ener. **33**, 700 (2006)
2. M. Chadwick, M. Herman, P. Oblozinsky, M. Dunn, Y. Danon, A. Kahler, D. Smith, B. Pritychenko, G. Arbanas, R. Arcillas et al., Nucl. Data Sheets **112**, 2887 (2011)

3. R. Rachamin, C. Wemple, E. Fridman, Ann. Nucl. Ener. **55**, 194 (2006)
4. G. Kessedjian, G. Barreau, M. Aïche, B. Jurado, A. Bidaud, S. Czajkowski, D. Dassié, B. Haas, L. Mathieu, L. Tassan-Got, J.N. Wilson, F.-J. Hambsch, S. Oberstedt, I. AlMahamid, J. Floyd, W. Lukens, D. Shuh, Phys. Rev. C 85, 044613
5. Y. Naitou, Y. Watanabe, S. Hirayama, M. Hayashi, A. Prokofiev, A. Hjalmarsson, S. Pomp, P. Andersson, R. Bevilacqua, C. Gustavsson, M. Österlund, V. Simutkin, H. Sjöstrand, M. Tesinsky, U. Tippawan, Journ. of Kor. Phys. Soc. **59**, No. 2 (2011)
6. K. Asai, N. Naoi, T. Iguchi, K. Watanabe, J. Kawarabayashi, T. Nishitani, J. Nucl. Sci. and Tech. **43**, No 4 (2006)
7. R. Beyer, E. Grosse, K. Heidel, J. Hutsch, A.R. Junghans, J. Klug, D. Légrády, R. Nolte, S. Röttger, M. Sobiella, A. Wagner, Nucl. Instr. and Meth. A **575** (2007)
8. A. Donzella, M. Barbui, F. Bocci, G. Bonomi, M. Cinausero, D. Fabris, A. Fontana, E. Giroletti, M. Lunardon, S. Moretto, G. Nebbia, M.M. Necchi, S. Pesente, G. Prete, V. Rizzi, G. Viesti, A. Zenoni, Nucl. Instr. and Meth. A **613** (2010)
9. J. Hassard, J.H. Liu, R. Mongkolnavin, D. Colling, Nucl. Instr. and Meth. A **416** (1998)
10. C. Cazzaniga, M.Rebai, M. Tardocchi, G.Croci, M. Nocente, S.Ansell, C.D. Frost, G. Gorini, Prog. Theor. Exp. Phys., 073H01 (2015)
11. R. Babut, V. Gressier, J. Instr. in Nucl. Sci. and Tech. **2** P01005 (2007)
12. J. Taforeau, I. Schaefer, S. Higueret, D. Husson, L. Lebreton, Prog. Nucl. Sci. and Tech. Vol 4 (2014)
13. M.A. Kovash, B. Daub, J. French, V. Henzl, K. Shoniyozov, J.L. Matthews, Z. Miller, H. Yang, Proc. of conf. Adv. in Nucl. Inst. Meas. Meth. and Appl. (2011)

EPJ Web of Conferences **111**, 11002 (2016)
DOI: 10.1051/epjconf/201611111002

Neutron Induced Capture and Fission Processes on ^{238}U nucleus

Cristiana Oprea[1] and Alexandru Oprea

Frank Laboratory for Neutron Physics, Joint Institute for Nuclear Research, 141980 Dubna, Moscow Region, Russian Federation

Abstract. Nuclear data on Uranium isotopes are of crucial interest for new generation of nuclear reactors. Processes of interest are the nuclear reactions induced by neutrons and in this work mainly the capture and the fission process on ^{238}U will be analyzed in a wide energy interval. For slow and resonant neutrons the many levels Breit – Wigner formalism is necessary. In the case of fast and very fast neutrons up to 200 MeV the nuclear reaction mechanism implemented in Talys will be used. The present evaluations are necessary in order to obtain the field of neutrons in the design of nuclear reactors and they are compared with experimental data from literature obtained from capture and (n,xn) processes.

1 Introduction

In the middle of XX century the scientists had already demonstrated that the classical fossil hydrocarbon fuel (gas, coal, oil) will be finished in approximately one century taking into account the predicted temps of world economic development. The nuclear power based on the fission of the ^{235}U induced by slow neutrons as solution for the energetically sustainable development was considered until now. It is well known that this is not a long term solution, because the amount of ^{235}U in natural Uranium is about 0.7% and considering the energy demand of world economy, the planet reserves of ^{235}U will be finished practically in the same time with the fossil fuel [1]. A perspective direction of the development of nuclear energy is the possibility to transform the fissionable nuclei ^{238}U, ^{232}Th and others in nuclear fuels which requires the design and construction of new type of nuclear reactors. Then the energy demand of the world could be ensured for millions years even if the energy request of the planet would be increased by 50-100 times compared with nowadays demand [1,2].

In many countries national programs of research and development of subcritical systems and new generation nuclear reactors based on the fission of ^{238}U induced by fast neutrons are already started or in progress. The experimental, theoretical, computer simulations and numerical results provided by these programs have demonstrated the feasibility of the developed methods based on ^{238}U, natural Uranium and without using ^{235}U, Thorium and spent fuel elements from nuclear plants [3,4].

The fission and capture processes are regarded as concurrent processes in the neutron reactors. The neutron fields can be measured by threshold detectors based on (n,xn) reactions on ^{89}Y. The neutron

[1] Corresponding author: istina@nf.jinr.ru

multiplicity and mass distributions are important and necessary elements for calculation of fission reaction chain.

In the present work the cross sections of fission and capture processes induced by fast neutrons on ^{238}U and of the (n,xn) reactions on ^{89}Y nucleus, the mass distributions and neutron multiplicity are evaluated.

2 Theoretical backgrounds

Bohr and Wheeler [5] and Frenkel [6] were the first who have explained quantitatively the fission process in terms of liquid drop model. The authors regarded the nucleus like a classical liquid drop with volume energy and surface tension and on this basis the normal modes of vibration of the drop can be interpreted as excited states of nucleus in a potential barrier. The fission cross section is rising rapidly if the excitation energy is higher than the barrier height or the fission threshold, although the fission is not a threshold process in a conventional way because fission is possible at any excitation energy by barrier tunneling.

The neutron induced fission and (n,xn) cross sections, the mass distribution and neutron multiplicity obtained in the present work were calculated with Talys codes [7]. Fission cross section induced by neutrons with energy from 1 MeV up to 200 MeV was calculated in the frame of a phenomenological approach based on experimental data of fission barrier height and widths with axial asymmetry of the compound nucleus, provided by Talys database [7] and other nuclear data bases from the net.

For the cross sections of neutron capture process by ^{238}U and (n,xn) reactions on ^{89}Y nucleus, in the wide incident neutrons energy range from 2 MeV up to 200 MeV, it is necessary to consider all nuclear reactions mechanisms i.e. the statistical, pre-equilibrium and direct model. Compound processes, which are dominant in the low energy range of 10-20 MeV, by the statistical model of nuclear reactions and Hauser – Feshbach formalism are described [8]. The contribution to the cross section of pre-equilibrium processes is obtained according with the simple two-component exciton model [9]. For the direct processes the Distorted Wave Born Approximation (DWBA) was used [10].

In the cross sections evaluations a local type Wood – Saxon optical potential with all components (volume-central, surface-central, spin-orbit) was considered mainly extracted from experimental data. In the case of no experimental data Talys permits also to define global optical potential according with reference [11]. The mass distribution of the fission fragments and the neutron multiplicity for different energies of incident neutrons, in the cases of pre-neutrons and post-neutrons emission, were obtained using the GEF model (**GE**neral **F**ission model) created by Schmidt and Jurado [12]. This model is founded on some general theoretical ideas with rather general character. It is avoiding microscopic calculation with corresponding approximations and limitations. The GEF model is complemented by a wide volume of empirical information and it is able to give quantitative predictions of fission observables for a large number of fissionable systems.

3 Results and discussions

We have evaluated the cross sections of the $^{238}U(n,f)$ and $^{238}U(n,\gamma)^{239}U$ processes for incident neutrons energy from 1 MeV up to 200 MeV using the standard Talys input. The results are shown in the Figure 1.a and 1.b. The theoretical evaluation of neutron capture and fission reactions are compared with existing experimental results from Internet nuclear data bases [13]. Since the capture and fission cross sections are time consuming the step for the calculations is 10 MeV. From Figures 1.a and 1.b it results that at high neutrons energies the fission process is dominant comparing with the capture process of type $^{238}U(n,\gamma)^{239}U$. We note that the total gamma production is larger with one order of magnitude and it represents the contributions from all processes in which gamma quanta are involved in the exit channels.

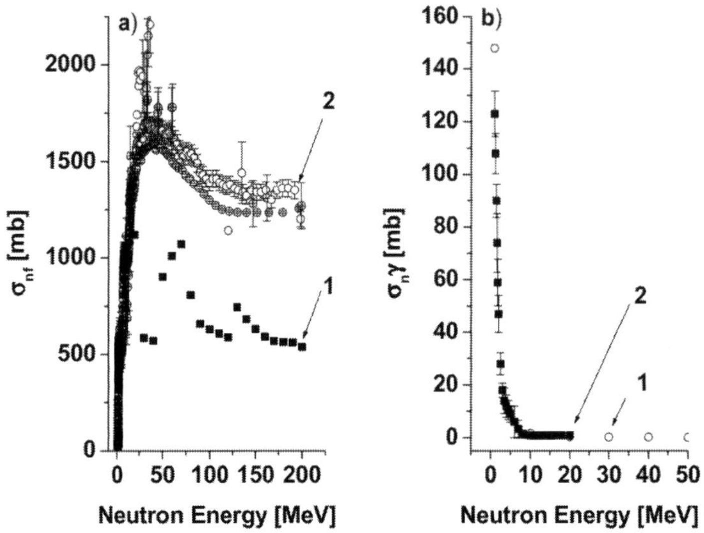

Figure 1. Cross section: a) Neutron induced fission of ^{238}U nucleus; b) Capture $^{238}U(n,\gamma)^{239}$; 1- Talys evaluation; 2- Experimental data.

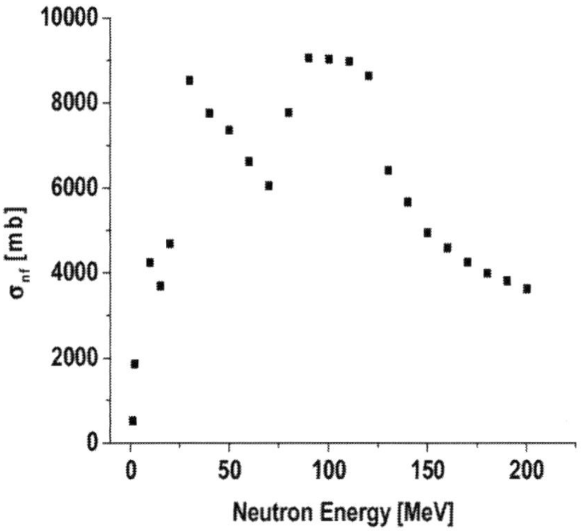

Figure 2. Total gamma production in the neutron capture by ^{238}U nucleus.

The total gamma production in the neutron capture by ^{238}U nucleus is represented in the Figure 2. The comparison of fission theoretical and experimental data can be considered satisfactorily (Figure 1.a). The capture experimental and theoretical data are in a better agreement than the fission data but it's necessary to extend the measurements higher than 20 MeV (Figure 1.b). The Figure 2 suggests that the measurements of the neutron capture of type $^{238}U(n,\gamma)^{239}U$ will be very difficult and affected by a large background.

Another interesting features of the fission process are the mass distribution of fission fragments and the neutron multiplicity at a given incident neutron energy before and after neutron emission by fission fragments. In Figures 3.a and 3.b we have the mass distribution at given incident neutron energies, 10 MeV and 100 MeV, respectively. One can observe that the mass distribution becomes more symmetrical with the increasing of the incident neutron energy.

The neutron multiplicity results are shown in the Figures 4.a and 4.b for 10 MeV and 100 MeV, respectively.

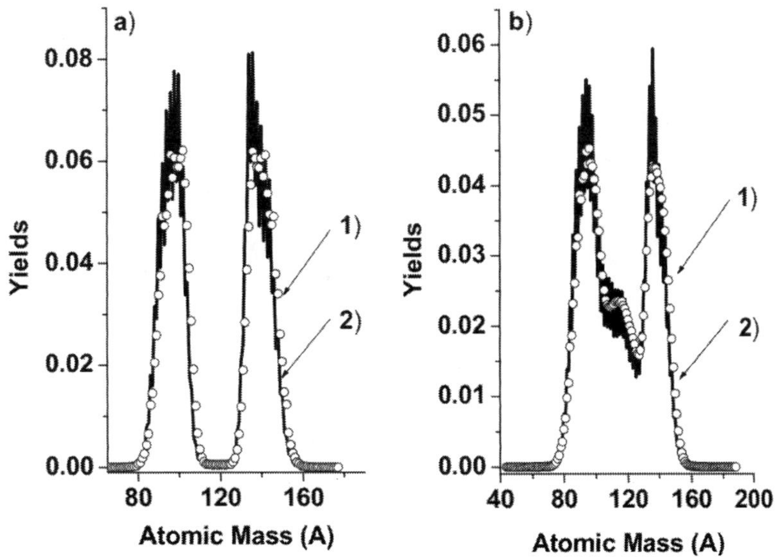

Figure 3. Mass distribution of fission fragment in $^{238}U(n,f)$ process at given incident neutron energy (E_n) before (1) and after (2) neutron emission; a) E_n = 10 MeV; b) E_n = 100 MeV.

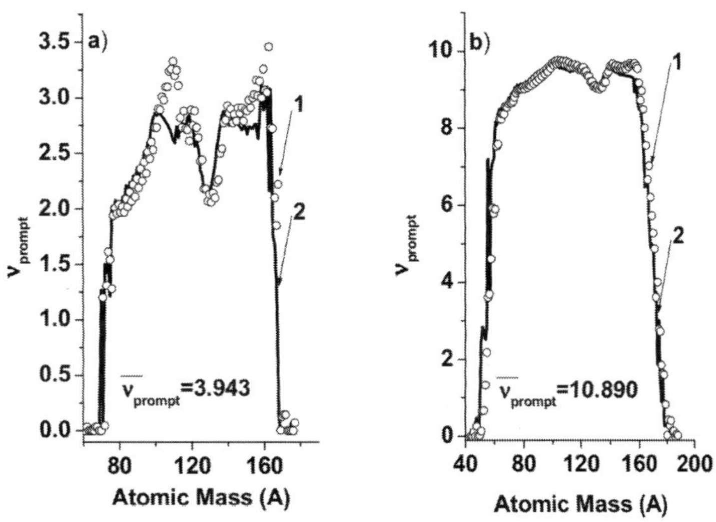

Figure 4. Average prompt neutron multiplicity as a function of mass at given incident neutron energy (E_n) before (1) and after (2) neutrons emission; a) – E_n=10 MeV; b) – 100 MeV.

The ^{89}Y nucleus has no isotope, and in many studies of new reactors, it is used in the design of threshold neutron detectors necessaries for the determination of fast neutrons energy spectrum. We have evaluated the cross sections of $^{89}Y(n,xn)$ reactions (x=1,2,3,4) with standard Talys parameters input. For $^{89}Y(n,2n)^{88}Y$ and $^{89}Y(n,2n)^{87}Y$ reactions the results are compared with experimental data and are shown in the Figure 5. In the case of production of ^{88}Y nucleus the agreement between theoretical and experimental data is very good (Figure 5.a). For ^{87}Y production in the $^{89}Y(n,3n)$ reaction our calculations from Figure 5.b are compared with two sets of experimental data. With the experimental data set noted by 2 in Figure 5.a there are serious differences as compared with theoretical one but with the data set noted with 3 there is a quit good agreement. The experimental data from Figures 5.a and 5.b are taken from [13, 14] and [13, 15, 16], respectively.

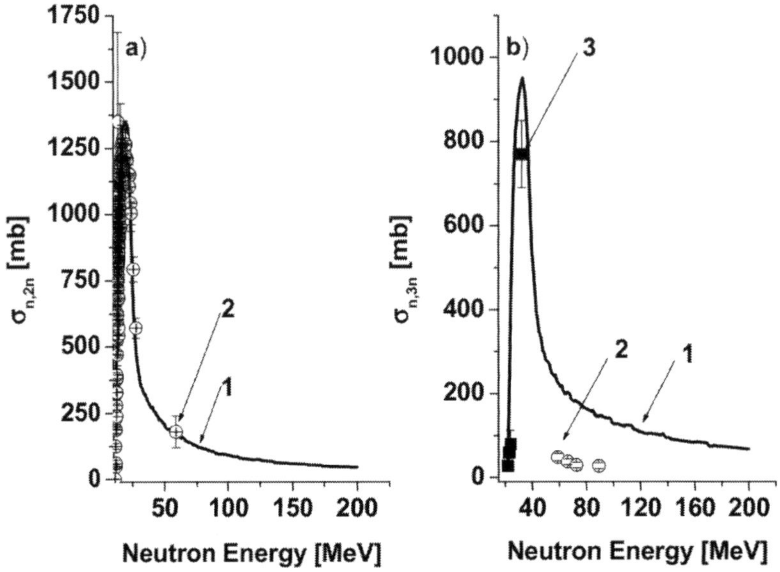

Figure 5. Production of ^{88}Y and ^{87}Y nuclei in: a) $^{89}Y(n,2n)^{89}Y$; b) $^{89}Y(n,3n)^{87}Y$; 1- Talys evaluation compared with experimental data – 2, 3.

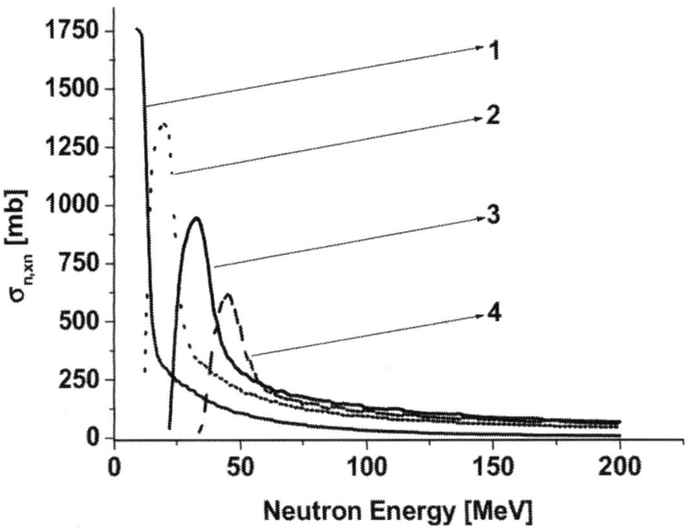

Figure 6. Talys evaluations of $^{89}Y(n,xn)$ reactions (x=1,2,3,4); Production of: 1 – ^{89}Y; 2 – ^{88}Y; 3 – ^{87}Y; 4 – ^{86}Y.

The quit good agreement between experimental and evaluated data of cross sections suggested to the authors to estimate (n,n') and (n,4n) cross sections where the experimental data are lacking or very poor. In Figure 6 the cross sections of production of $^{89,88,87,86}Y$ nuclei in (n,xn) reactions are represented. The cross sections calculations with standard Talys input had showed that in the low energy part of incident neutrons near the threshold, the compound processes are dominant first on discrete states of residual nuclei and step by step with the increasing of energy the discrete states are replaced by continuum ones. Going further on energy after the maximum is passed the contribution in the cross sections of compound processes is decreasing and gradually the pre-equilibrium and direct processes, respectively, become dominant in the cross sections.

4 Conclusions

We have evaluated the cross sections of neutron induced fission and capture on ^{238}U and of the (n,xn) reactions on ^{89}Y nucleus (x=1,2,3,4) from 1 MeV up to 200 MeV by using the Talys code. In the mentioned incident neutron energy interval the nuclear potential, density levels and other parameters (mainly extracted from experimental data practically for all nuclei) necessaries in the evaluations are locally defined. Talys software allows calculations in principle up to 1000 MeV with the so called global defined parameters. In the future the calculations should to be extended higher than 200 MeV up to some GeV as it is requested in the researches dedicated to the fast neutrons reactors of new generation.

On the other hand the energy range lower than 1 MeV, corresponding to the slow and resonant neutrons, is also of interest requesting treatment in the frame of other approaches and formalisms like Multilevel Breit- Wigner, R and/or S matrix.

The mentioned formalisms and computer codes dedicated to nuclear reactions are necessary to be combined with simulation software of the physical experiment as a new and interesting direction of development of the present research.

References

1. V.S. Barashenkov, Physics of Elementary Particles and Atomic Nuclei, 9, 5, 871 (1978) (in Russian)
2. Gh. Vladuca, *Nuclear Physics*, **2**, Bucharest University Press (1986) (in Romanian)
3. A. Wojciechowski, Y.C. Lim, I. Zhuk, K. Husak, S. Korneev, A. Potapenko, A. Safronova, V. Voronko, V. Sotnikov, M. Artiushenko, Scientific Report 2012 – 2013 of Laboratory of Information Technology – Joint Institute for Nuclear Research, Dubna, Russian Federation (LIT – JINR), 98 (2014) (http://lit.jinr.ru/Reports/SC_report_12-13/p98.pdf)
4. A. Wojciechowski, Y.C. Lim, V. Stepanenko, M. Bidus, I. Zhuk, K. Husak, S. Korneev, A. Potapenko, A. Safronova, V. Voronko, V. Sotnikov, M. Artiushenko, Scientific Report 2012 – 2013 of LIT – JINR Dubna, RF, 102 (2014)
5. N. Bohr, J.A. Wheeler, Phys. Rev. **56**, 426 (1939) (http://lit.jinr.ru/Reports/SC_report_12-13/p102.pdf)
6. Ya. I. Frenkel, Journal of Experimental and Theoretical Physics, **9**, 641 (1939)
7. A.J. Koning, S. Hilaire and M.C. Duijvestijn, .TALYS-1.0., Proceedings of the International Conference on Nuclear Data for Science and Technology, April 22-27, 2007, Nice, France, editors O. Bersillon, F. Gunsing, E. Bauge, R. Jacqmin and S. Leray, EDP Sciences, 211 (2008)
8. W. Hauser, H. Feshbach, Phys. Rev. **87**, 2, 366 (1952)
9. A.J. Koning, M.C. Duijvestijn, Nucl. Phys. **A744**, 15 (2004)
10. G.R. Satchler, *Direct Nuclear Reactions*, Oxford University Press, New York (1983)
11. S. Goriely, S. Hilaire, A.J. Koning, Phys. Rev. **C78**, 064307 (2008)
12. K. H. Schmidt, B. Jurado, *General Description of Fission Observables*, JEFF Report 24 / GEF Model, Data Bank NEA/DB/DOC 2014 (1) (http://www.khs-erzhausen.de/Preprints/db-doc2014-1.pdf)
13. Experimental Nuclear Reaction Data (EXFOR) (https://www-nds.iaea.org/exfor/exfor.htm)
14. S.M. Qaim, R. Woelfle, G. Stoecklin, Chemical Nuclear Data Conference, *Proceedings of the Conference*, 121, Canterbury (1971)
15. V. Wagner, O. Svoboda, J. Vrzalova, M. Schopar, B. Geier, A. Kugler, M. Honusek, Journal of Physics: Conference Series **366** 012047 (2012)
16. J. Vrzalova, O. Svoboda, A. Krasa, A. Kugler, M. Majerle, M. Suchopar, V. Wagner, NIMA, **726**, 84 (2013)

EPJ Web of Conferences **111**, 11003 (2016)
DOI: 10.1051/epjconf/201611111003

Impact of nuclear data on sodium-cooled fast reactor calculations

Alexander Aures[1,a], Friederike Bostelmann[1], Winfried Zwermann[1], and Kiril Velkov[1]

[1]*Gesellschaft für Anlagen- und Reaktorsicherheit (GRS) gGmbH, Boltzmannstr. 14,85748 Garching,Germany*

Abstract. Neutron transport and depletion calculations are performed in combination with various nuclear data libraries in order to assess the impact of nuclear data on safety-relevant parameters of sodium-cooled fast reactors. These calculations are supplemented by systematic uncertainty analyses with respect to nuclear data. Analysed quantities are the multiplication factor and nuclide densities as a function of burn-up and the Doppler and Na-void reactivity coefficients at begin of cycle. While ENDF/B-VII.0 / -VII.1 yield rather consistent results, larger discrepancies are observed between the JEFF libraries. While the newest evaluation, JEFF-3.2, agrees with the ENDF/B-VII libraries, the JEFF-3.1.2 library yields significant larger multiplication factors.

1 Introduction

Among the envisaged Gen-IV reactor concepts are the sodium-cooled fast reactors (SFR). However, there is not much operation experience with these concepts and only very few data is publicly available. For a reliable prediction of safety-relevant parameters, it is necessary to correctly assess the applicability and quality of simulation tools and nuclear data libraries for the analyses of these systems. On the basis of the SFR-FT benchmark specification [1], the OECD/NEA Working Group SFR-FT [2] performed a parametric study of various SFR concepts: large size cores with oxide and carbide fuel and medium size cores with oxide and metallic fuel. The present contribution focuses on the large oxide core (MOX-3600) and the medium size metallic core (MET-1000). The objective of this study is to assess the impact of nuclear data on safety-relevant parameters such as the multiplication factor, nuclide densities, and the Doppler and Na-void reactivity coefficients. Neutron transport and depletion calculations with different nuclear data libraries are performed, and systematic uncertainty analyses with respect to nuclear data are carried out.

2 Models and calculation methods

For both cores considered, a model of a sub-assembly is built that corresponds to the lowest axial section of a fuel assembly of the inner core region. Detailed data of the geometry and the material is provided in the SFR-FT benchmark specification [1]. The coolant is sodium and the fuel consists mainly of U-238 and Plutonium and corresponds to the equilibrium state. The entire inventory of

[a] Corresponding author: alexander.aures@grs.de

fission products is represented by a reactivity equivalent part of the molybdenum inventory. The fuel assemblies of the two core concepts are hexagonal and consist of 271 fuel pins.

2.1 Calculation tools and nuclear data libraries

The calculations in this study are performed with various neutron transport and depletion codes, and – wherever possible – with different nuclear data libraries. The TRITON/NEWT and the TRITON/KENO sequence of the SCALE 6.1 code system [3] are applied with the 238-group ENDF/B-VII.0 library. Continuous-energy Monte Carlo calculations are undertaken with MCNP-6 [4] and Serpent 2 (2.1.24) [5] along with the nuclear data libraries ENDF/B-VII.1 (provided with MCNP-6), ENDF/B-VII.0 and JEFF-3.1.1 (both provided with Serpent), and JEFF-3.1.2 and JEFF-3.2 (provided by OECD/NEA data bank).

Systematic uncertainty analysis with respect to uncertainties in the nuclear data is performed with the sampling-based XSUSA methodology *(Cross Section Uncertainty and Sensitivity Analysis)* and with the perturbation theory-based TSUNAMI module of the SCALE 6.1 code system. Details of both are given in Ref. [3, 6].

3 Results

3.1 Depletion calculations

Depletion calculations are carried out for a sub-assembly of both the large oxide core (MOX-3600) and the ABR metallic core (MET-1000). The results are obtained by various stochastic and deterministic neutron transport and depletion codes, and by the application of different nuclear data libraries either in continuous-energy or in multi-group representation.

The fuel assembly of the large oxide core is depleted for 410 days with a thermal power of 50.4 MW/t HM. Figure 1 shows the multiplication factor as a function of burn-up.

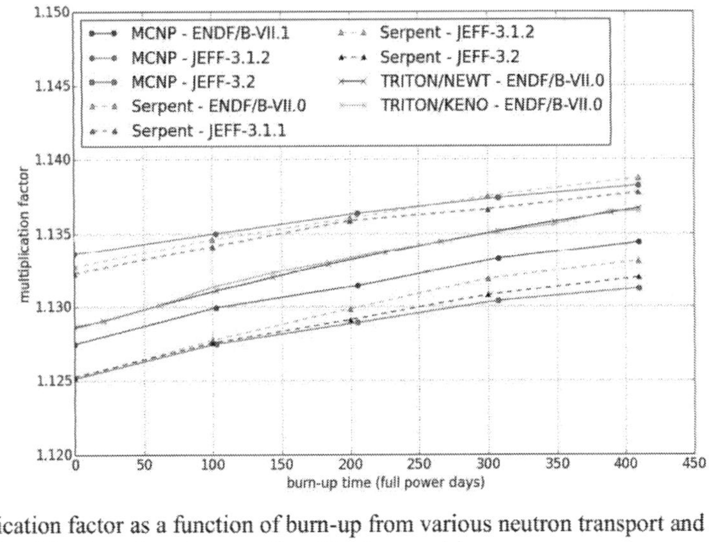

Figure 1. Multiplication factor as a function of burn-up from various neutron transport and depletion codes and different nuclear data libraries of a sub-assembly of the large oxide core.

The multiplication factors calculated by MCNP with JEFF-3.1.2 and ENDF/B-VII.1 shows a discrepancy of about 480 pcm at BOC (begin of cycle) and 300 pcm at EOC (end of cycle). This is in qualitative agreement with the results of other benchmark participants who obtained differences with the same sign and similar magnitude for whole core calculations when using ENDF/B-VII and JEFF-3.1 data [2]. Applying the newest version of the JEFF library, JEFF-3.2, differences of -184 pcm at BOC and of -247 pcm at EOC to ENDF/B-VII.1 are obtained. Between JEFF-3.1.2 and JEFF-3.2, the differences are significant large, -641 pcm at BOC and -514 pcm at EOC. Similar results are obtained with Serpent. In case of JEFF-3.2, the multiplication factors calculated with Serpent and MCNP are consistent at BOC. However, with increasing burn-up, Serpent yields slightly larger multiplication factors. The results of the TRITON/NEWT and the TRITON/KENO sequences with application of the 238-group ENDF/B-VII.0 library are virtually identical. This is expected, since both codes use the same nuclear data library.

Moreover, the nuclide densities of several actinides obtained from the depletion calculations of the large oxide core are analysed at EOC. Figure 2 shows the main findings, namely the maximum relative deviations from the mean values of the nuclide densities.

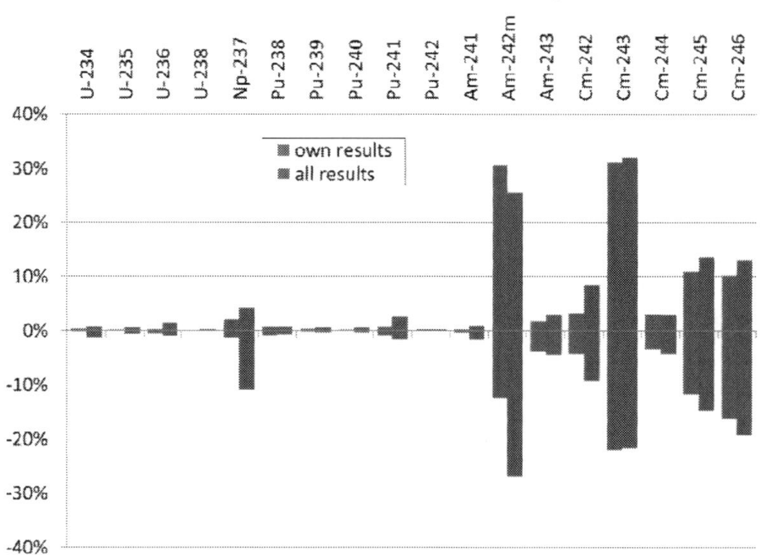

Figure 2. Range of results for actinide inventories at the end of the burn-up cycle obtained from own calculations only, and from own calculations and all available benchmark results.

The results denoted by *own results* are determined from the own TRITON/NEWT, MCNP (ENDF/B-VII.1, JEFF-3.1.2) and Serpent (ENDF/B-VII.0, JEFF-3.1.1 / -3.1.2 / -3.2) calculations. TRITON/KENO results are excluded from the evaluation because they are almost identical to the TRITON/NEWT results. Additionally, the results are supplemented by corresponding values obtained from the results of the benchmark contributions (denoted by *all results*). While the bandwidths of the results are small for most actinides, in particular Pu-239, they exhibit large values for Np-237 and some isotopes of the Minor Actinides Am and Cm.

The sub-assembly of the ABR metallic core is depleted for 328.5 days with a thermal power of 75.4 MW/t HM. Figure 3 shows the multiplication factors obtained with the various code systems and nuclear data libraries as a function of burn-up. Similar to the results of the large oxide core, a large discrepancy is observed between the multiplication factors obtained by MCNP with JEFF-3.1.2 and JEFF-3.2. The difference is about 850 pcm at BOC and 800 pcm at EOC. The difference between JEFF-3.2 and ENDF/B-VII.1 is about 200 pcm at BOC and 20 pcm at EOC. The multiplication factors calculated by the TRITON/NEWT sequence using the 238-group ENDF/B-VII.0 library are smaller than the continuous-energy Monte Carlo results.

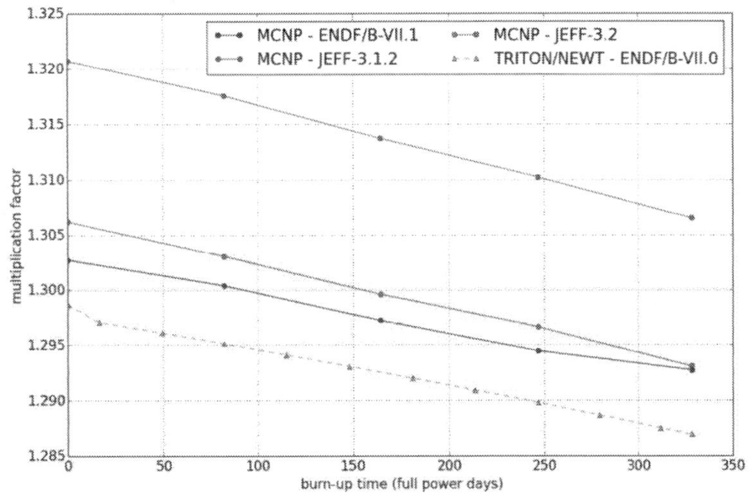

Figure 3. Multiplication factor as a function of burn-up from various neutron transport and depletion codes and different nuclear data libraries of a sub-assembly of the ABR metallic core.

3.2 Safety-relevant parameters and their uncertainty with respect to nuclear data

Safety-relevant parameters and their uncertainties are determined for the large oxide core and the ABR metallic core at BOC. For this, the temperature of all materials is set to 600 K. For the assessment of the Doppler reactivity, the temperature is changed to 1200 K, and for the sodium void reactivity, the sodium density is changed to 1% of the nominal density. Results of the safety-relevant parameters calculated with MCNP and different nuclear data libraries are presented in the Tables 1-2.

Table 1: MCNP results for the large oxide core fuel assembly at BOC.

	ENDF/B-VII.0	ENDF/B-VII.1	JEFF-3.1.2	JEFF-3.2
k-eff nominal	1.13712	1.13901	1.14505	1.13677
k-eff high fuel temp.	1.12823	1.13016	1.13620	1.12804
k-eff Na void	1.17258	1.17481	1.18222	1.17361
Doppler reactivity (pcm)	-693	-688	-680	-681
Na void reactivity (pcm)	2659	2675	2746	2761

Table 2: MCNP results for the ABR metallic core fuel assembly at BOC.

	ENDF/B-VII.0	ENDF/B-VII.1	JEFF-3.1.2	JEFF-3.2
k-eff nominal	1.30246	1.30418	1.32281	1.30791
k-eff high fuel temp.	1.29839	1.30020	1.31883	1.30399
k-eff Na void	1.40546	1.40718	1.43291	1.41628
Doppler reactivity (pcm)	-241	-235	-228	-230
Na void reactivity (pcm)	5627	5612	5809	5850

The differences in the Doppler reactivity are moderate for both fuel assemblies. However, significant differences can be observed between the multiplication factors and also for the sodium void reactivities. While ENDF/B-VII.0 and ENDF/B-VII.1 yield rather consistent results with a small and almost constant deviation in the multiplication factors for both cases, the deviations concerning the JEFF libraries are much larger. JEFF-3.1.2 yields substantially larger multiplication factors than the other libraries, in particular for the metallic core.

Figure 4 shows results from the systematic uncertainty analyses with respect to nuclear data. On the left, uncertainty values of the multiplication factors, the Doppler and the Na-void reactivity are plotted for both systems for BOC. On the right, uncertainty values of the multiplication factor at BOC / EOC and of the actinide nuclide densities at EOC of the large oxide core (MOX-3600) are shown. The uncertainty of the multiplication factors is significantly larger than those found for light water reactor systems. The uncertainty of the nuclide densities is moderate compared to the bandwidths of the results from the various depletion sequences.

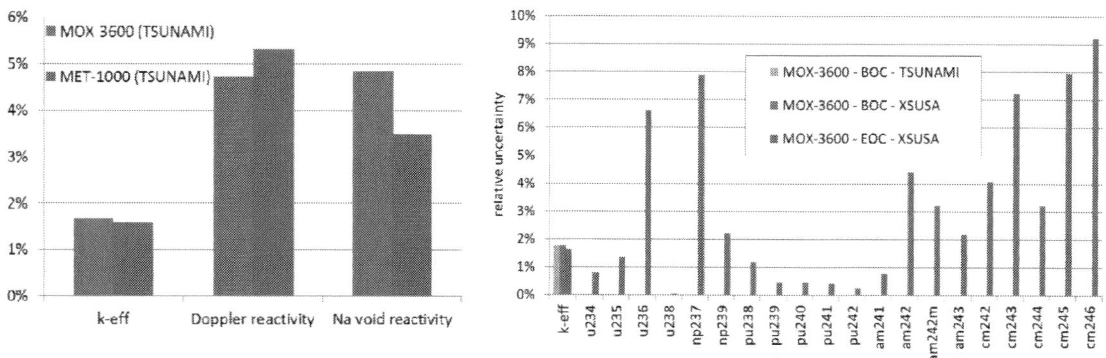

Figure 4. Uncertainties (1 σ) of the multipl. factor, Doppler and Na-void reactivity for both systems at BOC (left), and of the multipl. factor and nuclide densities for the large oxide core at BOC / EOC (right).

4 Summary

In this study, the impact of nuclear data on safety-relevant parameters such as multiplication factor, nuclide densities, the Doppler reactivity coefficient and the Na-void reactivity coefficient is studied. Neutron transport and depletion calculations as well as systematic uncertainty analyses are performed for an axial section of a fuel assembly of both the large oxide core and the ABR metallic core. The application of different nuclear data libraries leads to large discrepancies in the multiplication factors as a function of burn-up for both fuel assemblies. The bandwidths of the nuclide densities of the actinides obtained from the various depletion sequences are moderate. Only for Np-237 and for several Am and Cm isotopes, larger bandwidths are observed. The analysis of the safety-parameters showed that the application of different nuclear data libraries results in significant discrepancies for the multiplication factors and sodium void reactivities. The uncertainty of the multiplication factor determined by the systematic uncertainty analysis turned out to be significantly larger than for LWR systems.

Acknowledgments

This work is supported by the German Federal Ministry for Economic Affairs and Energy.

References

1. D. Blanchet, L. Buiron, N. E. Stauff, T. K. Kim, T. Taiwo, *AEN – WPRS Sodium Fast Reactor Core Definitions (Version 1.2 – September 19th)*, 2011, available at OECD/NEA Data Bank.
2. *Sodium Fast Reactor Core Feed-back and Transient Response Task Force (SFR-FT Task Force)*, https://www.oecd-nea.org/download/wprs-sfr/index.html (2014)
3. *SCALE: A Comprehensive Modeling and Simulation Suite for Nuclear Safety Analysis and Design, Version 6.1*, ORNL/TM-2005/39, Oak Ridge National Laboratory, Oak Ridge, Tennessee, June 2011. Available from Radiation Safety Information Computational Center at Oak Ridge National Laboratory as CCC-785.

4. J.T. Goorley et al., *Initial MCNP6 Release Overview - MCNP6 version 1.0*, Los Alamos National Laboratory, LA-UR-13-22934 (2013).
5. J. Leppänen, M. Pusa, T. Vittanen, V. Valtavirta, T. Kaltiaisenaho, *The Serpent Monte Carlo code: Status, development and applications in 2013*, Ann. Nuc. Energy **82**, 142-150, (2015)
6. W. Zwermann, A. Aures, L. Gallner, V. Hannstein, B. Krzykacz-Hausmann, K. Velkov, J.S. Martinez, *Nuclear Data Uncertainty and Sensitivity Analysis with XSUSA for Fuel Assembly Depletion Calculations*, Nucl. Eng. and Tech. **46**, No. 3, (2014)

EPJ Web of Conferences **111**, 11004 (2016)
DOI: 10.1051/epjconf/201611111004

ANITA-2000 activation code package - updating of the decay data libraries and validation on the experimental data of the 14 MeV Frascati Neutron Generator

Manuela Frisoni[a]

ENEA,CR "E.Clementel", Via Martiri di Monte Sole4, 40129 Bologna, Italy

Abstract. ANITA-2000 is a code package for the activation characterization of materials exposed to neutron irradiation released by ENEA to OECD-NEADB and ORNL-RSICC. The main component of the package is the activation code ANITA-4M that computes the radioactive inventory of a material exposed to neutron irradiation. The code requires the decay data library (file fl1) containing the quantities describing the decay properties of the unstable nuclides and the library (file fl2) containing the gamma ray spectra emitted by the radioactive nuclei. The fl1 and fl2 files of the ANITA-2000 code package, originally based on the evaluated nuclear data library FENDL/D-2.0, were recently updated on the basis of the JEFF-3.1.1 Radioactive Decay Data Library. This paper presents the results of the validation of the new fl1 decay data library through the comparison of the ANITA-4M calculated values with the measured electron and photon decay heats and activities of fusion material samples irradiated at the 14 MeV Frascati Neutron Generator (FNG) of the ENEA-Frascati Research Centre. Twelve material samples were considered, namely: Mo, Cu, Hf, Mg, Ni, Cd, Sn, Re, Ti, W, Ag and Al. The ratios between calculated and experimental values (C/E) are shown and discussed in this paper.

1 Introduction

The evaluation of the neutron induced material activation plays an important role for the development of future fusion power plants for issues related to safety, engineering design and radioactive waste management. The availability of reliable nuclear data and activation codes is required. ANITA-2000 [1,2] is a code package for the activation characterization of materials exposed to neutron irradiation released by ENEA to OECD-NEADB [3] and ORNL-RSICC [4]. It has been widely validated and used in the past in ENEA-Bologna for radioactive inventories and source terms calculations for fusion plants safety analysis [5,6,7].The original package contains: a) the activation code ANITA-4M that traces back to the ANITA code (Analysis of Neutron Induced Transmutation and Activation [8]), b) two activation cross section libraries based respectively on EAF-99 [9] and on FENDL/A-2.0 [10], c) the decay data library (file fl1) containing the quantities describing the decay properties of the unstable nuclides and d) the library (file fl2) containing the gamma ray spectra emitted by the radioactive nuclei. The fl1 and fl2 files of the ANITA-2000 code package, originally based on the

[a] Corresponding author: manuela.frisoni@enea.it

evaluated nuclear data library FENDL/D-2.0 [11], were recently updated on the basis of the JEFF-3.1.1 Radioactive Decay Data Library [12]. The ANITA-4M code actually uses a neutron cross section library based on the data of the European Activation File EAF-2010 [13] provided in the package EASY-2010 [14]. This work presents the results of the validation of the new fl1 decay data library through the comparison of the ANITA-4M calculated values with the measured electron and photon decay heats and activities of fusion material samples irradiated at the 14 MeV Frascati Neutron Generator (FNG) of the ENEA-Frascati Research Centre. Twelve material samples were considered, namely: Mo, Cu, Hf, Mg, Ni, Cd, Sn, Re, Ti, W, Ag and Al. The ratios between calculated and experimental values (C/E) are shown and discussed.

2 Updated decay data libraries

The Decay, Hazard and Clearance data library (file fl1) provides the information describing the decay properties of the unstable nuclides. It contains, for each nuclide, the decay mode, the branching ratios, the decay constant (s^{-1}), the total energy (MeV) released in the decay and the energy (MeV) released in the form of gamma or X-rays. The decay data of the new fl1 file were taken from the JEFF-3.1.1 Radioactive Decay Data library (MF=8, MT=457 section in the ENDF-6 [15] standard format notation).

The fl1 file provides also for each radionuclide the clearance level C_L (Bq/g). This quantity allows to establish if a radioactive material can be potentially moved out of the originating facility and recycled. The safe handling of radioactive waste is recognized as crucial to ensure protection of human health and the environment. IAEA publish regulations on these issues and Ref.[16] provides information on suggested clearance level values for a set of important radionuclides. The clearance level data contained in the new fl1 file have been updated by including the information contained in Ref.[16] and Ref.[17]. The updated fl1 file actually contains data for 3433 nuclides.

The gamma library (file fl2) contains the gamma ray spectra emitted by the radioactive nuclei in the ORNL-SCALE 18-γ energy group structure. It was updated on the basis of the gamma radiation spectra contained in the JEFF-3.1.1 Radioactive Decay Data library. The data given in the fl2 gamma library are used in ANITA-4M to compute the intensity and the energy distribution of the gamma-rays emitted by the irradiated material. The production of the new fl1 and fl2 files required the development of a series of proper interface codes in order to process the ENDF6 format JEFF-3.1.1 evaluated decay data, to calculate the clearance level quantities and to convert these data into the specific format required by the ANITA-4M code.

3 The FNG experiment

The availability of reliable nuclear data to be used in the neutron transport and activation calculations is highly recommended. To this end, suitable experimental activities (benchmark experiments) are necessary. One of those experiments was carried out at the ENEA Research Centre of Frascati at the 14 MeV Frascati Neutron Generator (FNG) based on the T(d,n)α fusion reaction [18]. The FNG group used a neutron reflector that forms an irradiation cavity where the neutron spectrum simulates that existing in the first wall of a typical fusion device. More details on the experimental set-up are given in Ref. [19]. Twelve high purity samples of different materials (i.e. Mo, Cu, Hf, Mg, Ni, Cd, Sn, Re, Ti, W, Ag, Al) were irradiated at a fixed position close to the FNG neutron target. The irradiation time ranged between 160 s and 1600 s . For each sample the experimental results were provided for various cooling times ranging from about 130 s to 1440 s. The experiment allowed to measure simultaneously the total activity and the electron and photon decay heats for each irradiated sample. The FNG experimental results enable to validate both the cross section and the decay databases used by the activation codes.

4 Calculation-experiment comparison

The validation of the updated decay data library (file fl1) has been performed through the comparison of the ANITA-4M results with the FNG experimental data. The physical quantities involved in the comparison are, for each one of the material samples, the electron and photon decay heats (kW) and the activity (Bq). The ANITA-4M calculations were performed by using the neutron activation cross sections based on EAF-2010. The assessment refers to Mo, Cu, Hf, Mg, Ni, Cd, Sn, Re, Ti, W, Ag, and Al samples. The details of the experimental set-up (material composition, sample mass, irradiation time, neutron flux, etc.) provided by FNG team were used in the ANITA-4M input files in order to model each experiment. The ratios between calculated (C) and experimental (E) values (with the error bars obtained by using the error propagation theory and considering the errors of both the C and E values at 1σ) for material samples electron and photon decay heat and activity are shown in Figs.1,2,3.

For each material the C/E values have been calculated as a mean over the cooling times. The overall set of detailed experimental data and calculation results is given in Ref. [20].

For sake of comparison, decay heat and activity calculations were also performed by using the EASY-2010 code package (FISPACT-2010 code + EAF-2010 libraries). The comparison between the ANITA-4M and FISPACT-2010 results is shown in Figs.4,5,6.

5 Results discussion

5.1 Comparison between experimental results and ANITA-4M calculations (new fl1 file and EAF-2010 neutron cross section activation library)

1. Electron decay heat:

- The discrepancies between the calculated mean electron decay heat C and the experimental E values are inside 20% for Cu, Mg, Cd, Sn, Ti, Al and up to 30% for Mo, W and Ag. Materials with discrepancies greater than 30% are: Hf (42%), Ni (70%) and Re (31%).
- The mean electron decay heat C/E values show agreement within error bars for Mo, Cu, Mg, Sn, Ti, W and Al.

2. Photon decay heat:

- The calculated photon decay heat values are lower than the corresponding experimental values for all the material samples.
- The discrepancies between the calculated mean photon decay heat C and the experimental E values are inside 20% for Mo, Mg , Al and up to 30% for Cu, Cd and Sn. The materials with discrepancies greater than 30% are: Hf (60%), Ni (40%) , Re (55%); Ti (40%), W (40%) and Ag (35%).
- The photon decay heat C/E values show agreement within the error bars only for Mo, Mg and Al.

3. Activity:

- The calculated activity values show in general an underestimation of the experimental values except for Mo and Ni.
- The mean activity values for Mo, Mg and Al show a good agreement between the experimental and calculated results (discrepancies $\leq$ 10%). The discrepancies are inside 20% for Cu, Sn, Ti, W and Ag. The largest discrepancies are for Hf (55%) , Ni (79%) , Cd (35%) and Re (30%).

5.2 Comparison between ANITA-4M and FISPACT-2010 calculations

For the electron decay heats, the ANITA-4M and FISPACT-2010 calculated results differ by less than 5% for all the materials except for Hf (13%), Ni(30%), Cd (12%) and Sn (26%). For Sn, ANITA-4M overestimates the experimental values whereas, on the contrary, FISPACT-2010 gives an underestimation.

For the photon decay heats, the ANITA-4M and FISPACT-2010 calculated results differ by less than 5% for all the materials, except for Hf (10%) and Ni (16%).

For the activity, the ANITA-4M and FISPACT-2010 calculated results differ by less than 2% for all the materials, except for Hf (7%).

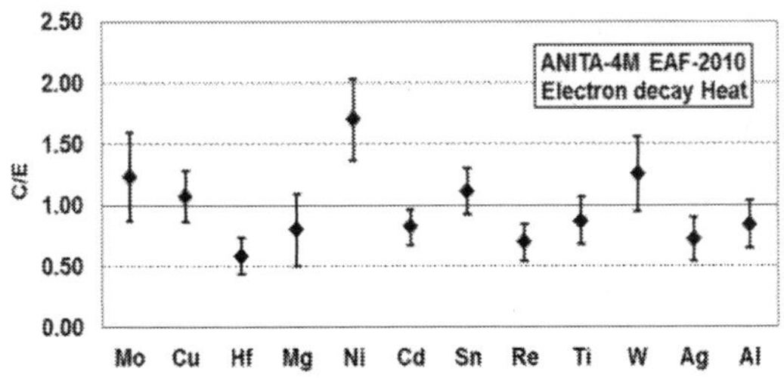

Figure 1. Material electron decay heat: calculated (C) and experimental (E) comparison.

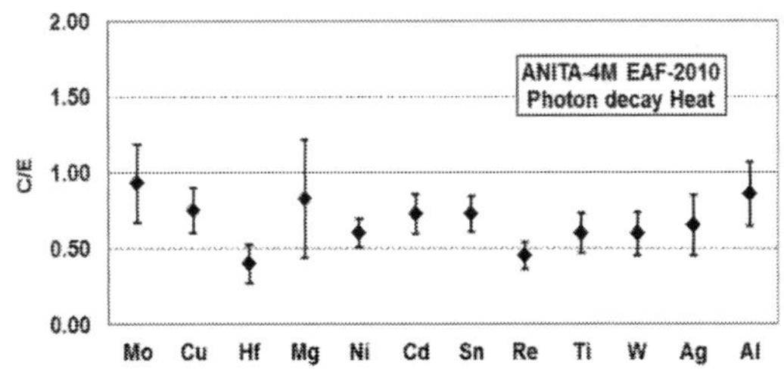

Figure 2. Material photon decay heat: calculated (C) and experimental (E) comparison.

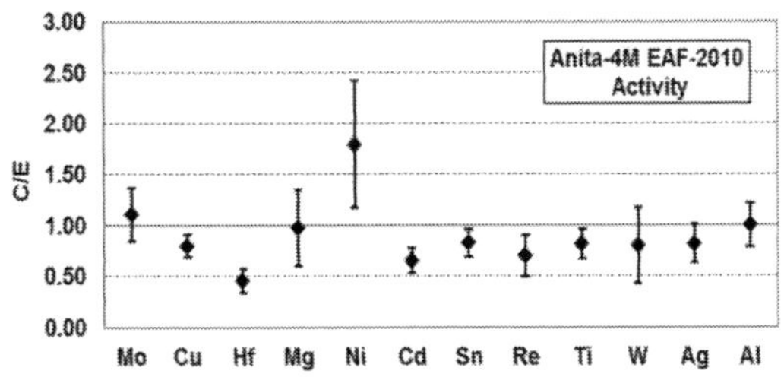

Figure 3. Material activity: calculated (C) and experimental (E) comparison.

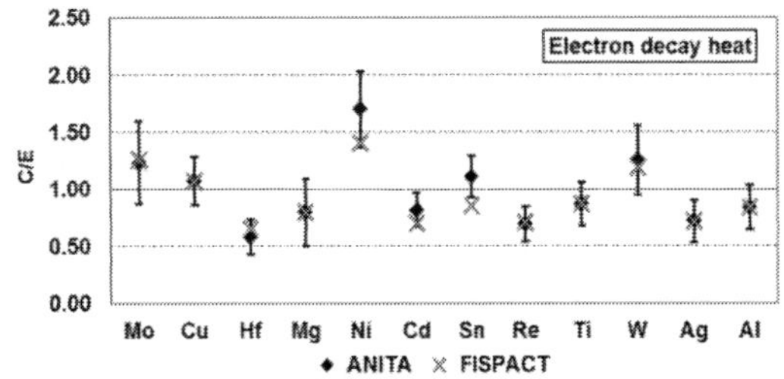

Figure 4. Material electron decay heat: ANITA-4M and FISPACT-2010 comparison.

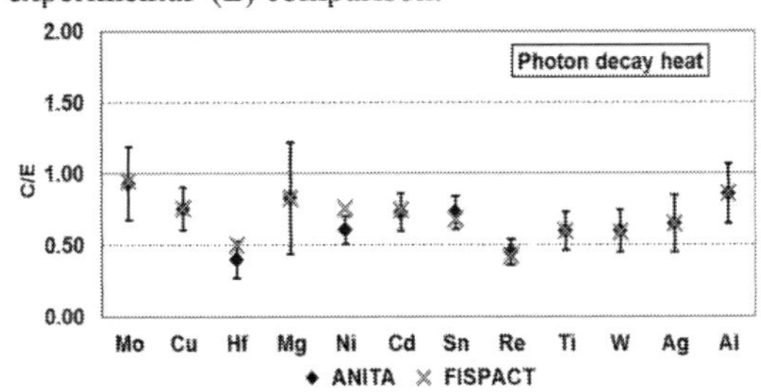

Figure 5. Material photon decay heat: ANITA-4M and FISPACT-2010 comparison.

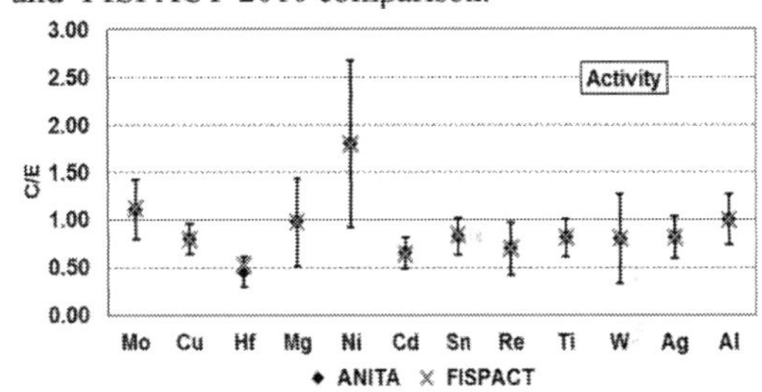

Figure 6. Material activity: ANITA-4M and FISPACT-2010 comparison.

6 Conclusion

The validation of the updated decay data library (file fl1) based on JEFF-3.1.1 evaluated data was performed through the comparison between ANITA-4M calculated values and the measured electron and photon decay heats and activities of twelve fusion material samples irradiated at the 14 MeV Frascati Neutron Generator (FNG) of the ENEA-Frascati Research Centre.

For sake of comparison the FISPACT-2010 code has been also used for decay heat and activity calculations.

The ANITA-4M and FISPACT-2010 results are quite comparable confirming once again that both are reliable tools for the activation calculations in the frame of the fusion field.

The ANITA-4M and FISPACT-2010 calculations were performed by using the same neutron activation cross sections based on EAF-2010 data. The differences between the ANITA-4M and FISPACT-2010 results can be explained as due to differences in the decay data (e.g. life-times, beta and gamma decay energies, etc.) and/or a different numerical treatment of the nuclear chains.

The calculation/experiment comparison presented in this paper shows that the decay data and the neutron activation cross sections also of the most recent evaluations need a further assessment and improvement.

References

1. D.G. Cepraga, G. Cambi, M. Frisoni, ANITA-2000 Activation Code Package. Part I - Manual, ENEA Report ERG-FUS/TN-SIC TR 16/2000 (2000)
2. D.G. Cepraga, G. Cambi, M. Frisoni, ANITA-2000 Activation Code Package. Part II : Code validation, ENEA Report FUS-TN-SA-SE-R-020 (2001)
3. D.G. Cepraga, G. Cambi, M. Frisoni, G.C. Panini, ANITA-2000, OECD NEA Data Bank NEA-1638, (2000)
4. D.G. Cepraga, G. Cambi, M. Frisoni, G.C. Panini, ANITA-2000, RSICC CCC-693 (2002)
5. G. Cambi, D.G. Cepraga, M. Frisoni, Experimental fusion material photon and electron decay heat measurements: Its use for activation codes validation, Fus. Eng. and Des. **75–79**, pp. 1253–1256 (2005)
6. D.G. Cepraga ,G. Cambi, F. Carloni, M. Frisoni, D. Ene, Neutronics and activation calculation for ITER generic site safety report, Fus. Eng. and Des. **63-64** pp. 193–197 (2002)
7. G. Cambi, D.G. Cepraga, M. Frisoni, Material activation assessment for waste analysis of the EU design of RC/RTO ITER, Fus. Eng. and Des. **58–59**, pp. 949–954 (2001)
8. C. Ponti and S. Stramaccia, ANITA: Analysis of Neutron Induced Transmutation and Activation, EUR 12622 EN report (1989)
9. J.-Ch. Sublet, J. Kopecky and R.A.Forrest, The European Activation File: EAF-99 cross section library, UKAEA FUS 408 (1998)
10. A.B. Pashchenko, H.Wienke, J. Kopecky, J.-Ch. Sublet and R.A. Forrest, FENDL/A-2.0 Neutron Activation Cross Section Data Library for Fusion Applications, report IAEA-NDS-173, Rev. 1(1998).
11. R.A. Forrest and F.M. Mann, FENDL/D-2.0, Decay Data Library for Fusion Applications, summary documentation by A. B. Pashchenko and H. Wienke, report IAEA-NDS-178 (1997)
12. M.A. Kellet, The JEFF-3.1.1 Decay Data Library, JEFDOC-1188 (2007)
13. J.-Ch. Sublet, L.W. Packer, J. Kopecky, R.A. Forrest, A.J. Koning and D.A. Rochman, The European Activation File: EAF-2010 neutron-induced cross section library, CCFE-R (10) 05
14. EASY-2010 OECD NEA Data Bank NEA-1564/13 (2011)
15. ENDF-6 Formats Manual Data Formats and Procedures for the Evaluated Nuclear Data Files ENDF/B-VI and ENDF/B-VII Written by the Members of the Cross Sections Evaluation Working Group, Edited by M. Herman and A. Trkov (2010)
16. Application of the Concepts of Exclusion, Exemption and Clearance, IAEA Safety Standard Series No. RS-G-1.7, IAEA Vienna (2004)
17. IAEA, Clearance levels for radionuclides in solid materials - Application of exemption principles, IAEA-TECDOC-855, IAEA, Vienna, (1996)
18. M. Martone, M. Angelone, M .Pillon, The 14 MeV Frascati neutron generator, J.of Nucl.Mat. **212-215,** pp. 1661-1664 (1994)
19. M. Pillon, M. Angelone, R.A. Forrest, Development of a Spectrometer to Measure Photon and Electron Decay Heat from Radionuclides, J. of Nucl. Sci. and Techn., Supplement 2, pp. 649-652 (2002)
20. M. Frisoni, Updating of the Libraries Included in the ANITA-2000 Code Package on the Basis of the JEFF-3.1.1 Radioactive Decay Data Library, ENEA Report ADPFISS-LP1-025 (2014)

EPJ Web of Conferences **111**, 11005 (2016)
DOI: 10.1051/epjconf/201611111005

Validation of the BUGJEFF311.BOLIB, BUGENDF70.BOLIB and BUGLE-B7 broad-group libraries on the PCA-Replica (H_2O/Fe) neutron shielding benchmark experiment

Massimo Pescarini[a], Roberto Orsi and Manuela Frisoni

ENEA, CR "E. Clementel", Via Martiri di Monte Sole 4, 40129 Bologna, Italy

Abstract. The PCA-Replica 12/13 (H_2O/Fe) neutron shielding benchmark experiment was analysed using the TORT-3.2 3D S_N code. PCA-Replica reproduces a PWR ex-core radial geometry with alternate layers of water and steel including a pressure vessel simulator. Three broad-group coupled neutron/photon working cross section libraries in FIDO-ANISN format with the same energy group structure (47 n + 20 γ) and based on different nuclear data were alternatively used: the ENEA BUGJEFF311.BOLIB (JEFF-3.1.1) and BUGENDF70.BOLIB (ENDF/B-VII.0) libraries and the ORNL BUGLE-B7 (ENDF/B-VII.0) library. Dosimeter cross sections derived from the IAEA IRDF-2002 dosimetry file were employed. The calculated reaction rates for the Rh-103(n,n')Rh-103m, In-115(n,n')In-115m and S-32(n,p)P-32 threshold activation dosimeters and the calculated neutron spectra are compared with the corresponding experimental results.

1 Introduction

ENEA-Bologna analysed the PCA-Replica 12/13 [1] water/iron (H_2O/Fe) low-flux engineering neutron shielding benchmark experiment included in the OECD-NEADB/ORNL-RSICC NEA-1517 SINBAD database. Three-dimensional (3D) fixed source transport calculations were performed through the TORT-3.2 [2,3] discrete ordinates (S_N) code. PCA-Replica reproduces a PWR ex-core radial geometry and is particularly suitable to test the cross section libraries specifically dedicated to LWR shielding and pressure vessel dosimetry applications. Two ENEA-Bologna freely released (at OECD-NEADB and ORNL-RSICC) broad-group libraries named NEA-1866 BUGJEFF311.BOLIB [4] and NEA-1872 BUGENDF70.BOLIB [5], respectively based on the OECD-NEADB JEFF-3.1.1 [6,7] and US ENDF/B-VII.0 [8] evaluated nuclear data libraries, were alternatively used together with the ORNL BUGLE-B7 [9] similar library based on ENDF/B-VII.0 data. All these libraries, dedicated to the previously cited applications, include parameterized sets of problem-dependent self-shielded neutron cross sections specifically prepared for BWR and PWR applications, adopt the FIDO-ANISN format and have the same broad-group coupled neutron/photon energy group structure (47 n + 20 γ) as the older ORNL BUGLE-96 [10] library. The ENEA libraries were obtained through problem-dependent cross section collapsing from their ENEA fine-group mother libraries with a nuclear data processing methodology similar to that adopted to generate BUGLE-96, using in particular the same compositional, geometrical and temperature input data to perform the cross section collapsing. BUGJEFF311.BOLIB and BUGENDF70.BOLIB were in fact respectively obtained from the NEA-

[a] Corresponding author: massimo.pescarini@enea.it

1869 VITJEFF311.BOLIB [11] and NEA-1870 VITENDF70.BOLIB [12] fine-group libraries in AMPX format for nuclear fission applications through problem-dependent cross section collapsing with the ENEA-Bologna 2007 revision [13] of the ORNL SCAMPI [14] nuclear data processing system. Both previous libraries are based on the Bondarenko [15] self-shielding factor method, were generated through the LANL NJOY-99.259 nuclear data processing system and have the same fine-group energy structure (199 n + 42 γ) as the ORNL VITAMIN-B6 [10] and VITAMIN-B7 [9] similar libraries from which BUGLE-96 and BUGLE-B7 were respectively obtained. It is underlined that, differently from the ENEA libraries, VITAMIN-B7 and BUGLE-B7 were generated through the recent ORNL AMPX-6.1 nuclear data processing system not yet freely distributed. In the present work the PCA-Replica calculated neutron dose and spectrum integral results, obtained using the three cited working libraries, are compared with the corresponding experimental data.

2 The PCA-Replica experimental facility and measurements

The PCA-Replica (Winfrith, UK, 1984) experimental facility duplicated exactly the ex-core radial geometry of the ORNL PCA (Pool Critical Assembly) similar experiment (Oak Ridge, US, 1981), simulating the ex-core radial geometry of a PWR. PCA-Replica reproduces the 12/13 configuration of PCA with a first water gap of about 12 cm between the core and a thermal shield (TS) simulator and a second water gap of about 13 cm between TS and a PWR pressure vessel (RPV) simulator. The PCA low-flux reactor neutron source was replaced in PCA-Replica with a neutron source emitted by a thin fission plate with the same rectangular cross-sectional area of the PCA reactor source. This simpler source configuration could more easily be calibrated with a high degree of accuracy with respect to the PCA reactor source. The parallelepiped fission plate (63.5 × 40.2 × 0.6 cm) was made of highly enriched uranium (93.0 w% in U-235) alloyed with aluminium. It was irradiated by the NESTOR reactor (30 kW at the maximum power) through a graphite thermal column (total thickness = 43.91 cm) in the ASPIS shielding facility. Beyond the fission plate (0.6 cm thick), the PCA-Replica shielding array was arranged in a large parallelepiped steel tank (square section; side 180.0 cm) filled with water and surrounded by a thick concrete shield. After the first water gap (12.1 cm thick) there was the stainless steel TS simulator (5.9 cm thick), the second water gap (12.7 cm thick) and the mild steel RPV simulator (thickness T = 22.5 cm) tightly connected with a void box made of a thin layer (0.3 cm thick) of aluminium, simulating the air cavity (29.58 cm thick) between the RPV and the biological shield in a PWR. All these components were perfectly orthogonally aligned and centred along the Z horizontal axis passing through the centroid of the fission plate. Threshold activation neutron dosimeters and neutron spectrometers were located in all or in part of the ten measurement spatial positions along the Z axis. The Rh-103(n,n')Rh-103m, In-115(n,n')In-115m and S-32(n,p)P-32 dosimeters employed are respectively characterized by the effective energy thresholds 0.69, 1.30 and 2.70 MeV, by the median energies 1.9, 2.4 and 3.9 MeV and by the 90% response energy ranges 0.53 - 5.4 MeV, 1.0 - 5.6 MeV and 2.2 - 7.4 MeV. The spectral measurements were performed only in two positions (1/4 T RPV and void box) with two kinds of spectrometer. The spherical hydrogen-filled proportional counters of type SP-2 were used in combination, to cover the energy range from 50.0 keV to 1.2 MeV. The neutron fluxes between 1.0 and 10.0 MeV were determined with a spherical 3.5 ml organic liquid (NE213) scintillator.

3 Transport calculations

The whole PCA-Replica experimental array was reproduced in Cartesian (X,Y,Z) geometry (see Figure 1) with the TORT-3.2 code to assure a detailed description of the spatial inhomogeneity of the neutron source emitted by the fission plate. The NEA-1708 ADEFTA-4.1 [16] program calculated the isotopic atomic densities from the isotopic abundances reported in the BNL-NNDC database. The automatic spatial mesh generation and the graphical verification of the geometrical model were performed through the NEA-1678/09 BOT3P-5.3 [17,18] pre/post-processor system.

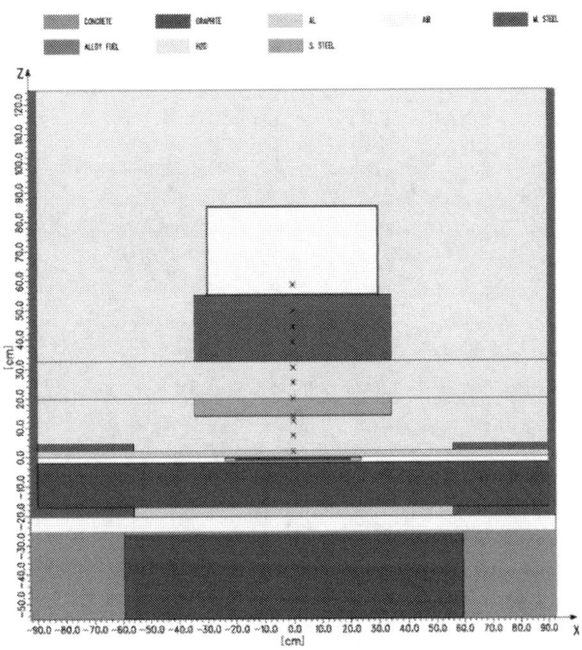

Figure 1. PCA-Replica - Geometrical and compositional model horizontal section at Y = 0. cm, dosimeter locations "×", TORT-3.2 (X,Y,Z) with 65X×63Y×182Z spatial meshes.

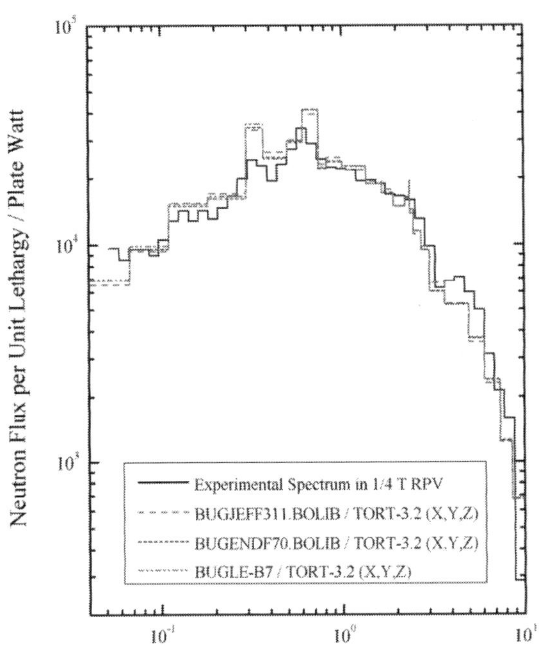

Figure 2. PCA-Replica - Comparison of experimental and calculated neutron spectra in mild steel, in the 1/4 T RPV measurement position at Z = 39.01 cm.

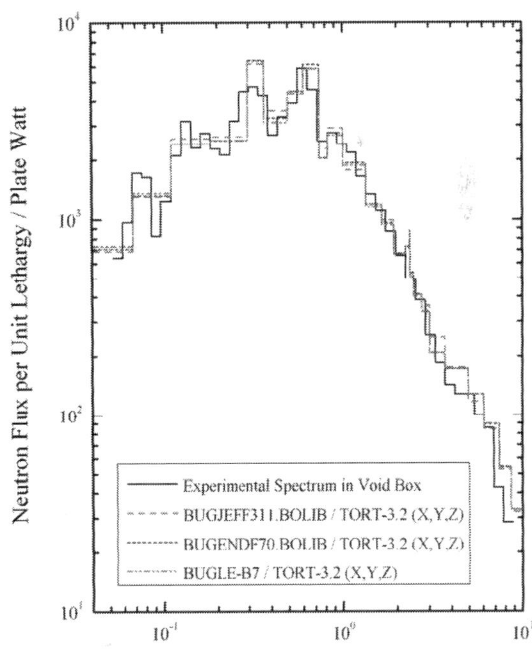

Figure 3. PCA-Replica - Comparison of experimental and calculated neutron spectra in air, in the void box measurement position at Z = 58.61 cm.

Table 1. PCA-Replica - Summary of Experimental (E) and Calculated (C) Rh-103(n,n'), In-115(n,n') and S-32(n,p) reaction rates[a] per NESTOR reactor Watt along the horizontal axis Z.

Dos. Pos.	Distance from Fission Plate on the Axis Z [cm]	Experimental Reaction Rate[a] ± Random Error (1σ) (E)	BUGJEFF311. BOLIB Calculation		BUGENDF70. BOLIB Calculation		BUGLE-B7 Calculation		Reference Location
			Calculated Reaction Rate[a] (C)	C/E[b]	Calculated Reaction Rate[a] (C)	C/E[b]	Calculated Reaction Rate[a] (C)	C/E[b]	
Rh-103(n,n')Rh-103m Systematic Error = ± 3.0% in All Dosimeter Measurement Positions									
1	1.91	1.69E-20 ± 3.0%	1.81E-20	1.09	1.81E-20	1.09	1.82E-20	1.10	12 cm Water Gap 1
2	7.41	3.78E-21 ± 3.0%	3.43E-21	0.93	3.44E-21	0.93	3.46E-21	0.93	
3	12.41	1.40E-21 ± 3.0%	1.30E-21	0.95	1.31E-21	0.95	1.32E-21	0.96	
4	14.01	1.27E-21 ± 3.0%	1.15E-21	0.93	1.16E-21	0.93	1.17E-21	0.94	
5	19.91	4.23E-22 ± 3.0%	4.18E-22	1.01	4.17E-22	1.00	4.16E-22	1.00	13 cm Water Gap 2
6	25.41	1.15E-22 ± 4.0%	1.02E-22	0.91	1.02E-22	0.91	1.03E-22	0.91	
7	30.41	4.73E-23 ± 4.0%	4.12E-23	0.89	4.13E-23	0.89	4.15E-23	0.89	
8	39.01	2.07E-23 ± 1.0%	2.02E-23	1.02	2.02E-23	1.01	2.02 E-23	1.02	1/4 T RPV
9	49.61	5.53E-24 ± 1.9%	5.62E-24	1.06	5.59E-24	1.05	5.56E-24	1.05	3/4 T RPV
10	58.61	1.80E-24 ± 1.6%	1.65E-24	0.96	1.66E-24	0.96	1.62E-24	0.94	Void Box
In-115(n,n')In-115m Systematic Error = ± 2.0% in All Dosimeter Measurement Positions									
8	39.01	3.93E-24 ± 0.9%	3.89E-24	1.03	3.87E-24	1.03	3.88E-24	1.03	1/4 T RPV
9	49.61	8.23E-25 ± 1.4%	7.80E-25	0.99	7.76E-25	0.98	7.70E-25	0.97	3/4 T RPV
10	58.61	2.31E-25 ± 1.5%	2.15E-25	0.97	2.16E-25	0.97	2.12E-25	0.95	Void Box
S-32(n,p)P-32 Systematic Error = ± 4.0% in All Dosimeter Measurement Positions									
8	39.01	1.08E-24 ± 1.5%	9.86E-25	0.95	9.78E-25	0.94	9.83E-25	0.95	1/4 T RPV
9	49.61	1.46E-25 ± 1.9%	1.38E-25	0.98	1.35E-25	0.97	1.36E-25	0.97	3/4 T RPV
10	58.61	3.73E-26 ± 1.3%	3.63E-26	1.01	3.57E-26	1.00	3.57E-26	1.00	Void Box

Note: Total experimental error (1σ confidence level) = $[(\text{random error})^2 + (\text{systematic error})^2]^{1/2}$ (see [1]).

([a]) Reaction rates are in units of reactions × s^{-1}× $atom^{-1}$ × NESTOR $Watt^{-1}$.

([b]) Experimental results contain a contribution from the NESTOR core background. Calculated results refer only to the neutrons produced in the fission plate per 1 Watt of NESTOR power. The E values, in the C/E ratios, are reduced by 4% in the RPV and void box and by 2% in the water gaps (see [1]).

A parallelepiped geometry (185.08 × 180.0 × 180.0 cm, respectively along the X, Y and Z axis) was described with a 65X×63Y×182Z fine spatial mesh grid. Volumetric meshes with sides always inferior to 0.5 cm were introduced to get the best result accuracy in the measurement positions on the Z axis. Group-organized files of macroscopic cross sections, requested by TORT-3.2, were prepared through the GIP [4] program. Fixed source transport calculations with one source (outer) iteration were performed using fully symmetrical discrete ordinates directional quadrature sets for the flux solution. The P_3-S_8 approximation, adopted in all the calculations, is the most widely used option in fixed source calculations for LWR safety analyses. P_N corresponds to the order of the expansion in Legendre polynomials of the scattering cross section matrix and S_N represents the order of the flux angular discretization. The same value (1.0E-03) for the point-wise flux convergence criterion was employed. The vacuum boundary condition was selected at the left, right, inside, outside, bottom and top geometrical boundaries. The same set of flat weighting dosimeter cross sections for Rh-103(n,n'), In-115(n,n') and S-32(n,p), derived from the IAEA International Reactor Dosimetry File IRDF-2002, was used in all the calculations. The volumetric neutron sources used in the calculations with the three libraries were got using their own U-235 fission neutron spectra whereas the common value of $\overline{\nu}$ (U-235) = 2.437 [1] for the average number of neutrons produced per U-235 thermal fission was used in all the calculations to assure a consistent result intercomparison. The value of 6.74E-04 fission plate Watts per NESTOR Watt [1] was given to the source multiplier parameter for the neutron source. The calculated/experimental dosimetric results are reported in Table 1. The calculated/experimental spectra in the 1/4 T RPV and void box positions are respectively shown in Figures 2 and 3.

4 Conclusion

The deviations of the dosimetric results from the corresponding experimental ones are within the desired ± 10-15% target accuracy. Given that the total experimental uncertainty ranges from about 2% to a maximum of 5% (1σ), then the calculated results exhibit very good statistical consistency with the corresponding measurements. The results obtained using BUGENDF70.BOLIB and BUGLE-B7 are very similar and this seems to indicate an equivalent performance of the NJOY-99.259/SCAMPI and AMPX-6.1 nuclear data processing calculation chains starting from the same ENDF/B-VII.0 data.

References

1. J. Butler et al., UKAEA, AEE Winfrith Report AEEW-R 1736 (January 1984)
2. W.A. Rhoades, D.B. Simpson, TORT Version 3, ORNL/TM-13221 (October 1997)
3. DOORS3.1, ORNL, RSIC CCC-650 (August 1996)
4. M. Pescarini, V. Sinitsa , R.Orsi, M. Frisoni, ENEA Int. Report UTFISSM-P9H6-2 Rev.1 (2013)
5. M. Pescarini, V. Sinitsa , R.Orsi, M. Frisoni, ENEA Int. Report UTFISSM-P9H6-8 (2013)
6. *The JEFF-3.1.1 Nuclear Data Library*, JEFF Report 22, OECD-NEA Data Bank (2009)
7. *The JEFF-3.1 Nuclear Data Library*, JEFF Report 21, OECD-NEA Data Bank (2006)
8. M.B. Chadwick et al., Nucl. Data Sheets, **107**, 12, pp. 2931-3060, (December 2006)
9. J.M. Risner et al., ORNL/TM-2011/12, NUREG/CR-7045 (September 2011)
10. J.E. White et al., ORNL-6795/R1, NUREG/CR-6214, Revision 1 (January 1995)
11. M. Pescarini, V. Sinitsa, R. Orsi, ENEA Int. Report UTFISSM-P9H6-003 Rev.1 (2011)
12. M. Pescarini, V. Sinitsa, R. Orsi, M. Frisoni, ENEA Int. Report UTFISSM-P9H6-005 (2012)
13. V. Sinitsa, M. Pescarini, ENEA Int. Report FPN-P9H6-006 (2007)
14. SCAMPI, ORNL, RSIC PSR-352 (September 1995)
15. I.I. Bondarenko et al., *Group constants for nuclear reactors calculations* (Cons. Bureau, 1964)
16. R. Orsi, ENEA Int. Report FPN-P9H6-10 (2008)
17. R. Orsi, ENEA Int. Report FPN-P9H6-11 (2008)
18. R. Orsi, Nuc. Sci. and Eng., Technical Note, **146**, pp. 248-255 (2004)

EPJ Web of Conferences **111**, 11006 (2016)
DOI: 10.1051/epjconf/201611111006

Toward Reanalysis of the Tight-Pitch HCLWR-PROTEUS Phase II Experiments

Grégory Perret[1,a], Efstathios Vlassopoulos[1,b], Mathieu Hursin[1,2], and Andreas Pautz[1,2]

[1] *Paul Scherrer Institute, 5232 Villigen, Switzerland*
[2] *Ecole Polytechnique Fédérale de Lausanne, 1015 Lausanne, Switzerland*

Abstract. The HCLWR-Proteus Phase II experiments were conducted from 1985 to 1990 in the zero-power reactor Proteus at PSI in Switzerland. The experimental program was dedicated to the physics of high conversion light water reactors and in particular to the measurement of reactor parameters such as reaction rate traverses, spectral indices, absorber reactivity worths and void coefficients. The HCLWR experiments are especially interesting because they generated knowledge in the epithermal range of the neutron flux spectrum, for which little integral experimental data is available.
In an effort to assess the interest of this experimental data to validate modern nuclear data and improve their uncertainties, a preliminary re-analysis of selected configurations was conducted with Monte-Carlo codes (MCNP6/SERPENT2) and modern nuclear data libraries (ENDF/B-VII.0, JEFF-3.1.1 and JENDL-4.0). The spectral indices, flux spectra and sensitivity coefficients on k_∞ were calculated using cell models representative of the tight-pitch measurement configurations containing 11% PuO_2-UO_2 fuel rods in different moderation conditions (air, water and dowtherm).
Spectral index predictions using the three nuclear data libraries agreed within two standard deviations with the measured values. The only exception is the Pu-242-capture-to-Pu-239-fission ratio, which was overestimated with all libraries by more than four standard deviations, i.e. 13%, in the non-moderated configuration. In this configuration, Pu-242 captures are few since the flux spectrum in the Pu-242 capture resonance region (between 1eV and 1keV) is small making this spectral index hard to measure. Sensitivity coefficient predictions with both MCNP6 and SERPENT2 were in good agreement.

1 Introduction

New integral experiments dedicated to neutronic codes and data validation are scarce nowadays. This is due in part to the already enormous knowledge gathered along the years but also to the increasing costs and long-time commitment necessary to perform experiments. There are, however, at least two areas in which zero-power reactor neutronics experiments are still of value for modern codes and nuclear data libraries, namely for validation of reactor concepts different from the traditional Light Water Reactors and for more detailed investigations (e.g. within pin power distribution). The

[a] e-mail: gregory.perret@psi.ch
[b] Now at NAGRA, 5430 Wettingen, Switzerland

first trend is illustrated by the investigation of the GEN-IV reactor concepts; the second by the development of new set of high-fidelity codes, like in the framework of CASL, whose aim is to provide a framework for improved multi-physics (i.e. coupling neutronics, materials and thermal-hydraulic phenomena) reactor analysis and which require detailed measurements for its validation.

Ironically enough, several experiments performed in the distant past (read 1960s-90s for neutronics) were performed on reactor concept not much different than those considered in GEN-IV or studied for the transition from the Light Water Reactors technology to fast reactors. A striking example is the Proteus experimental program, which was from the 1970s to 2000s dedicated in turn to Gas-Cooled Fast Reactors, High Conversion Light Water Reactors and High Temperature Reactors [1]. A cost effective solution, for both research institutes and international organisation, is to recycle these experiments by modeling them, in so-far as the experimental details are still available, with the codes and libraries of the day and validate them. We started this endeavour for the GCFR-Proteus program at Wonder-2012 [2], which will result in safeguarding these data in the IRPhE database [3].

In this work, we report on our initial effort to use the HCLWR-Proteus data for validation purposes. Section 2 summarizes the HCLWR-Proteus experiments and the selected configurations. Section 3 describes the code and pin-cell models used in this work and illustrate their characteristics with flux spectra and sensitivity coefficients and profiles. Section 4 presents spectral index predictions and their comparisons with the measurement performed during the HCLWR-Proteus experiments.

2 Experiments

The HCLWR-Proteus Phase II experiments were conducted from 1985 to 1990 in the Proteus reactor at Paul Scherrer Institute (PSI) in Switzerland. Proteus is a zero-power reactor featuring a cylindrical central cavity that is driven critical by a surrounding graphite region fueled with 5 w% UO_2 fuel pins. In the HCLWR-Proteus experiments, the central cavity was loaded with a D_2O zone fueled with 5 w% UO_2 fuel pins, a buffer region containing metallic natural uranium rods in air and a central test zone filled with a HCLWR-representative fuel lattice.

The experimental program was dedicated to the physics of high conversion light water reactors and in particular to the measurement of reactor parameters such as reaction rate traverses, spectral indices, absorber reactivity worths and void coefficients. In this paper we focus on the interpretation of the spectral index measurements performed in cores 7, 8 and 9. In cores 7, 8 and 9, the test zone comprises a tight-pitch hexagonal lattice of fuel pins (pitch of 1.07cm; moderator-to-fuel ratio of 0.48; pitch-to-diameter of 1.12) loaded in light water (atmospheric conditions), air and dowtherm[1], respectively. The fuel is PuO_2-UO_2 with about 11% PuO_2 of which 72% is fissile. This set of core configurations is highly interesting for validation purposes as it features spectral conditions in both the epithermal and fast ranges (see section 3.1 and 3.2) - for which experimental data are scarce.

Reaction rate ratios were measured using fission chambers and foils. Fissions in U-235 (F5), Pu-239 (F9) and U-238 (F8) were respectively measured using U/Al, Pu/Al and U_{metal} or UO_2 foils in an instrumented pin at the core center. The absolute reaction rates were determined using inter-calibrated foils located together with absolutely calibrated fission chambers in a cavity 25cm above the core center. Captures in U-238 (C8) was determined using UO_2 foils in the instrumented rod. Because of the difficulty to manufacture and handle large enough Pu-241 deposits, fissions in Pu-241 (F1) required using Al catcher foils wrapped around the Pu-241 deposits. Measuring the activity of the Al-foils in the fission chamber and in the center of the core allowed determining the absolute rate of Pu-241 fissions[2]. Finally, captures in Pu-242 (C2) were measured by counting the 84keV

[1] Dowtherm is an organic liquid used to simulate water with lower density; its effective moderation is 0.28 as compared to 0.48 for light water.

[2] The same Al-foils were also measured without the Pu-241 deposit to correct for their activation by other radiations.

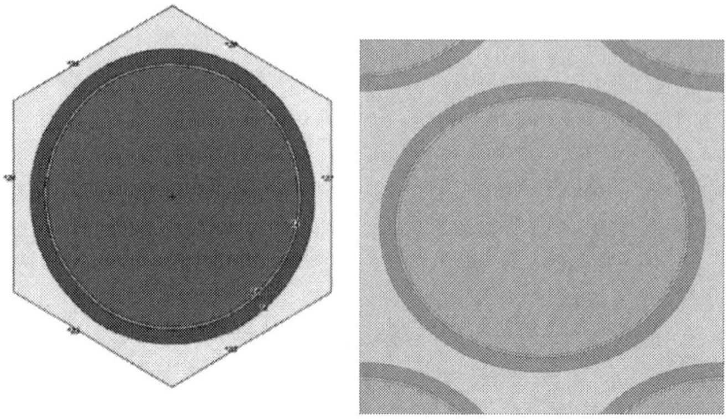

Figure 1. Pin-cell models - MCNP6 (left), Serpent (right).

γ-peak of Pu-243 emitted by a Pu-242 deposit irradiated in turn in a fission chamber located in the thermal column of the reactor and in the core center. The capture rate at the center of the core was determined using the relatively well known cross-section for capture in Pu-242 at the thermal energy[3]. Measurements of F1 and C2 were significantly harder than that of F5, F8, C8 and F9 and had higher uncertainties. Typical uncertainty on the measured reaction rate ratios are 1.5%-2% for C8/F9, F8/F9 and F5/F9, and 3% for F1/F9 and C2/F9 [4].

3 Codes and Models

In this work, we used the MCNP6 and SERPENT2 Monte-Carlo codes [5, 6] with the modern JEFF-3.1.1, ENDF/B-VII.0 and JENDL-4.0 nuclear data libraries [7–9]. The model presented here are pin cell models. Under the assumption that the center of the test zone of Proteus is representative of the neutronic conditions encountered in the infinite lattice, these models are sufficient to help characterize the experiments with flux spectrum and sensitivity profiles. In addition, they can be used to compare their predictions to the reaction rate ratio measured during the HCLWR-Proteus experiments, as the latter have been corrected for the low impact of the driver regions at the center of the core. Correction factors to account for leakage were calculated at the time of the experiments using diffusion theory and were less than 1% for the reaction rates measured in cores 7, 8 and 9, with the exception of C8/F9 in core 8 (1.8%), confirming that the center of the Proteus test zone is representative of an HCLWR [4].

Fig. 1 shows radial cut-away views of the MCNP6 and SERPENT2 pin cell models. The reference fuel material compositions were corrected to account for the α-decay of Pu-238 and β^- decay of Pu-241 in U-234 and Am-241, respectively. The boundary conditions are reflective in the radial direction. For the flux and reaction rate calculations, the height of the cell is adjusted to reach criticality and performed k-eigenvalue calculations with about 100 million particles.

3.1 Flux spectrum

The flux and reaction rates are tallied on the fuel region spreading ±2.5cm around the median plan, where the spectrum is not significantly distorted by the axial leakage. For the flux tallying, we used

[3]Although the Pu-242 thermal capture cross section should be well-known, the value taken from an outdated library may introduce some bias in C2 assessment.

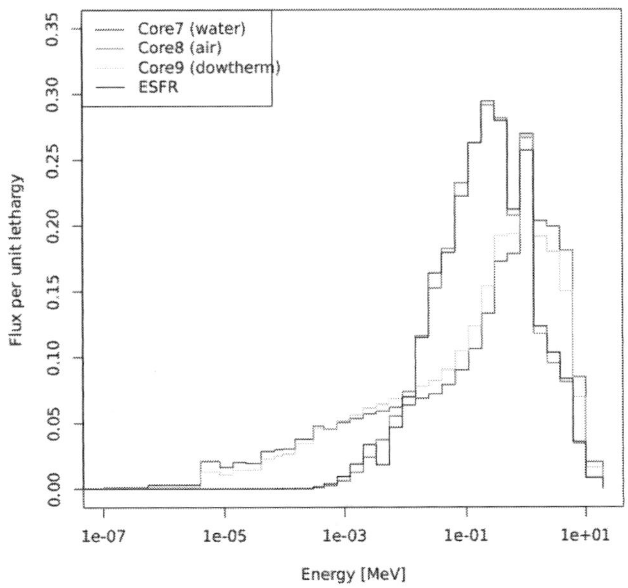

Figure 2. Flux spectra for the three studied configurations and the European sodium fast reactor.

the ECCO-33 energy group typically employed for fast reactor systems [10]. Typical uncertainty is lower than 0.1% for the bins in which the flux is not negligible.

Fig. 2 shows the flux spectra per unit lethargy for the three selected configurations and for the European sodium fast reactor. Calculations were performed with MCNP6 and ENDF/B-VII.0. The reference HCLWR core moderated with water at room temperature and pressure (core 7) has a softer spectrum than the sodium fast reactor, especially in the epithermal range from 10eV to 10keV. The flux spectra of the voided HCLWR core (core 8) is similar to that of the sodium reactor - the main visible difference being the first sodium scattering resonance at 2.85keV. Finally, the core moderated with dowtherm (core 9) is, as expected, in between the spectra of cores 7 and 8.

We calculated the same spectra with SERPENT2 and with JEFF-3.1.1 and JENDL-4.0 and obtained similar results. Fig. 3 shows, as an example, the flux spectra of the reference HCLWR core obtained with the three libraries. The maximum deviations are about 2%, i.e. negligible. Differences in the spectral index predictions come therefore mainly from the tabulated cross section and not from the flux calculation.

3.2 Sensitivity analysis

Sensitivity coefficients and profiles are required to assess how representative an experiment, like HCLWR-Proteus, is of a design case, e.g. Reduced Moderation Boiling Water Reactors or fast reactors [11], and to reduce the a priori uncertainty on the prediction of the reactor parameter for the design case [12]. These quantities also provides additional physical insight on the modeled experiment.

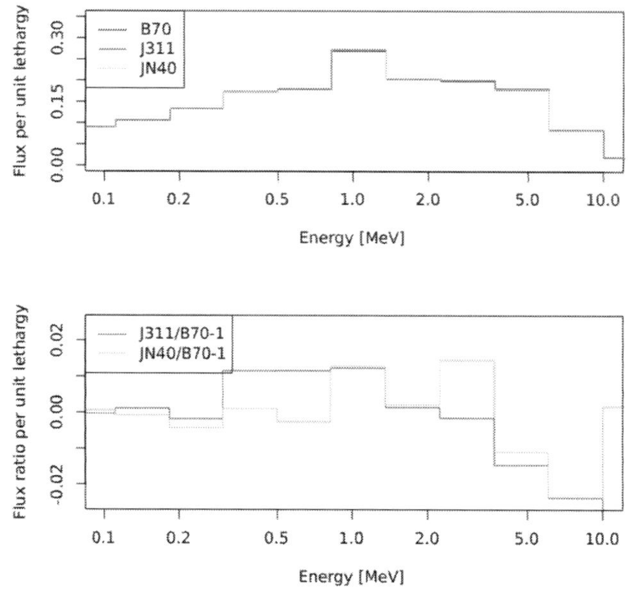

Figure 3. Flux spectra for different libraries (top) and differences (bottom).

Table 1. Sensitivity coefficients of k_∞ integrated on the whole spectra.

Isotope	Reaction	Core 7 (Water)	Core 8 (Voided)
U-238	Total	-0.146 (1.5%)	-0.337 (1.1%)
	Inelastic	-0.0379 (1.9%)	-0.128 (0.9%)
	Capture	-0.182 (0.2%)	-0.302 (0.1%)
	Fission	0.0639 (0.4%)	0.0937 (0.4%)
Pu-239	Total	0.212 (0.5%)	0.388 (0.3%)
	Inelastic	-0.011 (13%)	-0.038 (7%)
	Capture	-0.155 (0.2%)	-0.045 (1.2%)
	Fission	0.368 (0.3%)	0.438 (0.2%)

We calculated sensitivity coefficient relative to the infinite multiplication factor (k_∞). In these calculations, the pin cell axial boundary conditions are reflective and the sensitivity coefficients are calculated on all regions of the model.

Tab 1 shows, for illustrative purposes, the dominant sensitivity coefficients for U-238 and Pu-239 reactions obtained with MCNP6 and JEFF-3.1.1 in the water-moderated and voided configurations and their relative 1σ uncertainty. The coefficients show the total impact on k_∞, i.e. integrated over the whole energy domain, of the isotope-reaction pairs. We obtained the same sensitivity coefficients, i.e. within 2σ for the non negligible isotope-reaction pairs, when using SERPENT2 and JEFF-3.1.1.

We can notice the enhanced importance of U-238 reactions - especially inelastic and capture reactions - in the voided configuration. Regarding Pu-239 reactions, as expected, the impact of the

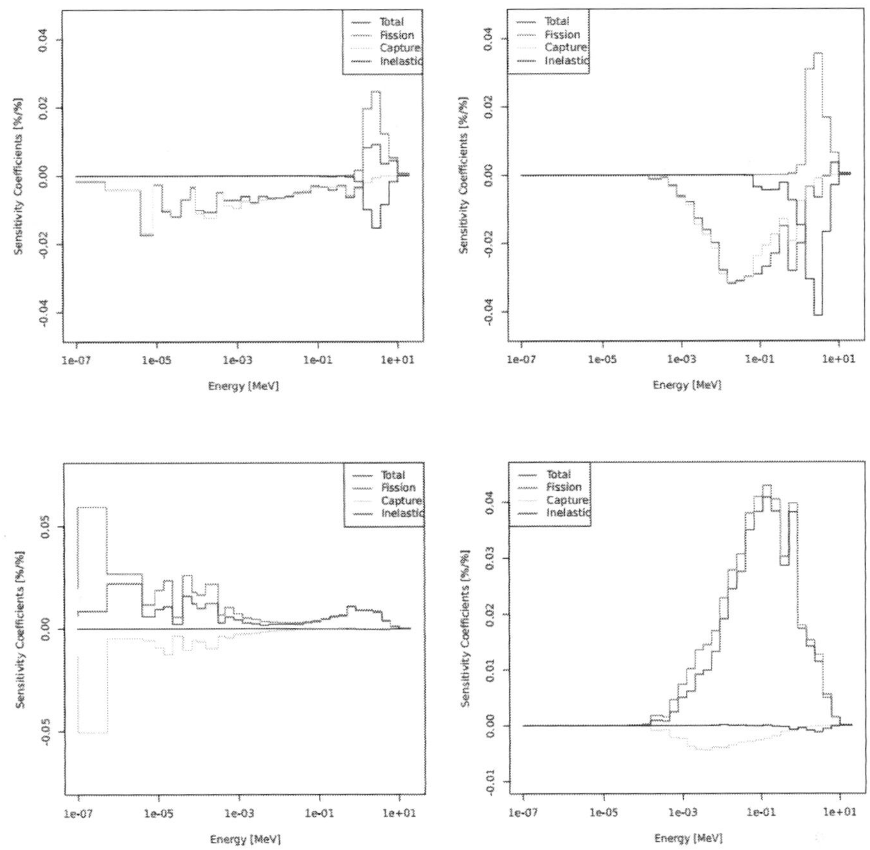

Figure 4. Sensitivity profiles for U-238 (top) and Pu-239 (bottom) in water-moderated (left) and voided (right) configurations.

capture cross section on k_∞ is strongly reduced in the voided configuration, and that of fission enhanced, pointing towards a better utilization of the Pu-239 resources in fast spectra.

More insight can be gained by plotting the sensitivity profiles, i.e. sensitivity coefficient as a function of energy. Fig. 4 shows the sensitivity profiles for k_∞ calculated with MCNP6 and JEFF-3.1.1 in the water-moderated (core 7) and voided configurations (core 8). The main reactions, i.e. total, inelastic scattering, fission and capture reactions, are plotted for U-238 and Pu-239.

The importance of the inelastic scattering for U-238 in fast spectra is illustrated when comparing the two top plots, i.e. water-moderated configuration on the left and voided on the right. The impact of the inelastic scattering of U-238 on k_∞ is largely enhanced above 1MeV and more than compensate the sensitivity coefficient for fission in U-238, rendering the total sensitivity coefficient positive above 2MeV. The total sensitivity coefficient for U-238 is only positive above 10 MeV in the moderated case. These graphs also shows the large compensating effects for U-238 reactions.

Finally, Pu-239 sensitivity profiles, illustrate, without surprise, the effect of the thermal fissions in the moderated case (left) as compared to the voided configuration (right). They also shows how the sensitivity of k_∞ to Pu-239 reaction is only governed by fissions in fast spectra and the role of

Table 2. Comparison of measured and calculated spectral indices.

Core 7 (Water)			
Index	B70	J311	JN40
C8/F9	1.007±0.018	1.007±0.018	1.004±0.018
F8/F9	1.024±0.019	1.005±0.019	1.016±0.019
F5/F9	1.017±0.015	1.016±0.015	1.012±0.015
F1/F9	0.988±0.030	0.988±0.030	0.982±0.029
C2/F9	1.044±0.031	1.020±0.030	1.025±0.031
Core 8 (Air)			
Index	B70	J311	JN40
C8/F9	1.021±0.016	1.018±0.016	1.006±0.016
F8/F9	1.002±0.018	0.983±0.018	0.994±0.018
F5/F9	1.023±0.013	1.025±0.013	1.001±0.013
F1/F9	1.011±0.030	1.022±0.031	0.994±0.030
C2/F9	1.133±0.034	1.230±0.037	1.167±0.035
Core 9 (Dowtherm)			
Index	B70	J311	JN40
C8/F9	0.989±0.017	0.992±0.017	0.986±0.017
F8/F9	1.017±0.018	0.999±0.018	1.013±0.018
F5/F9	1.023±0.015	1.024±0.015	1.018±0.015
F1/F9	0.987±0.049	0.990±0.050	0.981±0.049
C2/F9	1.038±0.031	1.035±0.031	1.036±0.031

capture in moderated configuration - which impaired the effective burning of plutonium. We ran the same sensitivity profiles with SERPENT2 and JEFF-3.1.1 and found the same results (within 2σ for the non negligible isotope-reaction pairs).

4 Spectral Index Results

Tab. 2 lists the calculated-to-measured ratio of the spectral indices detailed in section 2 and their 1σ uncertainty. For the moderated cases (cores 7 and 9) all predictions agree with the measurements within 2σ. Slightly better agreement is obtained with JEFF-3.1.1. The spectral indices for the voided configuration (core 8) are also correctly predicted but for the Pu-242 capture rate. The C2/F9 ratio is overpredicted by 13% to 23% depending on the library (i.e. by 4 to 6σ). The best predictions for the non moderated core are obtained, on average, with JENDL-4.0.

The difference in predictions for the Pu-242 capture rate are directly related to the difference in the cross sections in the unresolved resonance and fast ranges (see Fig. 5 reproduced from [13]), in which the flux is non negligible (see Fig. 2). The source of the bias between predictions and the experimental value can be due to error in the measurement or post-processing of the measured value, or in the evaluations of the cross section.

Possible errors in the measurement stems from: the very low reaction rate due to the small Pu-242 capture cross section above 1keV and the small amount of Pu-242 in the deposit (0.2 mg/cm^2 on a 0.1-mm thick aluminum foil), the difficulties to measure the low energy (84keV) gamma-ray lines[4] emitted by Pu-243 (which has a short half-life of 4.96 hours), and the calibration in the thermal

[4]due to self absorption and background of other radiations

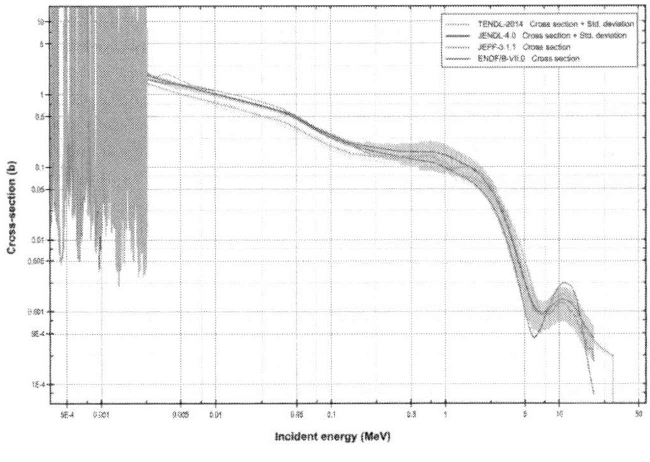

Figure 5. Pu-242 capture cross section above 1keV for different libraries.

column (see section 2), which uses the Pu-242 thermal cross-section of an outdated library. Further investigation of these sources of bias requires unfortunately a more detailed measurement report than the one currently available.

Regarding the nuclear data cross sections, Fig. 5 also shows that the uncertainty on the cross section, for example in JENDL-4.0 or TENDL-2014, is not negligible and that a bias may still exists today - as illustrated by the lower values of TENDL-2014 from 1 keV to 1 MeV.

5 Conclusion

We presented a first re-analysis of the HCLWR-Proteus experiments with the Monte-Carlo codes MCNP6 and SERPENT2 and the modern nuclear data libraries ENDF/B-VII.0, JEFF-3.1.1 and JENDL-4.0. We focused on the tight pitch core configurations moderated with water or dowtherm and voided, and calculated the spectral indices measured in the center of the core.

The calculation models are simple pin cells, as they are representative of the spectral conditions encountered in the center of core in the selected HCLWR-Proteus core configurations. Results showed an excellent agreement (within 1-2σ) for the C8/F9, F8/F9, F5/F9 and F1/F9 indices[5] in the three core configurations. The Pu-242-capture-to-Pu-239-fission index is well predicted only in the moderated configurations. In the voided configuration, in which the flux spectra is spanning from 1keV to 1MeV, the index is over-predicted by 13% to 23% (4 to 6 σ) depending on the library. In addition to possible biases in the evaluated cross sections of the selected libraries, errors in the measurements cannot be rejected. Errors could stems from the low measurement signal (few Pu-242 captures in the unresolved resonance range and high background) or the measurement method (comparison to thermal spectra measurement requiring the use of outdated Pu-242 thermal cross sections), and will require further investigations.

Finally, we calculated sensitivity coefficients and profiles on k_∞ with MCNP6 and SERPENT2 and the JEFF-3.1.1 library. Results are consistent between the codes, illustrate the physics of the water-moderated and voided core configurations and will serve to determine if the experimental results could help design HCLWR or fast reactors.

[5]C stands for capture, F for fission and 5, 8, 9 and 1 for U-235, U-238, Pu-239 and Pu-241, respectively.

References

[1] G. Perret, *http://www.psi.ch/erp/proteus* (2015)

[2] G. Perret, K.A. Jordan, Tech. rep., Paul Scherrer Institut, Villigen, 5232, Switzerland (2012)

[3] K. Jordan, G. Perret, J. Bess, *Benchmark Evaluation of PROTEUS Gas-Cooled Reactor Experiments* (2014)

[4] H. Berger, R. Böhme, R. Chawla, H. Hager, O. Köberl, R. Seiler, T. Williams, Tech. rep., Paul Scherrer Institute, Villigen, 5232, Switzerland (1993)

[5] T. Goorley, Tech. rep., LA-UR-14-24680 - Los Alamos National Laboratory (2014)

[6] J. Leppänen, *http://montecarlo.vtt.fi/index.htm* (2015)

[7] A. Santamarina, D. Bernard, P. Blaise, M. Coste, A. Courcelle, T. Huynh, C. Jouanne, P. Leconte, O. Litaize, S. Mengelle et al., *The JEFF-3.1.1 Nuclear Data Library - JEFF Report 22* (2009), `www.nea.fr/html/dbdata/nds_jefreports/`

[8] M. Chadwick, P. Obložinský, M. Herman, N. Greene, R. McKnight, D. Smith, P. Young, R. MacFarlane, G. Hale, S. Frankle et al., Nuclear Data Sheets **107**, 2931 (2006), evaluated Nuclear Data File ENDF/B-VII.0

[9] K. Shibata, O. Iwamoto, T. Nakagawa, N. Iwamoto, Journal of Nuclear Science and Technology **48**, 1 (2012)

[10] E. Sartori, Tech. rep., NEA DAta Bank, Gif-sur-Yvette, France (1990)

[11] T. Iwamura, S. Uchikawa, T. Okubo, T. Kugo, H. Akie, Y. Nakano, T. Nakatsuka, Nuclear Engineering and Design **236**, 1599 (2006)

[12] O. Leray, C. Vaglio-Gaudard, J.P. Hudelot, A. Santamarina, G. Noguere, *Sensitivity and Uncertainty Analysis applied to the JHR Reactivity Prediction*, in *Proc. of the PHYSOR 2012: Advances in Reactor Physics - Linking Research, Industry and Education* (American Nuclear Society, Knoxville, USA, 2012), 2010

[13] N. Soppera, M. Bossant, E. Dupont, Nuclear Data Sheets **120**, 294 (2014)

AUTHOR INDEX

AUTHOR INDEX

AUTHOR INDEX

AUTHOR INDEX